Regional Hydrological Response to Climate Change

The GeoJournal Library

Volume 38

The titles published in this series are listed at the end of this volume.

Regional Hydrological Response to Climate Change

edited by

J. A. A. JONES
Institute of Earth Studies,
University of Wales,
Aberystwyth, U.K.

CHANGMING LIU
Institute of Geography and United Research Centre for Water Problems,
Academia Sinica,
Beijing, China

MING-KO WOO
Department of Geography,
McMaster University,
Hamilton, Ontario, Canada

and

HSIANG-TE KUNG
Department of Geography,
University of Memphis,
Tennessee, U.S.A.

KLUWER ACADEMIC PUBLISHERS
DORDRECHT / BOSTON / LONDON

A C.I.P. Catalogue record for this book is available from the Library of Congress.

ISBN 0-7923-4329-8

Published by Kluwer Academic Publishers,
P.O. Box 17, 3300 AA Dordrecht, The Netherlands.

Kluwer Academic Publishers incorporates
the publishing programmes of
D. Reidel, Martinus Nijhoff, Dr W. Junk and MTP Press.

Sold and distributed in the U.S.A. and Canada
by Kluwer Academic Publishers,
101 Philip Drive, Norwell, MA 02061, U.S.A.

In all other countries, sold and distributed
by Kluwer Academic Publishers Group,
P.O. Box 322, 3300 AH Dordrecht, The Netherlands.

Printed on acid-free paper

Printed in the Netherlands

To our colleague Ming-Ko Woo, who suffered the most horrendous accident during the compilation of this volume and has continued to show great fortitude and perseverance. All strength to his continuing recovery.

CONTENTS

Conclusions

PREFACE

This volume arises from the work of the International Geographical Union Working Group on Regional Hydrological Response to Climate Change and Global Warming under the chairmanship of Professor Changming Liu (1992-96). The book consists mostly of peer-reviewed papers delivered at the Working Group's first three scientific meetings held in Washington, D.C. (1992), Lhasa, Tibet (1993) and Moscow (1995). These have been supplemented by a few additional chapters that have been specifically commissioned in order to give a well-rounded coverage of the global and scientific aspects of the topic.

As editors, we have sought to balance state-of-the-art reviews of methodology and regional research with detailed studies of specific countries and river basins. In the spirit of the IGU, we have devoted particular effort to encouraging contributions from scientists in the non-English-speaking world. These chapters provide valuable evidence of recent climatic change and predictions of future hydrological impacts from parts of the world where little detailed work has been conducted hitherto. They provide much valuable information that is new and interesting to an international audience and is otherwise very difficult or impossible to acquire.

It is hoped that the present volume will be not only a record of current achievements, but also a stimulus to further hydrological research as the detail and spatial resolution of Global Climate Models improves. One notable aspect that emerges from a number of the contributions is that many, though by no means all, recent hydrological trends are in line with global warming predictions.

J.A.A. Jones (UK)
Changming Liu (China)
Ming-Ko Woo (Canada)
Hsiang-Te Kung (USA)

May, 1996

LIST OF CONTRIBUTORS

Elena M. Aizen
Institute for Computational Earth System Science
University of California Santa Barbara
Santa Barbara
California, CA 93106, USA.

Vladimir B. Aizen
Institute for Computational Earth System Science
University of California Santa Barbara
Santa Barbara
California, CA 93106, USA.

Naresh Akkur
Environmental and Social Systems Analysts Ltd.
1765 W 8th Avenue
Vancouver
B.C., V6L 1A8 Canada.

Brad Bass
Environmental Adaptation Research Group
4905 Dufferin St.
Downsview
Ontario, M3H 5T4 Canada.

B.C. Bates
CSIRO Division of Water Resources
Private Bag PO
Wembley
Western Australia 6014, Australia.

A.D. Capriolo
Biometeorological Research Center
National Scientific and Technical Research Council (CIBIOM -CONICET)
Serrano 669
Buenos Aires 1414, Argentina.

F.H.S. Chiew
CRC for Catchment Hydrology
Department of Civil and Environmental Engineering
University of Melbourne
Parkville,
Victoria 3052, Australia.

S. Daultrey
Department of Geography
University College Dublin
Belfield
Dublin 4, Ireland.

Deng Yuren
Department of Hydraulic Engineering
Chengdu University of Science and Technology
Chengdu 610065, China.

Fu Guobin
Institute of Geography and United Research Centre for Water Problems
Academia Sinica
Building 917 Datun Road
Anwai
Beijing 100101, China.

Malgorzata Gutry-Korycka
Faculty of Geography and Regional Studies
University of Warsaw ul. Krakowskie
Przedmiescie 30
00-927 Warsaw, Poland.

C.P. Holt
School of Environmental Science
Nene College
Boughton Green Road
Northampton, NN2 7AL, UK.

Hou Yuguang
Department of Hydraulic Engineering
Chengdu University of Science and Technology
Chengdu 610065, China.

J.A.A. Jones
Institute of Earth Studies
University of Wales
Aberystwyth, SY23 3DB, UK.

Kang Ersi
Lanzhou Institute of Glaciology and Geocryology
Academia Sinica
Lanzhou 730000, China.

I. Kayane
Institute of Geoscience
University of Tsukuba
Kariya 3-123
Ushiku 300-12
Ibaraki
Tsukuba, Japan.

G.N. Kravchenko
Institute of Geography
Russian Academy of Science
Staromonetny 29
Moscow 109017, Russia.

A.N. Krenke
Institute of Geography
Russian Academy of Science
Staromonetny 29
Moscow 109017, Russia.

Hsiang-Te Kung
Department of Geography
University of Memphis
Memphis,
Tennessee, TN 38152, USA.

Herbert Lang
Swiss Federal Institute of Technology (ETH)
CH-8057 Zurich, Switzerland.

L. Yu Lin
Department of Civil Engineering
Christian Brothers University
Memphis
Tennessee, TN 38104, USA.

Liu Changming
Institute of Geography and United Research Centre for Water Problems
Academia Sinica
Building 917 Datun Road
Anwai
Beijing 100101, China.

Liu Tianchou
Bureau of Hydrological Water Resource Survey of Tibet
Lhasa
Tibet, China.

S. Malasri
Department of Civil Engineering
Christian Brothers University
Memphis
Tennessee, TN 38104, USA.

T.A. McMahon
CRC for Catchment Hydrology
Department of Civil and Environmental Engineering
University of Melbourne
Parkville,
Victoria 3052, Australia.

K. Mori
Department of Geography
Mie University
1515 Kamihama
Tsu
Mie 514, Japan.

Atsumu Ohmura
Swiss Federal Institute of Technology (ETH)
CH-8057 Zurich, Switzerland.

R.M. Quintela
Biometeorological Research Center
National Scientific and Technical Research Council (CIBIOM -CONICET)
Serrano 669
Buenos Aires 1414, Argentina.

Joe M. Russo
ZedX Inc.
P.O. Box 404
Boalsburg
Pennsylvania, PA 16827-404, USA.

O.E. Scarpati
Biometeorological Research Center
National Scientific and Technical Research Council (CIBIOM -CONICET)
Serrano 669
Buenos Aires 1414, Argentina.

Lei Shi
SeaSpace Corporation
San Diego
California, CA 92126, USA.

Shi Yafeng
Lanzhou Institute of Glaciology and Geocryology
Academia Sinica
Lanzhou 730000, China.

L.B. Spescha
Agronomy Faculty
University of Buenos Aires
Buenos Aires, Argentina.

Q.J. Wang
CRC for Catchment Hydrology
Department of Civil and Environmental Engineering
University of Melbourne
Parkville,
Victoria 3052, Australia.

Piotr Werner
Faculty of Geography and Regional Studies
University of Warsaw ul. Krakowskie
Przedmiescie 30
00-927 Warsaw, Poland.

P.H. Whetton
CSIRO Division of Atmospheric Research
Private Bag 1
Mordialloc
Victoria 3195, Australia.

Ming-Ko Woo
Department of Geography
McMaster University
Hamilton, Ontario, L8S 4K1 Canada.

Masatoshi Yoshino
Institute of Geography
Aichi University
Toyohashi, 441 Japan.

John W. Zack
Meso Inc.
Troy
New York, N.Y. 12180, USA.

Lan-sheng Zhang
Department of Geography
Beijing Normal University
100875 Beijing, China.

1 PREDICTING THE HYDROLOGICAL EFFECTS OF CLIMATE CHANGE

J.A.A. JONES

Throughout the 1980s and 1990s, General Circulation Models (GCMs) have been predicting that global warming will cause significant changes in temperature and rainfall patterns and, by implication, in regional water balances, riverflows and resources. Transforming the 'climatic scenarios' produced by these GCMs into estimates of regional and ultimately catchment-scale water balances and runoff is a major concern of this book.

The evidence that rising concentrations of 'greenhouse gases' in the atmosphere will enhance the natural greenhouse effect and cause global warming over the coming century is still largely theoretical. Observational evidence indicates a reasonably clear rise in global mean temperature of 0.5°C this century, an overall rise in mean sea level of about 10 mm per decade, and most conclusively of all a rise in concentrations of the major greenhouse gas, CO_2, from 315 ppm when measurements began in 1958 to 360 ppm in 1990. But there is no proof that the temperature rise is linked to rising CO_2 levels; in fact in the period immediately following World War II when fossil fuel burning should have raised CO_2 levels markedly the world temperature was either falling or static. We could simply be seeing the results of natural factors that have pulled our climate out of the Little Ice Age that seems to have afflicted much of the northern mid-latitudes from c. 1350 to the beginning of the 19th century.

The reported sea level rise may also be explained by the thermal expansion of ocean waters caused by the natural warming trend, and perhaps by other factors, like the return of waters for long lost in the 'deadlock arc' of 'fossil' groundwater that is now being increasingly mined for human use. There is also a pressing need for better sea level data that is being addressed by the new global sea level monitoring network, GLOSS (Pugh, 1990).

There does, nevertheless, seem to be a broad feeling if not a consensus amongst scientists that the very marked warming of the last two decades owes more to fossil fuel burning and changes in atmospheric composition caused by human activities. Furthermore, when the observed CO_2 concentration curve is extrapolated, it suggests that concentrations will reach double the estimated pre-industrial baseline of 290 ppm by 2030-2050. The strongest evidence for global warming is that all the GCMs currently predict that this increase will cause marked warming with an average rise of about 2°C, increasing to perhaps 7°C in polar regions (largely through the demise of snowcover) and reducing to less than 1°C in equatorial regions. Although there has been a small but discernable trend towards less marked effects as these GCMs have been refined during

1

J.A.A. Jones et al. (eds.), Regional Hydrological Response to Climate Change, 1–20.
© 1996 *Kluwer Academic Publishers. Printed in the Netherlands.*

the 1990s (IPCC, 1990a and b, 1992, 1996a and b), the predicted rates of change are still some 4 times any that have occurred during the 20th century (cp. Wigley and Raper, 1992).

These GCM predictions are based on the *effective* doubling of CO_2 concentrations. It is conceivable that if other gases that act as more effective greenhouse gases increase significantly the date of effective doubling could be sooner. Perhaps of greatest concern is methane, CH_4, which is 30 times more effective than CO_2 and for which a major source is irrigated agriculture. Methane concentrations are increasing at 1% per annum, twice the rate of CO_2, and seem to be less controllable in principle. At the current state of technology, the burgeoning world population cannot feed itself without expanding irrigation, whereas it could survive by switching to alternative non-fossil fuel energy sources. Moreover, point emissions from urban and industrial sites are more controllable than non-point, agricultural sources.

1.1. Hydrological Consequences of Global Warming

There is an urgent need to predict the hydrological consequences of global warming. Despite large uncertainties in GCM predictions and even larger uncertainties in the results of feeding these GCM predictions into water balance and rainfall-runoff models, there is no doubt that there is the potential for climatic change during the next 50 years to have dramatic impacts upon regional water resources. There has been some professional reluctance amongst water managers to consider the implications whilst the acknowledged predictive errors remain so large, but Arnell et al. (1994) are in no doubt that the impacts of global warming on water resources may not only be very large, but that may also be felt within the time horizons of current planning.

This realisation comes at a time when countries in many parts of the world have only recently become aware of critical limitations to their water supplies even under the present climatic regime. Global surveys by the Worldwatch Institute of Washington (1984) and WMO/UNESCO (1991) focused concern on North Africa and the Middle East as some of the highest rates of population growth, at 3% or more per annum, threaten limited resources. Eleven of the 20 countries in the Middle East and North Africa already use over 50% of their renewable resources. Libya and most of the Arabian Peninsula now use over 100%, with the support of desalination and fossil groundwater. And population is due to double or even quadruple before stabilizing (Falkenmark, 1989). Southern Asia and China are somewhat better served, but with approaching 20% of overall renewable resources currently in use they already have intraregional problems (see Liu and Fu, Section II).

In many parts of sub-Saharan Africa, the incidence of drought appears to have been on the increase in recent decades and there are many amongst the Least Developed Countries whose economies are already highly sensitive to current climatic fluctuations. As the Brandt Report from the Independent Commission on International Issues (1980) said of these countries:

'Without irrigation and water management, they are afflicted by droughts, floods, soil erosion and creeping deserts, which reduce the long-term fertility of the land. Disasters, such as droughts, intensify the malnutrition and ill-health of their people and they are all affected by endemic diseases which undermine their vitality.'

Sir Crispin Tickell (1977, 1991), former British ambassador to the UN, has predicted that future climatic change will so strain agriculture and water resources in many of these Least Developed Countries that civil disobedience, political coups and massive international migration will occur.

Nevertheless, ironically, many parts of the Developed World may have to adapt more under global warming. National surveys in Canada and the USA in 1978 and 1981 both foresaw critical shortages before 2000. Even in England and Wales, the National Rivers Authority's water resources strategy indicates concern over approaching intraregional water shortages, without making any allowance for the probable effects of global warming (NRA, 1994). Many of these countries are facing a three-pronged problem:

1) They have largely come to regard their water resources as developable and as no real constraint to economic growth. They have therefore developed a culture of consumption in the sure knowledge that either God or the engineers will provide. To learn that this faith may be misplaced comes as a severe shock to the system. Unlike the LDCs, these countries have grown unused to hardship, and disruptions and shortages have the potential to create greater economic impact. Many are only just beginning, perhaps somewhat half-heartedly, to encourage frugal use and cut back on waste.

2) Many of the sophisticated water supply systems in these regions are already attuned to exploit *current* resources to the full. Some are overexploiting, as evidenced by the drastic lowering of groundwater levels at up to 1 m per annum in the Ogallala Aquifer beneath the American Great Plains.

3) Global warming is likely to affect water resources most in some mid-latitudes regions, as the subtropical high pressure belts shift northwards. They lie in the frontier zone between tropical and arctic circulations, where small changes in meridional temperature gradients, sea surface temperatures or the size of the Hadley Cells can result in major shifts in the tracks of rain-bearing depressions. Moreover, temperature changes here are predicted to be about twice as great as in the tropics. While saturation deficits remain large enough, warmer air and sea surface temperatures will encourage greater evaporation. In contrast, in many parts of the humid tropics, low saturation deficits will constrain increases in evaporation.

Mid-latitude glaciation and seasonal snowcovers are also at particular risk. Most alpine glaciers are likely to suffer more intense ablation. For some, this will be offset by increased precipitation. But for many areas, there is likely to be an increase in the ratio of liquid to solid precipitation, which will help reduce glacial accumulation and the duration of snowcover.

By no means all countries will be 'losers'. But positive changes in the annual water balance are not always for the better and can bring their own problems. Increased flood hazard may be one. This is also likely to present problems in some areas afflicted by negative changes, where a shift towards more localised convective rainfall perhaps falling on parched, cracked and hydrophobic ground may increase flood runoff. Increasing contrasts between the seasons may also complicate water management, even though total annual resources may increase. Such increased seasonality could be caused, for example, by higher summer evapotranspirational losses, by more complete shifts of depression tracks away from the area during summer, and by higher sea surface temperatures persisting into the early winter, feeding depressions with more moisture and so fostering higher winter rainfall.

1.1.1. REVIEWS OF GLOBAL AND REGIONAL RESPONSE IN THIS VOLUME

Kayane (Section I) begins the reviews with a critical look at what is and is not understood. He pays particular attention to the uncertainties, which are not always explicitly recognised. In terms of historical processes, there is still considerable uncertainty about the causes of the Pleistocene glaciations and on the causes and processes involved in Postglacial sea level rise. One might add here the debate as to the reasons why the CO_2 and methane concentration curves derived from analyses of polar ice cores so perfectly match the temperature curve derived from ocean sediments over the last 160 000 years. Orogeny and increased erosion in Central Asia may have been a mechanism causing decreased CO_2 in the atmosphere by locking it up in ocean sediments according to Raymo and Ruddiman (1992), but why should CO_2 concentrations increase at the beginning of an interglacial when the burgeoning biosphere should be a sink: is its effect masked by chemical release from warmer ocean surfaces, and to what extent is the temperature curve being driven by CO_2 and methane, or *vice versa*?

In the present century, there seems to be contradictory evidence which indicates decreasing global precipitation at the same time as warmer sea surface and rising air temperatures that should be increasing evaporation and rainfall. There are also significant gaps in our understanding of mankind's recent effect upon runoff and atmospheric exchange processes through land surface modification and water engineering.

Kayane notes the valuable role that lake level studies have had over the last two decades in illuminating the history of climate change. And he suggests that more extensive isotope analyses could shed much needed light on the souces and sinks within the hydrological cycle and especially upon anthropogenic interference.

The regional reviews by Woo (North America), Jones (Europe) and Liu and Fu (China) provide up-to-date syntheses of work, which are supplemented in many cases by their own investigations. There is general agreement that mid-latitude areas south of about 50-55° N are likely to experience a net drying effect, associated with the poleward extension of the subtropical high pressure belt. North of this, however, there should be notable increases in precipitation at least in North America and Europe, as depression tracks migrate northwards and warmer sea surface temperatures press polewards in the North Atlantic and North Pacific and increase zonal oceanic evaporation. Each of the reviewers pays special attention to questions of methodology and the data sources. Most, but by no means all, trends in recent instrumental records seem to suggest that changes are already underway that are consistent with global warming predictions.

1.2. Predicting hydrological response

Four broad approaches have been taken to predicting hydrological response:
     1) analysis of instrumental data and extrapolation of trends or patterns of variability,
     2) temperature sensitivity models, in which instrumental records are used typically to calculate the rates of change in regional parameters (especially annual precipitation) per degree of change in global temperature,
     3) use of palaeoclimatic reconstructions as analogues, and
     4) using GCM calculations of temperature, precipitation, and possibly soil moisture either directly as they are or indirectly via some transformation as input into water balance or more occasionally into conceptual rainfall-runoff models.
     Each of these approach has certain advantages and disadvantages, and some are better for answering certain questions than others.

1.2.1. ANALYSES OF INSTRUMENTAL RECORDS

Analyses of past records may be used simply to discover and extrapolate trends, to calculate temperature sensitivity rates, or in a more sophisticated way to downscale GCM predictions, which do not have sufficient spatial resolution or are regarded as not being sufficiently reliable for precipitation.

Proving a statistically significant trend in recent data is useful evidence of environmental change. Inferring climatic change from this evidence is most straightforward in the cases of temperature and precipitation. However, trends in evapotranspiration and runoff could also be related to changes in vegetation, landuse and land management. Often climatic and land surface factors are combined and difficult to disentangle. Indeed, even temperature and precipitation could be affected by non-climate related changes in land surface properties: the team at the University of East Anglia Climatic Research Unit (CRU) took great pains to avoid the effect of urbanisation in their calculation of the accepted 0.5°C warming this century (Warrick et al., 1990), and enhancement of local precipitation by urban effects is also well documented (Jones, 1996).

Extrapolating any trends into the future has its difficulties, as always. It assumes no sudden change in the causal factors, such as shifts in average depression tracks. Confidence limits are also frequently enlarged by having to base trend calculations upon relatively short runs of records. This is often the case for discharge measurements. It is less of a problem for precipitation records, and because of this stochastic hydrology has developed a range of techniques to predict future runoff from extrapolations based on the parameters of the precipitation record. Most such applications have been undertaken in the belief that climate will not change over the projected planning period: indeed, future climate change is one of the main factors that could invalidate planning prognoses made on this basis. But it is clearly feasible to adapt these methods to extrapolate *ongoing* climate change.

The other major problem for prognoses based on past data is that the past provides only a sample of an infinite population of events and sequences. One way around this problem is to fit a theoretical distribution to the data and to sample from this to assess the risk of extreme events. Again, this has mainly been used for stationary series, i.e. assuming no climatic trends. But Gregory *et al.* (1990) perturbed the parameters of the rainfall probability density function they had established for the UK in order to mimick expected climate change.

1.2.1.1. *Analyses of instrumental records in this volume*
Numerous examples of trend analyses from around the world can be found in the succeeding chapters of this volume, beginning with the worldwide analysis of discharge records from 142 catchments by Chiew and colleagues. Chiew *et al.* (Section I) explicitly selected natural rivers with catchment areas of at least 1000 km^2, to give a reasonable representation of a large area, and with over 50 years of record. They conclude that there is no clear evidence yet of sustained trends worldwide, although some individual regions do seem to show unusual concentrations of statistically significant trends.

Consistent negative regional trends in precipitation are found in China and Central Asia by Liu and Fu in Section II, Yoshino in Section III, Zhang in Section IV, and Liu in Section V. In Japan, Mori (Section V) found no trends in rainfall, but increases in potential evaporation have caused a decline in the calculated runoff. In Argentina, Quintela *et al.* (Section V) find a positive trend in precipitation in the north, but none elsewhere, which is not consistent with GCM predictions for trends under global warming.

Both Yoshino in Tibet and Daultrey in Ireland (Section III) look beyond the precipitation record and relate its variability to indices of shifts in atmospheric circulation. In both areas, the authors find significant links with El Niño/Southern Oscillation (ENSO) events. In Ireland, Daultrey also detects contributions from the North Atlantic Oscillation, and from sunspot numbers and the Quasi-Biennial Oscillation. Beyond possibly improving long-range weather forecasting, the particular value of establishing such linkages is that they might be a means of improving or downscaling GCM global warming scenarios, to give them local detail at the catchment scale based on predicted shifts in global circulation. Interestingly, the UK Hadley Centre has begun to incorporate ENSO in its transient GCM.

Relating precipitation to General Circulation patterns makes sound physical sense, provided suitable classification systems can be devised for the circulation patterns and provided links can be statistically proven. We will return to this topic under the heading of downscaling GCM output.

1.2.2. TEMPERATURE SENSITIVITY MODELS

Attempting to correlate changes in recorded precipitation, or any other hydrological component, with changes in global temperature is inherently less physically sound. Global temperature is one stage further removed from the hydrological endproduct than General Circulation patterns, and its effects will commonly be expressed through the medium of airflow patterns.

Such temperature sensitivity models were used notably by Groisman (1981), as reviewed by Jones (Section II). Groisman's work was essentially before the second generation of GCMs provided better regional predictions. It is a crude approach, but, nevertheless, it has its place, especially as an extra method for cross-checking GCM predictions and determining whether current trends are in line with them. Again, it might form a basis for downscaling.

A much sounder basis for the latter, however, appears to be establishing sensitivities to *local* temperature changes, rather than *global*. This is particularly justifiable on the grounds that GCM predictions of regional temperature changes are accepted as being considerably more reliable than regional precipitation.

1.2.3. PALAEOCLIMATIC ANALOGUES

Concern about the limitations of GCMs, particularly amongst Russian meteorologists led by M.I. Budyko and amongst certain Quaternary scientists elsewhere, have led to the promotion of palaeoclimatic reconstructions as analogues for future climate change. The Cretaceous warm period 65-220 Ma BP that was 10-20°C warmer, the Pliocene Optimum 3-4 Ma BP 4°C warmer, or the Hypsithermal, 6-9 ka BP, 1-2°C warmer provided possible analogues for a warmer globe.

Budyko (1989) argues that this approach is superior because it is using the real world as the model, incorporating all its subsystems, whereas computer models are simplified abstractions of reality. Geomorphologists have argued that sedimentary records of past flood events may be used to reconstruct the magnitude-frequency curves for hydrological events during such warm periods. If this were possible, then it could provide a more direct method than trying to use hydrological models to predict the response of a specific river from current GCM scenarios.

In practice, there are numerous problems with both the climatic and the geomorphic arguments (see Jones, Section II). One centres on the dictum that history never repeats itself. If the causes of two particular climatic changes are not identical, then there could be significant differences in airflow patterns affecting a specific area. This is especially relevant to climatically sensitive regions like Western Europe, where sea surface temperatures and planetary wave patterns play such a crucial role. Budyko (1989) does acknowledge that only the warming of the 1970s and 1980s may be fairly confidently ascribed to anthropogenic influences on the greenhouse effect. Newson and Lewin's (1990) argument that the flood-prone period around the demise of the Little Ice Age in the early 19th century could be a possible analogue for future global warming fails to appreciate the considerably different state of midlatitude circulation patterns and sea surface temperatures at that time.

A second problem centres on the reconstruction of the palaeoclimates and palaeohydrologies and the inexact nature of the signal in palaeoclimatic reconstructions (Huntley, 1990). Fairly reliable methods are available for reconstructions extending back in the order of centuries. Jones *et al.* (1984) offer a good example of reconstruction of water balances and inferred riverflows in Britain, based on tree-ring analyses. But beyond this, we have to rely largely upon rather inexact correlations between floral and faunal assemblages and general climatic characteristics to reconstruct temperature and humidity. For floods, we have an incomplete sedimentary record, some floods will have washed away the evidence of their predecessors, combined with a variety of competing formulae for inferring palaeovelocities from sediments and then the difficulty of reconstructing channel cross-sections in order to calculate discharges. In most cases, climatic and geomorphic reconstructions also commonly have to rely upon a limited sample of sites. These sites could be biassed towards locations that offer the best preservation.

Nevertheless, palaeo-reconstructions will continue for there own intrinsic value in helping to understand past environments, and methods are improving rapidly (Starkel *et al.*, 1990; Gregory *et al.*, 1995). They also offer yet another tool for cross-checking the GCMs.

1.2.4. USE OF GCM SCENARIOS

GCMs are three-dimensional mathematical computer models representing the physical processes in the atmosphere and the interactions between the atmosphere, the oceans and the land surface. They are developments of weather forecasting models that, because they are predicting so far ahead, have to incorporate a considerable amount of detail on surface features and feedbacks. Ideally these should include realistic models of ocean circulation and the dynamics of the cryosphere and biosphere, although progress with the latter is still at a very early stage. The respiration and population dynamics of marine phytoplankton are now thought to be crucial to CO_2 balance in the atmosphere, yet these are poorly monitored and understood.

Until recently, research has concentrated on *equilibrium* runs of these GCMs, i.e. a concentration of greenhouse gases equivalent to double the pre-industrial concentration is input into the GCM and the model runs until equilibrium is achieved. Some of the latest runs have improved upon this by modelling *transient* response, in which the gases are allowed to increase gradually according to an accepted rate. Because of the importance of feedback processes between atmosphere and ocean during the transient timescale, this requires parallel improvement in the modelling of the oceans: 50-70 m 'ocean slabs' that were adequate for equilibrium models must be replaced with submodels for three-dimensional oceanic circulation in Coupled Atmosphere-Ocean GCMs (AOGCMs). This is clearly scientifically superior, but there are still a few problems, like the ocean still lagging behind atmospheric change at the end of the simulation (cp. Jones, Section II).

Notwithstanding all the other methods and the arguments in their favour, GCMs are the embodiment of the highest possible aim in science, that is, to predict outcomes on the basis of the laws of physics. Their main detraction is that in order to achieve this they must simplify process and scale in both space and time. They are further hampered by gaps in our current knowledge of the total system. The third critical limitation is that reruns are extremely costly, not least in the amount of computer time that is required even with the current level of simplification. A transient run of the UK High Resolution model (UKHI, or in transient mode UKTR), which is an AOGCM, currently takes 3 months of computing to cover 150 years. Not long ago, it took 9 months. Lack of reruns means that statistics of variability or error measurements are lacking. We should not be too critical of this, since few other techniques can offer them either. But of course variability is of key interest for hydrology and water resources.

One encouraging aspect of GCM output during the 1990s has been a reasonable level of convergence in the patterns, and latterly in the amount, of warming predicted by most of the models, in spite of a wide range of approaches to simplification or 'parameterisation'. Convergence is not so apparent in precipitation predictions, as illustrated in Figures 1.1 and 1.2. A common and realistic solution to nonconvergence has been to calculate composite scenarios from a number of GCMs, as in the IPCC reports and in Warrick *et al.* (1990).

Until very recently, the scenarios used by hydrologists have tended to combine composite estimates of temperature change with a range of precipitation scenarios that brackets output from the various GCMs, and maybe adds a little extra for safety and symmetry. This is the type of approach recommended by the British Institution of Civil Engineers (Binnie, 1991) and used by the Institute of Hydrology in most of its work (e.g. Arnell, 1992). Holt and Jones (Section V) used the same scenarios.

Figure 1.3 illustrates the variety of approaches that have been taken to scenario construction, according to Viner and Hulme (1994). This includes a number of approaches to downscaling from GCM outputs at grid intervals of a few degrees of latitude and longitude to the scales normally required for hydrological or ecological applications.

1.2.4.1. *Upscaling and downscaling for GCM scenarios*

The need for *downscaling* current GCM output is widely recognised. But as Bass *et al.* (Section I) demonstrate in their detailed discussion of the problems, *upscaling* is also required to transform process-based models of soil-vegetation-atmosphere interactions (SVAT models) from the microscale, at which the processes operate and the models have been developed, to the mesoscale at which they might be incorporated in high resolution GCMs.

As computational power increases GCMs are being developed with higher spatial resolutions. Resolutions of up to 8 x $10°$ or 4 x $7.5°$ a decade ago have been replaced by 2.5 x $3.75°$. But it is a slow process and hydrological requirements are running ahead of it.

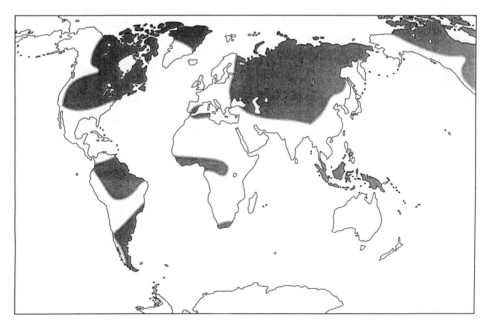

Figure 1.1. Areas of the world predicted to receive lower annual precipitation according to early American GCM equilibrium simulations. After Myers (1985).

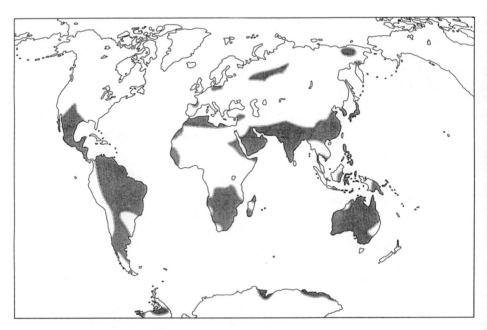

Figure 1.2. Areas of the world predicted to receive lower annual precipitation according to UKTR, a transient experiment incorporating the effects of sulphate aerosols. After The Hadley Centre (1995).

PROCESS-BASED	Advantages	Disadvantages
1. Single GCM	Attempt to simulate fully the 3-D atmosphere and ocean system	GCMs contain systematic errors in their simulation of present-day climate No assessment of uncertainty
2. Multiple GCMs	Allows inter-model differences to be compared	Assumes that the spatial pattern of change remains the same
3. Simple models	Can operate easily under a range of emissions scenarios Computationally efficient Allows the effects of uncertainty to be explored	Models the climate in one-dimension only
4. Nested HRLAMs	High spatial resolution of regions can be simulated Localised topography can be modelled	Driven by data from coarse resolution GCMs Development of HRLAMs still in its infancy
5. Composite model	Utilises GCM experiments and uses state-of-the-art high resolution one-dimensional models Allows pseudo-transient climate change scenarios to be constructed Climate response to a range of emissions and scenarios and climate sensitivities can be easily calculated	The method assumes that the spatial pattern of change does not change Assumes that different greenhouse gases produce the same climate change 'fingerprint'

LINKED	Advantages	Disadvantages
1. Down-scaling	Allows local climate detail to be obtained	Assumes constant relationships between climates at different space scales
2. Stochastic/synthetic	Allows a long time series of daily climate variables to be constructed	Requires good quality historical climate data sets Difficult to employ for variables other than temperature

EMPIRICALLY-BASED	Advantages	Disadvantages
1. Instumental analogues	Use historical records comprising of a number of variables	The climate will always be in a transitional state Definition of a warm period is difficult to determine
2. Palaeo-analogues	Long time-periods can be represented	Does not take into account different forcing mechanisms
3. Spatial analogues	Simple and easy to use Allows for easy perception and comparison	No account is made of spatial location Very simplistic

Figure 1.3. Methods of constructing climate change scenarios, according to Viner and Hulme (1994).

In recent years, the CRU has worked with the UK Hadley Centre to produce another type of scenario in which experimental results from the UKHI GCM are combined with a simple one-dimensional higher resolution model (STUGE), which requires far less computer processing time (Viner and Hulme, 1994). This combination enables more reruns and higher spatial resolution than would be possible for the full UKTR, which has a 20-level atmosphere and a 17-level ocean with an ocean circulation model and a thermodynamic sea ice model. Arnell and Reynard (1993) worked with transient scenarios calculated by combining equilibrium output from UKHI with the one-dimensional CRU's STUGE model interpolating to the 40 x 40 km grid interval of the Meteorological Office's Rainfall and Evaporation Calculation System (cf. Jones, Section II). Recent work has used STUGE to interpolate results to a 10 x 10 km grid, which has also been used to construct a baseline database of mean monthly climatological data for 1961-1990 in the UK (Viner and Hulme, 1994). The STUGE model can also be used to produce transient scenarios for alternative rates of emission/accumulation of greenhouse gases.

A major step forward in downscaling GCM output has been achieved in recent years by *nesting* higher resolution models within coarser global hosts. Publishable results have now been achieved in North America and Europe using this strategy. Dickinson *et al.* (1989) and Giorgi (1990) reported on the results of nesting one of these 'high resolution limited area models' (HRLAMs in Figure 1.3) within the National Centre for Atmospheric Research (NCAR) GCM, which proved sucessful. The 50 km or 0.5° resolution in the European Window nested within UKHI not only models important topographic controls like the Alps far better, it also provides output that is almost as detailed as the MORECS net, which is used on a daily basis by the UK water industry for operational decision-making. The Regional Model which operates within the European Window is capable for the first time of resolving the low pressure area in the lee of the Alps, which is so important to large-scale flow patterns, cyclogenesis and precipitation patterns over a wide area (Murphy *et al.*, 1993). Early work with the European Window showed some disparities between the predictions of the coarse resolution host and the Regional Model at the edge of the window, but this has since been improved (Bennetts, pers. comm.). Even higher resolution may be achievable by linking to larger scale Geographical Information Systems (GIS) (Parr and Eatherall, 1994). However, the success of all these local or regional higher resolution models is still dependent upon reliable initial conditions being supplied by the host GCM.

A third alternative method of downscaling that has proven popular is using regression analyses to provide local transfer functions. Typically, these have established a statistical relationship between local climate large-scale climate using current climatic data, which is then used as the transfer function to downscale GCM predictions. Both Karl *et al.* (1990) and Wigley *et al.* (1990) took this approach with output from the Oregon State University (OSU) GCM, respectively in North America and Europe. Palutikof et al. (1992) successfully tested regression equations to predict local temperature anomalies in the Mediterranean from mean temperature anomalies, using the CRU composite GCM scenarios (Warrick *et al.*, 1990).

1.2.4.2. *Approaches to upscaling and downscaling in this volume*

Bass *et al.* (Section I) begin with a thorough review of the work of the International Geosphere-Biosphere Programme's Biological Aspects of the Hydrological Cycle (IGBP-BAHC) project, giving detailed examples of the adopted approaches. Upscaling may be achieved simply weighting gridcell values according to the aggregated percentage of cover, or by developing statistical parameters to represent the subgrid variability. These approaches may be used to upscale ecological phenomena from the patch scale to mesoscale grids. Downscaling may be achieved by stochastic or dynamic approaches, or they suggest a combination of the two, downscaling to a certain level with a mesoscale atmospheric simulator and completing the work with a statistical interpolation technique to reach the desired level.

Three methods of statistical interpolation are described in Section III for precipitation. Kung *et al.* discuss the application of neural networks for 'learning' and reproducing local rainfall patterns. Gutry-Korycka and Wetner (Section III) take a more conventional approach and use established spatial statistical methods, kriging and distance-decay function interpolation, to produce maps of temperature and precipitation in Poland on a 6x6 km grid, based on macroscale scenarios from the Geophysical Fluid Dynamics Lab (GFDL) and Goddard Institute of Space Studies (GISS) GCMs. In contrast, Robinson proposes downscaling by using the relationships between local precipitation and General Circulation patterns, which are better predicted by GCMs. Robinson deals with the probabilities of rainfall events, but this method could be extended to provide rainfall parameters like intensity-duration-area statistics that are crucial to storm runoff calculations but are not yet available from GCMs.

1.2.4.3. *Methods used for calculating water balance and runoff in this volume*

Water balance and runoff calculations have mostly been undertaken using either lumped conceptual flow generation models or a combination of cruder evaporation loss equations and the hydrological continuity equation (Runoff = Precipitation - Evaporation). The resolution and reliability of the input data do not merit more sophisticated modelling at the present time.

Calculating evapotranspiration is one of the most difficult aspects of hydrological modelling at present, because of the lack of reliable GCM output on local windspeeds, advection, relative humidities and insolation. Many have turned to the simple statistically derived temperature and precipitation based formula of Turc (1954). This is simpler than its main temperature-based rival the Thornthwaite (1948) formula. Liu and Woo (Section V) find that it compares well with the physically-based Penman formula in their tests in China. They use the Turc formula in conjunction with standard water balance equations. Deng and Hou (Section V) test the Turc formula alongside three alternative methods of deriving water balance from temperature, including nonlinear regression, Langbein's empirical formula and a Grey Box model. They found that these techniques all gave quite comparable forecasting success rates and they extol their virtues for applications where few data are available.

Holt and Jones (Section V) take an alternative approach and use local transfer functions to calculate runoff directly from temperature and precipitation. Temperature changes are taken from the CRU composite equilibrium scenarios and used as input into local regression equations. These equations were calculated from the MORECS database, which includes temperature, precipitation, actual and potential evapotranspiration and 'effective rainfall'. MORECS calculations for effective rainfall include a two horizon soil moisture model. And this effective rainfall is taken as the best estimate of local runoff. Where possible, effective rainfall is compared with actual discharge records in the UK Surface Water Archive, and adjusted if necessary by a regression-based calibration function. This approach therefore combines evapotranspiration, water balance and runoff calculations in one simplified approach, which seems valid at the very least for sensitivity studies. It may have a broader validity, if it can be argued that transfer functions derived from local temperature, rainfall and runoff are more likely to subsume local climatic characteristics, like the amount of advected heat, sunshine durations and relative humidity, than globally-derived formulae which do not explicitly include these terms, especially in western Britain where advection and cloud cover are particularly important. After all, the Turc formula is based on just 254 locations covering a global range of climates and, like the rival Thornthwaite formula, it can produce nonsensical results. For example, in Turc's formula, if precipitation is zero then evaporation is zero whatever the temperature. In Thornthwaite's, if *mean* monthly temperature falls subzero, then there is no evaporation in that month. Turc may conceivably have some advantage over Thornthwaite in that it includes precipitation, which may bear some relationship to the crucial factor of atmospheric humidity.

Chiew *et al.* (Section V) are amongst the few who have used a hydrological simulation model approach (cp. Bultot *et al.*, 1988; Arnell and Reynard, 1993; see review in Jones, Section II). Chiew *et al.* use a 19 parameter conceptual daily rainfall-runoff model. The parameters are optimised with current data and then left at these settings and run for scenarios derived from GCM output and from a stochastic weather generator. The weather generator enables exploration of the effects of changes in the timing and frequency of events (cp. Bass *et al.*, Section I).

1.2.5. PREDICTING THE RESPONSE OF THE CRYOSPHERE

The general response of the ice and snow bodies to global warming has been extensively researched and modelled as an intrisic part of the work on climatic prediction. A major factor in the above average temperature rises expected in arctic regions is the effect of an anticipated reduction and demise of large sections of snow and ice cover on land and sea upon regional radiation balances. Meltwaters are also expected to be a major factor in sea level rise, although there is perhaps more disagreement on this point. There is general agreement that non-polar snow and ice is most clearly at risk, and this may in net terms be the second largest contributor to sea level rise after the thermal expansion of the oceanic water body (Warrick and Farmer, 1990).

There is less agreement about polar ice masses. It is possible that some increase in snowfall caused by warmer seas and less ice ice cover could increase ice accumulation. But higher temperatures around glacial margins will increase ablation. Certainly activity rates will increase. Modelling at the CRU suggests that the Greenland Icecap will undergo a net increase in ablation, whereas Antarctica could see net accumulation, perhaps almost balancing out the effect of Arctic icemelt on global sea level (Warrick and Farmer, 1990). However, the Antarctic high pressure belt is likely to block the intrusion of snow-laden depressions far into the continental interior and so limit accumulation to the continental margins.

The melting of mid-latitude ice bodies will also have a marked impact upon riverflows in many regions with important implications for water resources and flood risk. Meltwater is a major source of flooding on the Rhine (cp. Jones, Section II). More than a third of the world's irrigation water is snowmelt, and meltwater is a major source for 28 countries. Icemelt from alpineor warm-based glaciers comprises about 1/200th of world river runoff. 15 per cent of the discharge of the Columbia River as it crosses from Canada into the US is derived from the Columbia Icefield, which is also a major source of runoff for the Mackenzie and Sakatchewan Rivers. Spring meltwater from this and numerous other alpine ice bodies in the Rocky Mountains is a vital source of water for agriculture in the Canadian prairies. Large volumes of icemelt also serve agriculture along the major rivers of the Himalayas. Indeed, Pakistan's national glacier inventory was undertaken as part of the planning exercise for the Tarbela Dam in the 1970s, because of the strong dependence of flow in the Indus on glacier meltwater. And glacial meltwaters provide indispensable sources for hydropower in Norway and the Alps.

1.2.5.1. *Predictions in this volume*

Woo (Section II) reviews work in northern North America and reports on his own calculations. He predicts that snowfall will probably increase at high altitudes, but not at low altitudes. Some climatically marginal glaciers may disappear, but larger ice bodies could benefit from higher snowfall. Melt events will probably become more frequent and ablation on the lower slopes will increase the total amount of meltwater runoff for some time. But he finds some disagreement between predictions based on output from GISS and GFDL: the former suggests generally increased amounts of snowfall, but the latter suggests that increases during the winter will be counterbalanced by reduced falls in summertime.

Permafrost and freshwater ice covers are especially important for hydrology in Canada. The divide between continuous and discontinuous permafrost will clearly migrate northwards with major implications for hydrological processes. With the deepening of the active layer, infiltration capacities will rise and there will be a consequent increase in soil and ground water storage. This will reduce overland flow and alter the pattern of spring freshets and summer runoff.

Impacts on freshwater ice covers are also likely to affect runoff. Applying Stefan's heat balance equation, Woo predicts reductions of up to 1 m in river and lake ice thickness. This will reduce ice cover durations by a few weeks. To the present author this suggests two things. First, many spring floods are created by ice jams, so reduced ice thicknesses are likely to reduce the risk of this type of flooding, especially if this is accompanied by the reductions in peak discharges envisaged by Woo. Secondly, thinner ice and more winter snowfall are likely to increase the amount of white ice or slush ice formation. White ice is formed when the hydrostatic water level stands above the upper surface of the ice due to the weight of snowcover and cracking of the ice allows water to flood the overlying snow (Jones, 1969; Adams, 1981). It commonly accounts for half the ice cover in southern Ontario and been reported at up to 100% of the cover in Michigan. But it is a highly variable component: 11 years of measurements at Knob Lake in subarctic Quebec showed it varying from 13% to 57% of total ice. Jones correlated phases of white ice formation with large temperature fluctuations (diurnal ranges exceeding 12.5°C), which occur mainly in early and late winter.

Global warming is therefore likely to increase the potential for white ice formation in northern Canada, and perhaps shift the zone of optimum formation northwards. This could mean that overall ice thicknesses are not reduced as much as Stefan's equation suggests, if at all. It does not, however, alter the net hydrological effect so much. White ice tends to melt more rapidly than the original 'black' ice, because it is composed of smaller crystals, and it forms ice floes that have less compressive strength during breakup. So higher percentages of white ice should reduce the potential for ice jams. Jones (1969) also calculated that in lake basins it could contribute to lower meltwater discharges in a rather novel way: although Archimedes' principle should mean that ice formation will not displace water from the catchment, if the ice sheet is prevented from floating freely, for example by a snow overburden, then it may, provided the sub-ice stream network remains intact. Jones calculated that 14% of winter discharge from the Knob Lake basin could be thus generated, which would leave more lake storage available to dampen the spring flood wave.

Turning to Europe, Jones (Section II) reviews work which suggests a variety of responses. In the south, higher temperatures and a general tendency for lower snowfall should accelerate glacier retreat, reduce the amount and duration of snowcover, cause spring freshets to occur earlier, and to increase the risk of spring flooding in the initial stages. This initial increase in flood risk could be supported by increased rates of melting. Icemelt would be helped by lower albedoes as snowcover disappears earlier. But this would be followed after a few decades by a notable decline as glacial volumes shrink. However, altitude, location and the degree of glacierisation all affect the nature and timing of response. The simulation experiments of Bultot et al. (1994) show that some alpine basins may display reduced meltwater runoff from the outset. In Scandinavia, the expected long-term increases in precipitation will not necessarily mean more snowmelt runoff, as an increasing proportion of winter precipitation is expected to take the form of rain. Glaciers are still expected to retreat and snowcover duration to reduce.

Other predictions emerge from surveys of current precipitation trends in Central and Eastern Asia. The rising temperatures and reduced annual precipitation found by Yoshino (Section III) in Tibet suggest that glaciers and annual snowcover may be threatened by lower accumulation rates. This reduction in annual precipitation is supported by observations by Liu (Section V) on lake levels in Tibet, and by analysis of meteorological records by Zhang (Section III). Liu and Woo (Section II) and Zhang (Section III) all report that Chinese glaciers have been declining for most of this century. Zhang relates this particularly to the drying trend and points out that this this is a continuation of a trend that has lasted for the past 500 years. They all predict continued reductions in glacier mass balance and rising equilibrium lines.

In contrast, Krenke and Kravchenko (Section III) based their calculations of future trends upon analyses of current climatic records by Kovaneva (1982) which suggest that the Pamirs and Tianshan Mountains have cooler summers in years when global temperatures are higher. Their calculations of glacier mass balance suggest a lowering of the equilibrium lines and a reduction in meltwater discharges. Increases in the accumulation zone will out-balance some increase in melt rates on the warmer lower slopes.

Aizen and Aizen (Section III) demonstrate the sensitivity of cryospheric balances in Central Asia to General Circulation patterns and explore possible feedbacks in which Eurasian snowcover acts as a regulator of atmospheric circulation patterns. Snow-covered area reaches a maximum at the end of an epoch of strongly zonal winds and reaches a minimum at the end of an epoch of strong meridional flow.

Kang et al. (Section III) report on the development of a meltwater runoff model that incorporates a tank model for runoff and storage and takes a much more process-based view of meltwater runoff than generally hitherto. The model is the result of collaboration between the Langzhou Institute of Glaciology and Geocryology and the Swiss Federal Institute of Technology (ETH), Zurich. Tests of their model in a basin in the Tianshan suggests that runoff there is mainly controlled by the heat balance rather than precipitation receipts.

1.3. The structure of the book

The following chapters of this book cover four main aspects: 1) techniques for data synthesis and predictive modelling, 2) regional applications, 3) analyses of trends in recent instrumental records, and 4) critical reviews of regional predictions.

The chapters have been divided into 5 sections beginning with questions relating to the general sensitivity of the global hydrosphere to thermal and biospheric changes. This is followed by detailed analytical reviews of regional research work in North America, Europe and Eastern Asia, and by systematic sections on precipitation, the cryosphere, and water balance and management implications. Each section is prefaced by a bulleted summary of the main conclusions from each chapter. Broadly speaking, chapters on methodology and modelling precede those which analyse trends in recent data.

The final chapter draws together some of the main themes that emerge from this work and discusses the priorities for future research.

Acknowledgments

I would like to thank Dr David Carson, Director of the Hadley Centre for Climate Prediction and Research, and Dr David Bennetts, former Research Coordinator at the Centre, for permission to use output from UKHI/UKTR and for valuable discussions. Thanks also to Dr David Viner of the Climate Impacts LINK programme, University of East Anglia, for publications and for permission to reproduce Table 1.1.

References

Adams, W. P. (1981) Snow and ice on lakes, in D.M. Gray and D.H. Male (eds.) *Handbook of Snow: Principles, Processes, Management and Use*, Pergamon, Oxford, 437-74.

Arnell, N.W. (1992) Impacts of climatic change on river flow regimes in the UK, *Journal of Institution of Water and Environmental Management* **6**, 4, 432-442.

Arnell N.W., Jenkins, A., and George, D.G. (1994) *The Implications of Climate Change for the National Rivers Authority*, National Rivers Authority, Bristol, R & D Report 12, 94 pp.

Arnell N.W., and Reynard, N.S. (1993) *Impact of Climate Change on River Flow Regimes in the United Kingdom*, Institute of Hydrology, Wallingford, 129 pp.

Binnie, C.J.A. (ed.) (1991) *Policy Paper on: Securing Adequate Water Supplies in the United Kingdom in the 1990s and Beyond*, Institution of Civil Engineers, London.

Budyko, M.I. (1989) Climatic conditions of the future, in *Conference on Climate and Water*, Academy of Finland, Helsinki, vol. 1, 9-30.

Bultot, F., Coppens, A., Dupriez, G.L., Gellens, D., and Meulenberghs, F. (1988) Repercussions of a CO_2-doubling on the water cycle and on the water balance: a case study from Belgium, *J. Hydrology* **99**, 319- 347.

Bultot, F., Gellens, D., Spreafisco, B., and Schedler, B. (1994) Effects of climate change on snow accumulation and melting in the Broye catchment (Switzerland), *Climatic Change* **28**, 339-363.

Dickinson, R.E., Errico, R.M., Giorgi, F., and Bates, G.T. (1989) A regional climate model for the western U.S., *Climatic Change* **15**, 383-422.

Falkenmark, M. (1989) The massive water scarcity now threatening Africa - why isn't it being addressed? *Ambio* **18**, 2, 112-8.

Giorgi, F. (1990) Simulation of regional climate using a limited area model nested in a general circulation model, *J. Climate* **3**, 941-963.

Gregory, K.J., Starkel, L., Baker,V.R. (1995) *Global Continental Palaeohydrology*, Wiley, Chichester, 352 pp.

Gregory, J.M., Jones, P.D., and Wigley, T.M.L. (1990) *Climatic Change and its Potential Effect on UK Water Resources*, Report to the Water Research Centre, Medmenham, UK, Parts 1 and 2.

Groisman, P.Ya. (1991) Data on present-day precipitation changes in the extratropical part of the Northern Hemisphere, in M.E.Schlesinger, (ed.) *Greenhouse-gas-induced Climatic Change: a Critical Appraisal of Simulations and Observations*, Elsevier, Amsterdam, 297-310.

Hadley Centre (1995) *Modelling climate change, 1860-2050*, The Hadley Centre for Climate Prediction and Research, Bracknell, UK, 13 pp.

Huntley, B. (1990) Lessons from climates of the past, in J. Leggett (ed.) *Global Warming - The Greenpeace Report*, Oxford University Press, Oxford.

Independent Commission on International Issues (Brandt Commission) (1980) *North-South: a Programme for Survival*, Pan Books, London.

Intergovernmental Panel on Climate Change (IPCC) (1990a) *Climate Change. The IPCC Scientific Assessment*, J.T. Houghton, G.J. Jenkins and J J Ephraums (eds.), Cambridge University Press, Cambridge.

Intergovernmental Panel on Climate Change (IPCC) (1990b) *Climate Change. The IPCC Impacts Assessment*, W.J. McTegart, G.W. Sheldon, D.C. Griffiths (eds.), Australian Government Publishing Service, Canberra.

Intergovernmental Panel on Climate Change (IPCC) (1992) *Climate Change 1992. The Supplementary Report to the IPCC Assessment*, J.T. Houghton, B.A. Callander, and S.K. Varney (eds.), Cambridge University Press, Cambridge.

Intergovernmental Panel on Climate Change (IPCC) (1996a) *Climate Change 1995: The Science of Climate Change,* J.T. Houghton, L.G. Meiro Filho, B.A. Callander, N. Harris, A. Kattenburg, K. Maskell (eds.), Cambridge University Press, Cambridge, 564 pp.

Intergovernmental Panel on Climate Change (IPCC) (1996b) *Climate Change 1995: Impacts, Adaptations and Mitigation of Climate Change: Scientific-Technical Analyses,* R.T. Watson, M.C. Zinyowera, R.H. Moss (eds.), Cambridge University Press, Cambridge, 890 pp.

Jones, J.A.A. (1969) The growth and significance of white ice at Knob Lake, Quebec, *Canadian Geographer* **13**, 4, 354-72.

Jones, J.A.A. (1996) *Global Hydrology: Processes, Resources and Environmental Management*, Addison Wesley Longman, Harlow.

Jones, P.D., Briffa, K.R., and Pilcher, J.R. (1984) Riverflow reconstruction from tree rings in southern Britain, *J. Climatology* **4**, 461-472.

Karl, T.R., Wei-Chung Wang, Schlesinger, M.E., Knight, R.W., and Portman, D. (1990) A method of relating GCM simulated climate to the observed local climate. Part 1, Seasonal statistics, *J. Climate* **3**, 1053-1079.

Kovaneva, N. (1982) Statistical studies of the regularities in the contemporary changes of the surface air temperature and pressure fields, *Voprosy Hydrologii Sushi*, Hydrometeoizdat., Leningrad, 200-210. (In Russian.)

Murphy, J.M., Cookmartin, G., Jones, R.G., and Noguer, M. (1993) Regional climate, in *The Hadley Centre Progress Report 1990-1992*, Bracknell, The Hadley Centre for Climate Prediction and Research, 24-26.

Myers, N. (ed.) (1985) *The Gaia Atlas of Planet Management*, Pan Books, London, 272 pp.

National Rivers Authority (NRA) (1994) *Water - Nature's Precious Resource. an Environmentally Sustainable Water Resources Development Strategy for England and Wales*, HMSO, London, 93 pp.

Newson, M.D., and Lewin, J. (1990) Climatic change, river flow extremes and fluvial erosion - scenarios for England and Wales, *Progress in Physical Geography* **15**, 1, 1-17.

Palutikof, J.P., Guo, X., Wigley, T.M.L., and Gregory, J.M. (1992) *Regional Changes in Climate in the Mediterranean Basin due to Global Greenhouse Gas Warming*, Climatic Research Unit, University of East Anglia, Norwich, 162 pp.

Parr, T., and Eartherall, A. (1994) *Demonstrating Climate Change Impacts in the UK: The DoE Core Model Programme*, Institute of Terrestrial Ecology/NERC, Huntingdon, 18 pp.

Parry, M.L., Carter, T.R., and Porter, J.H. (1989) The greenhouse effect and the future of UK agriculture, *J. Royal Agricultural Society of England*, 120-131.

Parry, M.L., and Read, N.J. (eds.) (1988) *The Impact of Climatic Variability on UK Industry*, AIR Report No. 1, Atmospheric Impacts Research Group, University of Birmingham, UK, 71 pp.

Pugh, D.T. (1990) Is there a sea-level problem? *Proceedings of the Institution of Civil Engineers* Part 1, **88**, 347-366.

Raymo, M.E., and Ruddiman, W.F. (1992) Tectonic forcing of late Cenozoic climate, *Nature* **359**, 117-22.

Starkel, L., Gregory, K.J., and Thornes, J.B. (eds.) (1990) *Temperate Palaeohydrology*, Wiley, Chichester, UK, 548 pp.

Thornthwaite, C.W. (1948) An approach towards a rational classification of climate, *Geographical Review* **3**, 55-94.

Tickell, Sir C. (1977) *Climate Change and World Affairs*, Harvard Studies in International Affairs No. 37, Cambridge, Mass.

Tickell, Sir C. (1991) The human species: a suicidal success? *Geographical Journal* **159**, 2, 219-26.

Turc, L. (1954) Calcul du bihan de l'eau evaluation en fonction des precipitations et des temperatures, *International Association of Scientific Hydrology*, Pub. No. 38, 188-202.

Viner, D., and Hulme, M. (1994) *The Climate Impacts LINK Project*, University of East Anglia, Climatic Research Unit, Report to Department of the Environment, Norwich, 24 pp.

Warrick, R.A., Barrow, E.M., Wigley, T.M.L. (1990) *The Greenhouse Effect and its Implications for the European Community*, Commission of the European Communities, Brussels, EUR 12707 EN.

Warrick, R.A., and Farmer, G. (1990) The greenhouse effect, climatic change and rising sea level: implications for development, *Transactions of the Institute of British Geographers* **15**, 1, 5-20.

Wigley, T.M.L., and Raper, S.C.B. (1992) Implications for climate and sea level of revised IPCC emissions scenarios, *Nature* **357**, 293-300.

Wigley, T.M.L., Jones, P.D., Briffa, K.R., and Smith, G. (1990) Obtaining sub-grid scale information from coarse resolution GCM output, *J. Geophysical Research* **95**, 1943-1953.

WMO/UNESCO (1991) *Report on Water Resources Assessment*, WMO, Geneva, 64 pp.

Worldwatch Institute (1984) *Water - the Next Resource Crisis?* Worldwatch Institute, Washington, D.C.

J.A.A. Jones
Institute of Earth Studies
University of Wales, Aberystwyth, SY23 3DB, UK.

SECTION I

SENSITIVITY OF THE GLOBAL HYDROSPHERE

SUMMARY

Global water dynamics

■ Global precipitation seems to be decreasing, but evidence of warmer sea surface temperatures suggests that there should be more evaporation and therefore more precipitation.

■ Certain regions, especially deserts, are highly sensitive to small changes in the global cycle.

■ The effect of human alteration of surface hydrology on the global cycle is probably currently comparable to that predicted for CO_2 changes in the near future. Land surface modification alters evapotranspiration by about 200 mm per annum, and double this in the humid tropics. 13% of global discharge is currently diverted by mankind.

■ Lack of knowledge in the above areas and lack of adequate models, in comparison with the well-developed level of physically-based modelling in the realm of rainfall-runoff processes, seriously hampers global predictions.

■ Prediction and diagnosis are complementary. There is an urgent need to analyse observational data and to infer causes from effects (diagnosis), to improve the prevalent predictive approach.

■ The causes of the Pleistocene glaciations remain an area of considerable uncertainty and a high priority field for study.

■ More needs to be learnt about Postglacial sea level rise, especially the effects of 'negative heat' supply from meltwaters.

■ Lake level studies have proven a valuable source for reconstructing global patterns in Postglacial water balances.

■ The scientist's role in assisting global policymaking may be to present an optimal strategy, which properly recognises the extent of anthropogenic compared with natural forcing factors. The Indian monsoon and African drought are taken as examples.

Biospheric aspects of the hydrological cycle

■ The current IGBP-BAHC research programme focuses on upscaling from ecological-scale observation and modelling as well as downscaling from GCM output.

■ Upscaling algorithms are beginning to address two major problems: including the spatial heterogeneity of the landsurface (the LSH) and parameterising biological processes which occur on short timescales.

■ Classification of General Circulation patterns is a major problem in downscaling.

■ Prototype models for upscaling and downscaling are discussed, including extending Soil-Vegetation-Atmosphere Transfer (SVAT) models.

■ Subgrid-scale processes may be aggregated by (1) weighting according to the percentage of grid covered or (2) parameterising the subgrid-scale variance.

■ Remote sensing may provide an alternative to using aggregating rules, although the HAPEX-MOBILHY experiment suggests that simple rules of aggregation may be sufficient.

■ Mountain areas present particular problems for regional SVAT models because of nonlinear increases in variance and lack of ecological studies in many areas.

■ From the series of algorithms being developed and tested by BAHC, details are given of the Regional Hydro-Ecological Simulation System and the Mountain-Micro-Climate Simulator (MT-CLIM).

■ Calibration and validation of SVAT models requires weather data with higher spatial resolution than normally available.

■ Downscaling algorithms are required for this and for increasing the resolution of GCM output.

■ Both stochastic and dynamic approaches may be taken.

■ An effective solution may be to combine approaches by using a dynamic mesoscale model like the Mesoscale Atmospheric Simulator System (MASS) to downscale to a 10 km grid and refine this with stochastic interpolation methods.

■ The Weather Generator project aims to create high resolution data from coarse-scale inputs, e.g. downscaling from 1000 km to 10 or even 1 km.

■ The Grey-theory Prediction Model (GPM), a stochastic generator which seeks to reproduce an element of uncertainty, has been developed to extrapolate temperature and precipitation in Canada and tests have proved successful.

Changes in global riverflow

■ No clear evidence is found of any global trend in peak discharges or annual volumes.

■ However, records are short, variances are high and we may expect some lag time for effects to work through the system.

■ Some individual regions do seem to show slight evidence of trends:

    1. 20% of North American basins pass 4 or more tests, with NE basins showing negative trends and SE basins showing positive in both peak flows and total volumes.

    2. In Europe 25% show some evidence of trend in peak discharges and 10% in total flows, but negative only.

    3. Little change is evident in South America, though there are positive trends in West Africa and 15% of East Australian basins show trends in both peaks and totals.

2 AN INTRODUCTION TO GLOBAL WATER DYNAMICS

I. KAYANE

Abstract

The role of the hydrological cycle in the Earth system is discussed through a review of papers related to paleohydrological processes in the glacial-interglacial cycle, the postglacial period, and the recent age of global warming, with emphasis on anthropogenic interactions with the hydrological cycle. The importance of solving appropriate "inverse problems" related to the hydrological cycle through field work is stressed.

2.1. The Hydrological Cycle as an Unknown Part of the Global Earth System

Water's role in the Earth system may be grouped into three processes; physically informing climate and shaping topography by transporting heat and sediments respectively, chemically in weathering rocks, and biologically in supporting soil-vegetation ecosystems. These processes are eternal in the sense that the necessary energy depends solely on solar energy and gravity. Climate, topography and ecosystems are the products of hydrological processes. On the other hand, if climate changes, either naturally or anthropogenically, the global hydrological processes will also be changed. This results in turn in changing the water-related erosional, biogeochemical and geophysical processes. Feedback mechanisms between climate and the hydrological cycle within the Earth system are nonlinear and intricate.

Irrespective of the principal role of water in the Earth system, many aspects of hydrological processes remain only poorly understood. One example is the global annual precipitation. Figure 2.1 shows plots of average annual precipitation on land and ocean estimated by different authors (USSR Comm. IHD, 1978). Years on the horizontal axis indicate the year when quoted papers were published. Though there is some scatter, it seems that the same figures have been quoted repeatedly by different authors. Since the global precipitation must equal the global evaporation, a new estimate of the former must change the latter and vice versa. Global warming has become a matter of urgent concern, yet global wetting or drying is very difficult to discuss due to the lack of reliable data. Precipitation fluctuations over global land areas for the last 100 years have been analyzed (Diaz et al., 1989; Houghton et al., 1990), but not for over the ocean.

J.A.A. Jones et al. (eds.), Regional Hydrological Response to Climate Change, 25–38.
© 1996 *Kluwer Academic Publishers. Printed in the Netherlands.*

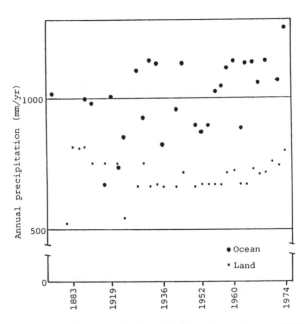

Figure 2.1. Variation in global precipitation estimates (Kayane, 1992).

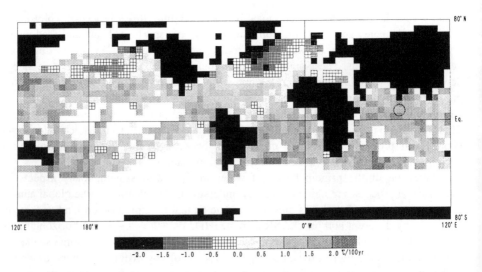

Figure 2.2. Global distribution of changes in SST, in °C/100 years, for the period 1930-1989 (Kayane *et al.*, 1995). The hatched square indicates the area with increased temperature, the meshed area the area with decreased temperature, the white area the region with insufficient available data (less than 480 months), and the black squares are regions with no data, which roughly correspond to continents.

Figure 2.2 shows the global distribution of the trends of SST (sea surface temperature) with time in °C/100 years for the period 1930-1989 (Kayane *et al.*, 1995), obtained by analyzing the SST data released from the United Kingdom Meteorological Office (UKMO/SST). There seems no doubt that SSTs in the tropical and subtropical ocean have increased considerably, though they have decreased in the North Atlantic and North Pacific Oceans poleward of 30°N, probably due to the increase in the atmospheric concentration of effective greenhouse gases since the eighteenth century (Roberts, 1994). According to the published data (JMA, 1994), however, global precipitation shows a slight decreasing trend, in contrast with the marked increasing trend of global temperature. Theoretically, the increase in evaporation expected from the warmer ocean and the observed decrease in global precipitation are mutually contradictory.

Has global precipitation been changing historically or on a geological timescale? Will anthropogenic changes in climate cause continental precipitation to increase or to decrease? What about global evaporation, which must be equal to global precipitation? To believe or not to believe the future prediction by GCMs is still a matter of belief, until full validation of the models has been achieved.

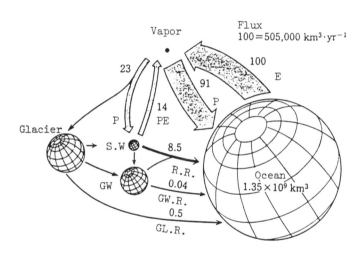

Figure 2.3. Global hydrological cycle (Kayane, 1992). Sphere = storage; arrow = flux.
Precipitation (P); evaporation (E); evapotranspiration (PE); surface waters (S.W.); groundwater
(GW); river runoff (R.R.); groundwater runoff (GW.R.); glacier runoff (GL.R.).

2.2. Scale of Anthropogenic Interactions with the Hydrological Cycle

Land surface hydrological processes occupy about 20% of the global hydrological cycle (Figure 2.3). Changes in the land-based branch of the hydrological cycle are affected by various causes, mostly through oceanic processes. A 10% change in land precipitation is equivalent to about 2% of the ocean evaporation. The precipitation in interior drainage basins, which occupy about 30% of the total land area, is equivalent to only 0.2% of the ocean evaporation (L'vovich, 1979). Therefore, desertification, for example, is the result of a very small change within the overall hydrological cycle in the Earth system. It is estimated that 13% of global runoff is now diverted by humans from temporary high flow to regulated flow by damming. Changes in evapotranspiration due to landuse change amount to 200 mm a^{-1} in the humid temperate zone (Swank and Douglass, 1974), and more than twice as much in the humid tropics (Bruijinzeel, 1990). Irrigation for cotton in central Asia has caused the inland lake area of the Aral Sea to decrease dramatically.

It might be possible to change the hydrological processes on land drastically, if we could use the present civil engineering technology to its maximum extent. However, effects and feedbacks from the anthropogenically induced changes on the global as well as the regional water cycle are not yet fully understood.

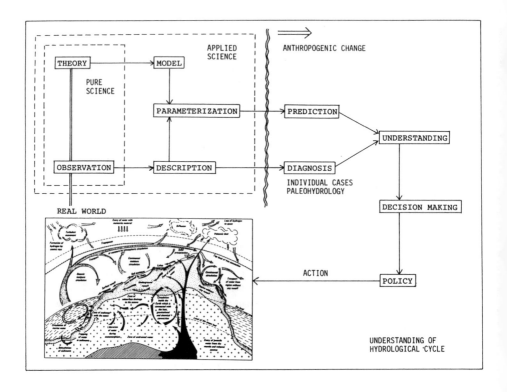

Figure 2.4. Hydrologic research and decision making (Kayane, 1992).

Figure 2.4 attempts to relate scientific research to decision making processes (Kayane, 1992). Many hydrologic models have been built for the purpose of prediction of hydrological phenomena. However, the applicability of these models to hydrological processes in an anthropogenically changing world is not yet proven, because the parameters of most of the mathematical input-output hydrologic models are inversely identified from 'observed' input and 'observed' output. For such deterministic models, the following criticism might be made: Are the inversely identified parameters universally applicable? Is the presumed algorithm accurate enough to represent the modelled hydrological processes? A point to be stressed is that prediction by such deterministic models alone is not sufficient to reveal the real picture of hydrological processes.

As the medical doctor diagnoses an individual patient based on personal clinical data before preparing a prescription, a diagnosis based on observed data is necessary for scientific understanding of the hydrological cycle, because the deductive approach is not necessarily effective in elucidating individual regional cases in the evolving heterogeneous real world. We might say that the diagnosis is a scientific approach to find out causes from the results. In comparison with a conventional normal scientific approach to solve a 'normal problem' proceeding from the cause to the result, the diagnosis might be called a method to solve an 'inverse problem', to find out the cause from the result.

Prediction and diagnosis are not in conflict with each other, but mutually complementary. Data concerning the past hydrological cycle caused by natural climatic change can be obtained through analyses of paleohydrologic records. The accuracy of the diagnosis depends very much upon the availability of reliable hydrological data from both the present and past. The importance of observation and its appropriate interpretation will increase more and more as it becomes necessary to make appropriate decision making on global environmental problems in the heterogeneous real world.

2.3. Scale Problems in Global Water Dynamics

2.3.1. THE GLACIAL-INTERGLACIAL CYCLE

Paleoclimate reconstruction using stable isotope ratios (^{18}O) in deep ocean sediment cores and continental glaciers has revealed an almost complete picture of Quaternary climate history. The last interglacial period was recorded 130 ka BP in the middle latitudes of the Northern Hemisphere. An almost identical glacial-interglacial cycle with a period of ca. 100 ka could be traced back to 800 ka BP. However, the reconstructed climate does not exhibit such a marked glacial-interglacial cycle before 800 ka BP (Figure 2.5 from Raymo, 1992).

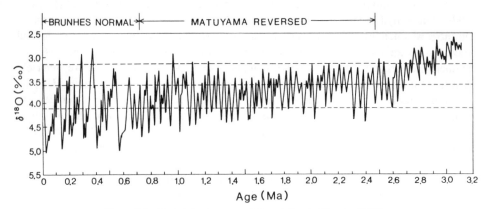

Figure 2.5. Oxygen isotope record of past climate (Raymo, 1992).

If past climate changes were interpreted as being due to the Milankovitch orbital forcing alone, we must seek a different reason for the abrupt change in climatic rhythm at ca. 800 ka BP. Was it caused by changes in the earth surface configuration due to continental uplift or sinking, or by mantle convective processes? An hypothesis called the 'Himalaya and Tibet (HT) effect hypothesis' was proposed by the present author (Kayane, 1993), which suggests that the appearance of the present glacial-interglacial climatic rhythm was initiated by the feedback effect caused by the uplift of the Himalaya mountains and Tibetan plateau. Whatever the reasons for the past climatic change, if a rhythm similar to that of the last glacial-interglacial cycle continues in future, we could extrapolate the past climate into future, provided the anthropogenic effect on climate is negligibly small. Is the anthropogenic effect large enough to disturb the natural climatic rhythm?

Stable isotope analyses have also revealed the history of Quaternary sea level changes. The present conversion factor for sea level reconstruction is a 0.1 per mil increase in ^{18}O content for a 10 m sea level decrease. Research into glacial-interglacial global water dynamics is based primarily upon the accurate evaluation of the past continental volume of ice caps. The ice caps imposed great stress on parts of the continental crust. The crustal deformation due to the stress affects global earth processes, such as crustal movement, volcanic activity, ocean circulation and their feedback mechanisms to climate. The re-evaluation of the conversion factor might be one of the highest research priorities for Quaternary paleohydrology.

2.3.2. THE POSTGLACIAL PERIOD

One of the most important issues in global water dynamics during the postglacial period is the cause of the rapid sea level rise at a rate of the order of 10 mm a^{-1}.

A partially imposed burden on the continental crust of the order of 10^6 kg m^{-2} has been removed within ca. 10 ka. In contrast, the ocean crust has been burdened with an additional weight caused by increasing sea water mass. The isostatic and hydro-isostatic adjustments of the continental and oceanic crusts have been continuing to the present day. Further research into paleohydrological response to ice cap melt, including the evolution of coastal topography responded to changing sea level (Pirazzoli, 1991), is needed in relation to newly elucidated boundary conditions.

Huge amounts of negative heat might be transported into the sea by the melt water from the ice caps. The outcome of this effect have seldom been discussed so far, although the start of a glacial was discussed by Kukla and Went (1992) as an historical analogue to the Younger Dryas during the last deglaciation. The residence time of the ocean bottom water is of the order of 1 to 10 ka. The present day ocean bottom water might have a memory three to four orders of magnitude larger than that of the atmosphere, extending back to the last glacial maximum. Has the feedback from the cold melt water mass to global climate, if it exists, been continuing?

The most direct influence of climatic change on human activities is through changes in the global pattern of precipitation. The past global distribution of precipitation is partly revealed by the analysis of closed lake level records by Street-Perrott and Harrison (1985), as follows.

At 18 ka BP, when the last glacial period had reached its maximum stage, the present Northern Hemisphere subtropical arid zone was occupied by a belt of enlarged lakes, suggesting a ca. 5°N equatorward shift of the westerly storm tracks in the Northern Hemisphere. In the tropics, however, an extensive dry zone had opened. There are slight indications of wetter conditions in Australia at 30°-38°S (Figure 2.6).

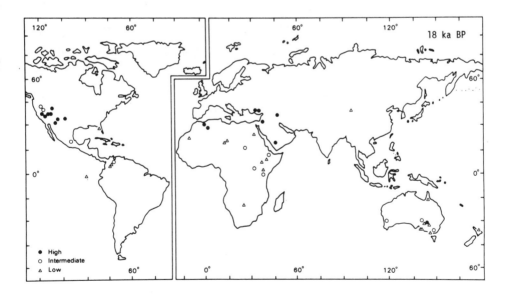

Figure 2.6. Global map of lake-level status at 18 000 years BP (Street-Perrott and Harrison, 1985).

Zonally-averaged temperatures at the last glacial maximum are estimated to be lower than the present day by ca. 6°C for 30°-60°N and 3°C for 30°-60°S (Roberts, 1989). Therefore, the equatorial shift in climatic zones might be larger in the Northern Hemisphere than in the Southern Hemisphere. As the present 'meteorological equator' with the highest zonally-averaged precipitation appears at ca. 10°N, the larger shift in the Northern Hemisphere might cause a southward shift of the meteorological equator. Then the shift, in turn, might cause a dramatic decrease or increase of precipitation in the equatorial zone.

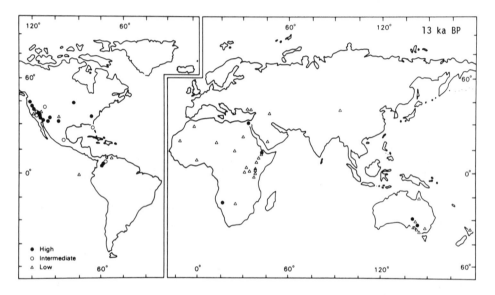

Figure 2.7. Global map of lake-level status at 13 000 years BP (Street-Perrott and Harrison, 1985).

It appears that 13 ka BP (Figure 2.7) was the most arid period in the Late Quaternary throughout Africa and western Europe. Australia also appears to have been very arid at this time. In contrast, North America was still experiencing relatively moist conditions, indicating the existence of a mean storm track at about 35°N. The minimum of atmospheric vapor convergence and runoff over Africa, western Eurasia, and tropical Australia at this period is one of the most intriguing features of the whole Late Quaternary record. It implies a strong suppression of summer monsoon precipitation and a radically different state of the Asian/Indian monsoon system from the present day.

The pattern of lakes status at 6 ka BP (Figure 2.8), when the warmest climate was recorded in the last 100 ka period, is almost completely opposite to that around 18 ka BP. A very wide and strongly developed belt of enlarged lakes occupied the tropics. This implies a significant northward shift in the mean position of the equatorial rain belt, as well as a greatly enhanced monsoonal transport of moisture into the tropical continents.

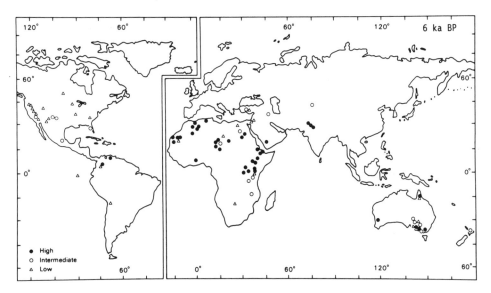

Figure 2.8. Global map of lake-level status at 6000 years BP (Street-Perrott and Harrison, 1985).

The above Figures by Street-Perrott and Harrison are a good example of world water dynamics during the glacial and interglacial cycle, but information concerning Asia is lacking. Figures 2.9 and 2.10 are the result of a comprehensive review on environmental change in south Asia made by Fujiwara (1990). The Indian subcontinent was cooler and drier in 18 ka and warmer and wetter in 8 ka than the present climate, respectively. The Indian summer monsoon was enhanced during the warmer period.

Mörner (1988) indicated that at around 13 ka, there was a rapid immigration and spread of a new fauna and flora around Scandinavia and in the North Sea. This is linked with a rapid deglaciation of the Fennoscandian ice cap, and a significant continental climatic amelioration. A similar pulse came at around 10 ka BP. Both events must represent significant intensifications of the warm water carried by the Gulf Stream to northwestern Europe. Reversed effects are noted at around 11 ka BP (Younger Dryas), 2.5 ka BP and during the so-called Little Ice Age, though the definition of Little Ice Age is controversial (Bradley and Jones, 1992), when pulses of southward expansions of Arctic water are recorded. He insists that the past climate could be understood in terms of variations within given budgets of energy, mass and momentum, or more precisely by the redistribution of heat and mass and the interchange of momentum that primarily drive the terrestrial climate. The role of the hydrosphere, especially that of the ocean, seems to be a crucial key to understanding the climatic change.

On the other hand, Wigley and Kelly (1990) report that evidence from the advances and retreats of alpine glaciers during the Holocene suggests that there were at least 14 century time-scale cool periods similar to the recent Little Ice Age. They examined the hypothesis that these cool periods were caused by reductions in solar irradiance.

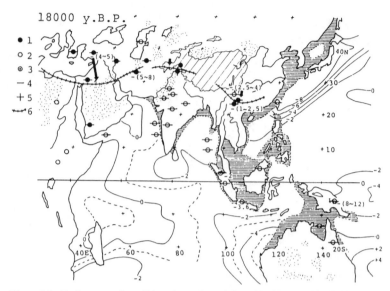

Figure 2.9. Environmental conditions in southern Asia at 18 000 years BP (Fujiwara, 1990).
1: Wetter than present, 2: Drier than present, 3: Same as present, 4: Cooler than present,
5: Warmer than present, 6: Winter position of polar front. SST as °C difference from present.

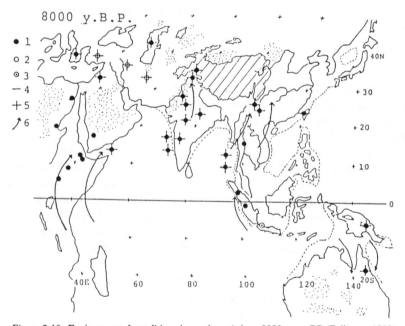

Figure 2.10. Environmental conditions in southern Asia at 8000 years BP (Fujiwara, 1990).
Legend: 1-5 are the same as in Figure 2.9. Enhanced SW monsoon.

A statistically significant correlation was found between the global glacial advance and retreat chronology and variations in atmospheric ^{14}C concentration.

More data have been accumulating over Asia, as one example is shown above, and North America than those in Figures 2.6 to 2.8. To test the validity of different hypotheses, comparison of the computer simulated past climate using GCMs with the paleohydrological evidence is an interesting topic, both for climatologists and hydrologists, especially if the comparison is made aiming at understanding 'paleohydrological processes'.

2.3.3. RECENT GLOBAL WARMING

The causes of climatic change are divided into two categories: natural and anthropogenic. Paleoclimatic analogues (Gleick, 1989; Epstein and Krishnamurthy, 1990) show great variations in the paleohydrological cycle in time and space. These variations originated in natural causes, although the real causal relation is not yet clearly understood. We can do nothing to counter the climatic change which is caused by natural causes. We have to accept it as a given natural condition for human activities.

In contrast, we could do something to counteract the climatic change caused by anthropogenic causes, provided global consensus is obtained. At this stage, a scientist's role in assisting global action to protect the global environment may be to present an optimal strategy. Indian monsoon rainfall and Sahelian drought are considered here as examples.

It is a well known fact that a close inverse correlation exists between the area of snow accumulation in Tibetan plateau and its surrounding area and the following year's summer monsoon rainfall in central India (Hahn and Shukla, 1976). A qualitative explanation is possible as follows: if the snow covers a wide area, a considerable amount of heat will be consumed to melt the snow, so that less is available to heat the continental air mass. This may act as a negative factor for the development of the Indian summer monsoon. It is also supposed that SST in an area is positively correlated to the Indian summer monsoon rainfall.

In accepting the above hypotheses, there remain a number of problems to be solved. Can we change the amount of snow accumulation anthropogenically? The globally averaged SST not only underwent a large fluctuation over the last century but these large changes, i.e. cooling about 1900 and warming in the 1930s, were sudden rather than gradual (Radok, 1987). Are anthropogenic effects large enough to change the SST voluntarily?

During the last two decades, most of the African continent has experienced extensive, severe and prolonged droughts. The area most affected has been the semiarid subtropical region of the Northern Hemisphere. Despite numerous claims that the Sahel drought was anthropogenically induced via direct human impact on the land surface, according to Nicholson (1989) at present no evidence exists to support such an hypothesis. A vast number of numerical simulations do, however, suggest that land surface changes, if sufficiently large and extensive, can influence large-scale climate, especially in the Sahel. Nicholson remarks that currently these are viewed as potential factors in perpetuating rather than initiating drought, the ultimate trigger being some change in the large-scale atmospheric dynamics.

For the understanding of anthropogenic effects on climate, therefore, a biogeophysical feedback mechanism (Kemp, 1989) needs to be investigated not only by computer simulation but also by field observation. The following two papers suggest the possibility of elucidating the feedback mechanism by analyzing environmental isotopes.

According to Gat and Matui (1991), an increase of 3 per mil in the "d" value of the isotopic composition of precipitation over the Amazon basin suggests that an isotopically fractionated evaporation flux contributes to the atmospheric water balance over the region. Their tentative conclusion is that 20-40% of the total evapotranspiration flux is accompanied by an isotopic fractionation from open water.

In the study by Ingraham and Taylor (1991), samples of surface water, shallow groundwater, and precipitation were collected for isotopic analysis along three east-west traverses, which began at the Californian coast. The results suggested partially closed hydrologic systems implying terrestrial recycling of water. A simple model estimated this amount to be about 20% across northern and central California.

2.4. Conclusions

1) Environmental isotope techniques are also valid to investigate paleohydrologic cycle, and future paleohydrological study should include study of paleohydrological processes.

2) Computer models have been developed to "simulate" the paleoclimate, but the simulation is based on many assumptions. It seems that the computer should be used as a tool to do "numerical experiments" rather than to simulate the real world.

3) The field work to reveal the biogeochemical and geophysical feedback mechanisms is probably one of the most urgent tasks needed to improve the accuracy of global water dynamics.

References

Bradley, R.S., and Jones, P.D. (eds.) (1992) *Climate since A.D.1500*, Routledge, London, 679pp.

Bruijinzeel, L. (1990) *Hydrology of moist tropical forests and effects of conservation: a state of knowledge review*, UNESCO IHP, 224 pp.

Diaz, H.F., Bradley, R.S., and J.K. Eischeid (1989) Precipitation fluctuations over global land areas since the late 1800's, *J. Geophysical Research* **94**, 1195-1210.

Epstein, S., and Krishnamurthy, R.V. (1990) Environmental information in the isotopic records in trees, *Philosophical Transactions of Royal Society of London* **A330**, 427- 439.

Fujiwara, K. (1990) Environmental change in tropical Asia. In, Environmental change in monsoon Asia, Series of Research Center for Regional Geography, Hiroshima University **20**,

1-63. (In Japanese.)

Gat, J.R., and Matui, E. (1991) Atmospheric water balance in the Amazon basin: An isotopic evapotranspiration model, *J. Geophysical Research* **96**, D7, 13179-13188.

Gleick, P.H. (1989) Climatic change, hydrology, and water resources, *Reviews in Geophysics* **27**, 329-344.

Hahn, D.G., and Shukla, J. (1976) An apparent relationship between Eurasian snow cover and Indianmonsoon rainfall, *J. Atmospheric Science* **33**, 2461-2462.

Houghton, J.T., Jenkins, G.J., and Ephraums, J.J. (eds.) (1990) *Climate Change: the IPCC Scientific Assessment*, Cambridge University Press, Cambridge, UK, 364 pp.

Ingraham, N.L., and Taylor, B. (1991) Light stable isotope systematics of large-scale hydrologic regimes in California and Nevada, *Water Resources Research* **27**, 77-90.

Japan Meteorological Agency (1994) Inter-annual variation of temperature and precipitation, *Tenki* **41**,482-483. (In Japanese.)

Kayane, I. (1992) Global hydrologic cycle, in J.S. Theon *et al.* (eds.) *The Global Role of Tropical Rainfall*, A. Deepak Publishing, 33-51.

Kayane, I. (1993) On the raison d'être of physical geography, *Geographical Review of Japan* **66**, 735-750. (In Japanese.)

Kayane, I., Nakagawa, K., and Edagawa, H. (1995) Hydrological consequences of global warming revealed by the century-long climatological data in Sri Lanka and southwest India, in *An Interim Report of IGBP Activities in Japan, 1990-1994*, Science Council of Japan, 65-81.

Kemp, D.D. (1989) *Global Environmental Issues: A Climatological Approach*, Routledge, London, 220 pp.

Kukla, G.J., and Went, E. (eds.) (1992) *Start of a Glacial*, Springer-Verlag, 352 pp.

L'vovich, M.I. (1979) *World Water Resources and Their Future*, American Geophysical Union, 415 pp.

Mörner, N-A. (1988) Terrestrial variations within given energy mass and momentum budgets: Paleoclimate, sea level, paleomagnetism, differential rotation and geodynamics, in F.R. Stephenson and A.W. Wolfendale (eds.) *Secular Solar and Geomagnetic Variations in the Last 10,000 Years*, Kluwer Academic Publishers, Dordrecht, Netherlands, 455-478.

Nicholson, A.E. (1989) African drought: Characteristics, causal theories and global teleconnections, in A. Berger *et al.* (eds.) *Understanding Climatic Change*, Geophysical Monograph, 52, American Geophysical Union, 79-100.

Pirazzoli, P.A. (1991) *World Atlas of Holocene Sea-level Change*, Elsevier, Amsterdam, 300pp.

Radok, U. (1987) *Toward Understanding Climate Change*, Westview Press, Boulder, USA, 200 pp.

Raymo, M. (1992) Global climate change: a three million years perspective, in G.L. Kukla and E. Went (eds.) *Start of a Glacial*, Springer-Verlag, 207-223.

Roberts, N. (1989) *The Holocene: An Environmental History*, Blackwell, Oxford, UK, 227 pp.

Roberts, N. (1994) *The Changing Global Environment*, Blackwell, Oxford, UK.

Street-Perrott, F.A., and Harrison, S.P. (1985) Lake levels and climate reconstruction, in A. Hecht (ed.) *Paleoclimate Analysis and Modeling*, John Wiley, Chichester, UK, 291-340.

Swank, W.T., and Douglass, J.E. (1974) Streamflow greatly reduced by converting deciduous hardwood stands to pine, *Science* **185**, 857-859.

USSR Committee for the IHD (1978) *World Water Balance and Water Resources of the Earth*, UNESCO, 663 pp.

Wigley, T.M.L., and Kelly, P.M. (1990) Holocene climatic change, ^{14}C wiggles and variations in solar irradiance, *Philosophical Transactions of Royal Society of London* **A330**, 547-560.

I. Kayane
Institute of Geoscience
University of Tsukuba, Tsukuba, Japan.

3 MODELLING THE BIOSPHERIC ASPECTS OF THE HYDROLOGICAL CYCLE

Upscaling Processes and Downscaling Weather Data

BRAD BASS, NARESH AKKUR, JOE RUSSO and JOHN ZACK

Abstract

The International Geosphere-Biosphere Program - Biological Aspects of the Hydrological Cycle (IGBP-BAHC) is concerned with the role of the biosphere in the hydrological cycle over a range of space and time scales. One crucial emphasis in this exercise involves 'upscaling' or aggregating the processes in the soil-vegetation-atmosphere interface from the patch scale of ecology to the mesoscale, larger regional scales and the continental scale. In time the upscaling must move from hours to decades and even longer time periods. A second emphasis involves the 'downscaling' of low-resolution climatic and meteorological data to high-resolution grids that are suitable for ecological and hydrological research. These two emphases are described along with prototype models for upscaling and downscaling. Other issues such as the incorporation of land surface heterogeneity in upscaling and the classification of atmospheric circulation patterns in downscaling are discussed along with other aspects related to these emphases are presented in greater detail. In particular the discussion of the downscaling component in BAHC summarizes the results of the most recent workshops (BAHC, 1993a and b).

3.1. Introduction

General circulation models (GCMs) provide a global simulation of all the major components of the climate system. The model simulations are sensitive to parameters, such as surface albedo, surface roughness, evaporation and soil moisture (Dorman and Sellars, 1989). These parameters are influenced by vegetation (Dickinson, 1992), snow cover and rainfall yet biospheric processes, the hydrologic cycle and their interaction are poorly described in GCMs. The Biospheric Aspects of the Hydrologic Cycle (BAHC) project has recently been established as a core project within the International Geosphere-Biosphere Programme (IGBP, 1990) to address the following general question over the full range of spatial and temporal scales, for different types of landscapes and ecosystems, and under existing and changing conditions: How do plant communities, ecosystems, and topography interact with the hydrologic cycle, and the

J.A.A. Jones et al. (eds.), Regional Hydrological Response to Climate Change, 39–62.

partitioning of energy at land surfaces? There are two major goals in the BAHC program:

1) To determine the biospheric controls of the hydrologic cycle through field measurements for the purpose of developing models of energy and water fluxes in the soil-vegetation-atmosphere system at a wide range of spatial and temporal scales;

2) To develop appropriate data bases that can be used to describe the interactions between the biosphere and the physical earth system and to validate model simulations of these interactions.

The role of the biota in the hydrologic cycle is described in Figures 3.1a and 3.1b (IGBP, 1993). Vegetation affects the exchange of energy, water, carbon and other substances with the atmosphere. Root growth modifies the soil affecting infiltration, percolation and drainage. While direct evaporation from the soil plays an important role in the hydrologic cycle, soil moisture also influences the control of transpiration from plants. The biological interaction with the hydrologic cycle can be represented by soil-vegetation-atmosphere transfer (SVAT) models. SVAT models, as described in the BAHC Operational Plan (IGBP, 1993) represent the critical interconnecting processes controlling energy, water and carbon transfer between soils, vegetation and the atmosphere. SVATs are used to model individual ecosystems over small homogeneous land surface areas, the patch scale (microscale), but they are also the basis of the surface parameterization in general circulation models (GCMs). The integration of energy, water, carbon and nutrient cycles is fairly complex, but SVAT models must represent land surface properties in a simple way if they are to be used at larger spatial scales over more heterogeneous landscapes. For example, the SVATs that are used to represent the land surface in GCMs are not as complex as the the patch-scale models due to the area represented by a grid cell. Adapting SVAT models to larger surface areas, a process of 'upscaling', often requires the use of a reduced number of parameters to represent complex land surface processes.

Moving to larger scales requires the inclusion of additional elements of the land surface and the atmosphere including the upper soil layers, the planetary boundary layer (PBL), surface rivers and reservoirs, and ground water resources. Investigation of the process dyamics in the PBL is a component of BAHC and will be based on short-term observation programs. These programs are designed to include more real-world complications such as land surface heterogeneities, discontinuities, and topography. In addition, the observation programs include the monitoring of several ecological characteristics such as nutrient cycling so as to improve the basis for large scale, integrative hydroecological modelling (IGBP, 1993).

Shuttleworth (1994) reviewed several regional observation programs. One such program designed to, the Hydrological Atmospheric Pilot Experiment - Modélisation du Bilan Hydrique (HAPEX-MOBILHY) was undertaken in southwest France in 1986. The 100-km square area included coniferous forest and small 'patches' of mixed agricultural vegetation. The program included year-long data collection supplemented by an intensive 3-month period of observation. The data include direct measurements of the energy exchanges, water exchanges and the near-surface climate of different patches. Additional data were collected by satellite and boundary-layer soundings. Additional

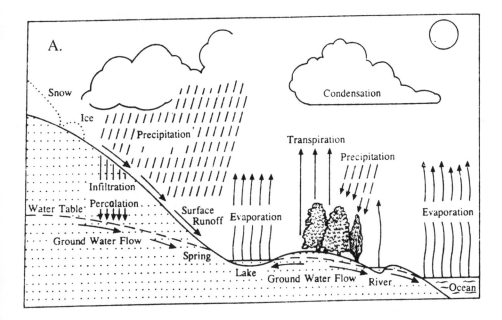

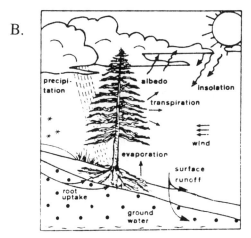

Figure 3.1. (a) The hydrologic cycle (IGBP, 1993). (b) Vegetation structure, soil properties and other biological characteristics that influence the hydrologic cycle (IGBP, 1993).

observations programs have been carried out in semi-arid regions in Spain in 1991 (the European Field Experiment in Desertification-threatened Area) and in Niger in 1992 (HAPEX-SAHEL).

A number of questions involving the differing spatial and temporal scales of processes cannot be answered by intensive short-term experiments. The spatial scales of processes ranges from the scale of the stomata and the regulation of evapotranspiration to continental hydrologic discharge (Sellers, 1991; Dickinson, 1992; Bolle, 1993). The temporal scales range from minutes to centuries or millennia for climate-related vegetation changes. As the spatial scale of the process changes, different data sources are required to characterize the process at different sites. At a time scale of seconds to days the metobolic activities of terrestrial plants govern the surface-atmosphere exchanges of latent heat, water and carbon dioxide. At a time scale of days to weeks the soil water balance and the exchange of trace gases assumes an operative importance. Changes in the amount of carbon stored in plants occurs at an annual time scale. Moving in the other direction extreme weather events can severely affect vegetation in hours or even minutes. Other climate related impacts occur on a range of scales of years to centuries. Many of the meteorological data required for this research cannot be obtained, or are not available at the correct scale, from existing observations or GCM scenarios. It is therefore necessary to 'downscale' from the variables and scales of the available data to an area or a site for which an observation record is available. Additional interpolation methods are required to extend site specific data over a grid at the   space-time resolutions required for ecohydrological modelling. If such a record does not exist for a site, then alternative methods, based on benchmarks, remote sensing and diagnostic models must be used to estimate the truth set that is needed for validation. One such diagnostic model, used to resolve subgrid estimates of flux values (heat, moisture, precipitation) has been tested over Europe (Hantel, 1987).

3.2. SVAT Modelling over Heterogeneous Landscapes

For hydrology and water resources, one of the more relevant components of BAHC is the understanding of the role of land surface heterogeneity (LSH) in modelling evaporation, areal precipitation and runoff at the regional scale. LSH incorporates many additional land surface processes which are not sufficiently understood nor represented in parameterizations in other models. A general representation of a complex landscape with the important phenomena, subsystems and processes involved is provided in Figure 3.2a (Bolle, 1993). Running et al. (1989, 1992) extended SVAT modelling to heterogeneous landscapes using leaf-area index (LAI) to describe plant behaviour and the control of the main processes in a specific biome type (Figure 3.2b). In addition, a set of site and meteorological data is used to describe the specific conditions of the area. Since these data can be derived from satellite imagery it is possible to apply this type of model over a wide variety of regions.

Within a GCM, limited landscape heterogeneity has only recently been incorporated into a grid cell through improved land surface models (Pitman et al., 1991). In the Bare Essentials of Surface Transfer (BEST) model, which is run with the Bureau of Meteorology Research Centre GCM, in Australia, the fraction of snow, bare soil and vegetation can be represented in any grid cell over a land surface. In BEST, the critical fraction is vegetation. Land surface models are qualitatively insensitive only to small variations in this fraction. Therefore a realistic estimate is required, and in BEST this

A

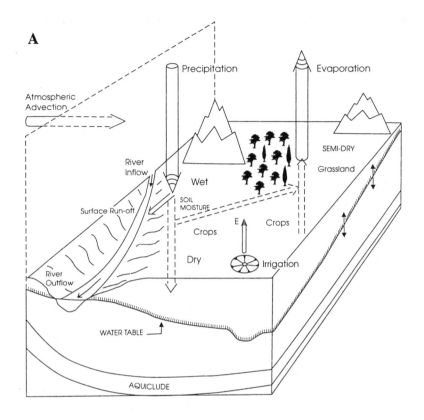

B

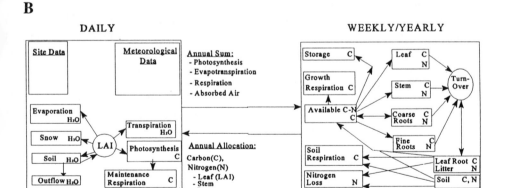

Figure 3.2. (a) Representation of landscape heterogeniety and complex processes (Bolle, 1993).

(b) A SVAT model integrating water, carbon and nitrogent cycles (Biome-BGC).

(Running, unpublished; IGBP, 1993).

has to be derived from data. The methods used to calculate this fraction and the snow depth are described in Pitman *et al.* (1991). The fraction of bare soil is calculated as a residual value.

Shuttleworth (1991) proposed a simple set of rules for aggregating SVAT model parameters. HAPEX-MOBILHY identified two characteristic length scales (10 km and 1 km), based on the size of coherent areas of land cover, each affecting the overlying atmospheric boundary layer in different ways (Shuttleworth, 1994). Using these results model-based aggregate descriptions were used to provide a synthetic benchmark of area-average energy, momentum and water vapour flow exchange. The benchmarks were used to evaluate a simple set of aggregation rules based on area-average weights in SVAT parameterizations. In dry conditions, this aggregation rule was used to estimate area-average fluxes to an accuracy of 10% at each of the characteristic length scales. The incorporation of sub-grid scale processes in GCMs is also a process of aggregation.

This is usually handled in one of two ways for the land surface (Pitman *et al.*, 1991). The first approach regards a grid cell as a combination of several sub-cells, and calculations are performed on each sub-cell. The results are aggregated by a weighting system based on the quantity of whole grid element that is estimated by each sub-cell. The drawback is that this approach is ineffective if the number of sub-cells is large. The second approach parameterizes the sub-grid scale variability, and this is used to adjust the transfer coefficients, but it is difficult to validate this parameterization.

Extending SVAT modelling to mountainous ecosystems presents additional problems due to the effects of mountain topography on the redistribution of water and energy (BAHC, 1994). The topography, elevation, and the ecology impose nonlinear effects on precipitation, runoff, infiltration and soil moisture patterns. The effects of elevation, slope, and climate on mountain ecosystems vary highly in space due to the extreme topography.

This produces steep ecological gradients whose complexity is difficult to capture within a modelling framework, except at very high spatial resolutions. In addition the knowledge of mountain ecosystems is limited, based only on a few regions in Europe and North America.

The state-of-the-art in using SVAT models over heterogeneous landscapes is illustrated by the Regional Hydro-Ecological Simulation System (Figure 3.3) which calculates the most important hydrological and hydroecological processes in a forested river basin. This system contains a SVAT model, represented on the left side of the Figure, and areal integration begins with the column 'GIS Processing'. The input data are daily maximum and minimum air temperature, total incoming solar radiation, daily total precipitation, LAI, topographic data and soil water holding capacity. Provision of the necessary data required the construction of a simple weather simulation model, or a prototype weather generator (Running *et al.*, 1987). The mountain micro-climate simulator (MT-CLIM) model has two modules, one for the derivation of daily humidity and incoming radiation at places where these data are unavailable, and a section for topographic extrapolation of point weather-station data (Figure 3.4). BAHC - Focus 4, the Weather Generator (WG), is coordinating a large-scale development of a series of such algorithms for different regions and for use with different coarse-scale data sources.

Parameter	Units	GIS	Models	Output
Climate	Radiation			
	Temperature			
	Humidity			
	Rainfall			
Vegetation	Biome type	Drainage area		
	Leaf area index		MT - CLIM	
		Stream network		Net photo-synthesis $(Mg\ ha^{-1}\ a^{-1})$
Topography	Elevation	Land unit extraction	FOREST - BGC	
	Slope	Topography - soils index		Evapo-transpiration $(mm\ a^{-1})$
	Aspect			
Soils	Texture	Registration	TOP Model	Discharge
	Depth			
	Water holding capacity			

Figure 3.3. A prototype for a regional 'upscaled' SVAT model (Running, 1992).

Topographic Climatology

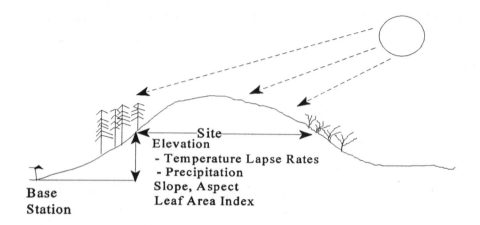

Diurnal Climatology

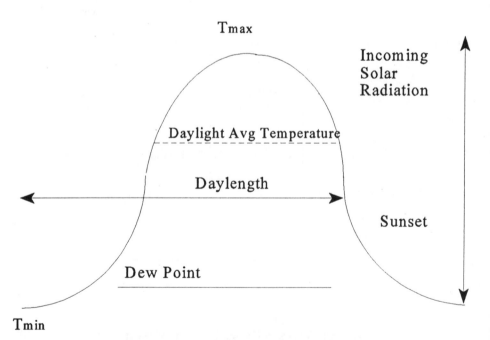

Figure 3.4. The Mountain Climate Generator (IGBP, 1993).

3.3. Downscaling Meteorological Data and Model Output

3.3.1. THE SIMULATION OF HIGH-RESOLUTION WEATHER INFORMATION

The primary goal of the WG project within BAHC is to develop the ability to create high-resolution meteorological information for ecological and hydrological modelling as well as related disciplines such as agriculture and water resources modelling. The WG will primarily consist of three components: a coarse-scale climate data source for input, a set of algorithms to downscale data from large scales to regional values or even specific sites, and the output variables required by the users. Daily weather results from the combination of forcings at several scales. For example, at the largest scale, forcings govern the general circulation defining the sequence of weather events in a region. Forcings at the mesoscale are associated with topographical and other surface charcteristics. These forcings alter the synoptic event in smaller areas and induce mesoscale circulations (Giorgi and Mearns, 1991; Hantel, 1987). The second component of the WG are the algorithms for projecting, or downscaling, coarse-scale climate or weather data down to smaller areas or individual sites. This is a problem of representing the climatic forcings on a large scale, delimited from 1000 km to global, down to the mesoscale, delimited from a few to several hundred kilometres (Giorgi and Mearns, 1991), but for the Weather Generator, this range has been narrowed to 1 - 10 km.

The third component 'Data Requirements' identifies (1) the minimum set of variables required for ecological and hydrological simulation, (2) the appropriate spatial and temporal resolution required for ecological and hydrological modelling, and (3) the minimum allowable uncertainty. In conjunction with a group of hydrologists and ecologists, this first component has been completed. In addition the sensitivity of models to each variable was assessed, and the uncertainty was explicitly defined. a standard definition of uncertainty has been discussed and accepted.

3.3.2. INPUT DATA AVAILABLE FOR DOWNSCALING

Climate and weather data or information are available at a variety of spatial and temporal resolutions. Figure 3.5 frames the spatial and temporal scales of the meteorological phenomena along with the scales of decisions (BAHC, 1993a). Data about these events are partly met by models, public station networks, and local monitoring programs. The foremost limitation of each data source is that their spatial and temporal scales are not appropriate for ecology, hydrology and related decision-oriented areas of inquiry which model processes or require decisions within the range of mesoscale events (kilometres to tens of kilometres).

A downscaling sub-group of the WG project has considered several data sets as potential candidates for the downscaling program (BAHC, 1993b). They have been classified according to the spatial resolution of the data source and the longest period in the future for which each source could provide data (Figure 3.6). It is also assumed that each source can provide data at the daily timestep. Figure 3.6 also indicates the most appropriate technique for downscaling a particular data source to surface observations. The most important source of data are station observations which offer the only set of standard on-site, surface data. They can be aggregated statistically to provide information for any period out to an indefinite number of years with the implicit assumption that the present climate state and trends are indicative of the future. Station data also represent the 'truth values' for judging the performance of any scheme or model that is used to generate high-resolution data. Finally, because the data can be transformed into a stochastic model, predictions made from station observations represent a baseline of accuracy against which to judge the ability of generated data to represent future states. That is, all other approaches must improve upon the climatological uncertainty embodied in station observations, at the chosen scale of analysis, or else the approach is not justified.

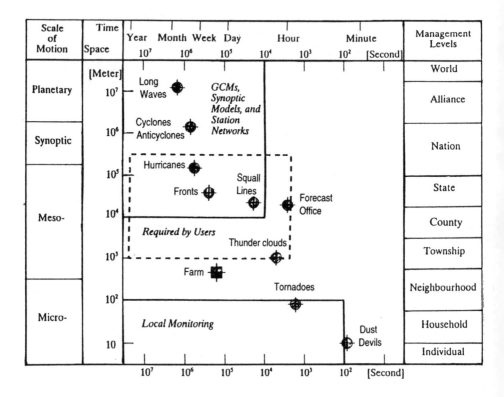

Figure 3.5. Scales of weather events and decision-making communities
(adapted from BAHC, 1993a). Scales of available data sources (——), as
compared to data required by users for management decisions (- - - -).

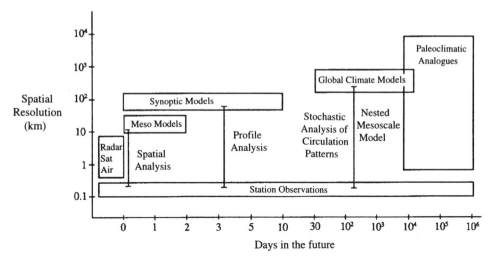

Figure 3.6. The spatial resolution of data sources and their range of prediction. Vertical bars
indicate an appropriate downscaling algorithm to be used in conjunction with each data source
(adapted from BAHC, 1993b).

Real time upper-air data remotely sensed by radar, profilers, rawinsondes, aircraft and
satellites complement station observations to give a three-dimensional, large-scale
description of the heat, momentum and moisture transfer in the lower atmosphere for
timesteps on the order of hours. While many of these sources could be further refined to
provide data on a scale useful for SVAT and hydrology models, they are not easily
accessible to research groups on a timely basis. Dynamic models simulate weather states
by solving the primitive set of equations, which represent mathematically the fluid-
dynamic properties of the atmosphere. Although they operate over a wide range of
spatial resolutions, there are always trade-offs between computer speed, model
complexity, forecast scale and prediction accuracy.

For example, mesoscale models have better formulation for the radiative, hydrologic,
convective and turbulent processes than the coarser-scale models, but they can only
provide forecasts out to 36 hours over a small spatial domain. Synoptic models can
forecast atmospheric states for a larger spatial domain and for a period further into the
future, but the accuracy is below that of the mesoscale models. However, the resolution
of mesoscale models varies between 80 and 10 km, while the spatial resolution of the
synoptic model ranges between 200 and 80 km. At the spatial resolution of 10 km,
mesoscale models can provide a source of weather data that can be directly input into
hydrological models, whereas synoptic models must be linked to other downscaling
approaches.

The largest-scale dynamic model, and that which sits at the coarsest resolution in
Figure 3.6, is the GCM. GCMs have less detail in their description of atmospheric
dynamics. Thus, reduction of complexity allows them to make 'predictions' in the order

of months out to 100 years when run in the transient mode. The spatial resolution is typically between 600 to 200 km, and must be linked with downscaling approaches for hydrological applications. In addition, even though the output is often treated as a prediction, it is only a plausible future scenario, and the daily values are not representative of specific days as is the case with the other dynamic models. Nevertheless, government policy makers and other decision makers view GCM output as a valuable guide in assessing future vulnerability to climate sensitive economic activities.

The problem with treating the GCMs in this manner is that the accuracy of the GCMs as a source of point data at the surface is unknown, whereas synoptic and mesoscale models out-perform climatology, and the evaluation is limited by a number of other factors. First, even the best spatial resolution is very coarse (200 km). Second, below the seasonal time scale, the credibility of the data is questionable. Third, and most critical, one would have to wait decades to obtain the necessary true values in order to evaluate GCM output of a future climatic state. Paleoclimatic analogues, such as tree rings, ice cores and pollen sediments are useful in providing qualitative guidelines for historical climate scenarios for periods ranging from tens of thousands to thousands of years, over regional or global domains. However, paleoclimatic analogues cannot provide data at specific space-time resolutions that are suitable for ecological and hydrological modelling.

At this time they are not considered as suitable inputs to any downscaling technique (BAHC, 1993b).

3.3.3. DOWNSCALING

There are two fundamental approaches to downscaling. In the first approach parameters of a daily stochastic weather model are conditioned on classified large scale circulation patterns or other variables in the free atmosphere. These linkages are used to construct a stochastic model to reproduce the spatial characteristics of meteorological variables over a number of sites in a small area. The values at each site can also be used as a basis for interpolating the meteorological variables to a high-resolution grid over the area.

For example, Bardossy and Plate (1992) link precipitiation, at sites in a small catchment, on the order of 5000-10 000 km to specific 500 or 700 hPa height anomalies, classified into 11 weather types. The precipitation is then simulated with a multivariate autoregressive AR(1) model. The model is able to reproduce both the probability of the occurrence of a rainfall event and the spatial distribution of rainfall amounts. The underlying assumption in these approaches is that the observed circulation patterns and the associated surface weather will appear under a global warming, but with a different frequency of occurrence. This approach requires a classification method that is consistent and meteorologically meaningful. In addition, the process has to be automated given the size of the data base.

Recently, computational intelligence techniques have been used to validate and check the extent of classification offered by some of the statistical clustering techniques. Akkur *et al.* (1992) used a rule induction algorithm on a k-means/Ward's classification to describe the influence of synoptic climatology on the ground-level ozone concentration at Brisbane, Australia. The rule induction method, employing a decision tree algorithm, gives an objective rule based interpretation of the synoptic categories. Neural networks, representing a connectionist approach, also offers a powerful tool for testing the extent of accuracy of weather pattern classification.

Subsequently, Akkur (1992) devised a two step integrated statistical-computational intelligence framework for synoptic classification problems wherein

- as a first step, a statistical clustering procedure (k-means/average linkage) is used to produce the weather pattern categories;
- as a validation step, train the categories using neural networks and rule induction algorithm.

Through the integration of the statistical and computational intelligence approaches, the advantages of each can be realised while counterbalancing their disadvantages.

Bardossy and Plate (1992) use expert meteorological experience to define their circulation patterns from daily 500 hPa height anomalies. Typical patterns are defined for large regions, such as Europe and North America, and these patterns are used to classify the 40-year record of daily 50 hPa height anomalies. Bardossy and Caspary (1990) automated the classification procedure for Germany using neural net technology. This approach made use of an existing classification which is based on daily records of European circulation patterns from 1881. However, a similar data base is not available on a global scale and the procedure was adapted to other regions by using a fuzzy rule base to match the daily anomalies with the typical patterns suggested by the meteorologists (Bardossy *et al.*, in press). On the German data, the fuzzy rule base was able to obtain a crisper delineation between patterns than either the neural net or the k-means clusturing.

The two main criticisms of this approach are the lack of a physical basis for the relationships between the meso and synoptic scales and limitations imposed by the range of data that were used to develop the stochastic models. Thus, the stochastic relationhips may not be applicable under a climate change scenario.

Dynamic approaches to downscaling may provide a means of meeting these two criticisms. One such technique alters the GCM resolution, creating a finer grid over those regions of interest, thus increasing the grid cell size over the rest of the globe. A second technique uses the three-dimensional subset of the GCM data as an initialization for a high-resolution limited area model, such as a mesoscale model. A third technique, profile analysis, uses the dynamic equations of the atmosphere as a constraint in resolving sub-grid scale fluxes within a larger grid cell. The dynamic approach, because it uses physically based models is not as strictly limited as is the stochastic approach, but it is difficult to obtain a large enough sample to be statistically meaningful.

The Mesoscale Atmospheric Simulation System (MASS) is one mesoscale model that is used in a downscaling mode on an operational basis. MASS was orginally developed as a tool for research in remote sensing applications, but it has evolved into a sophisticated simulation system with detailed representations of several bulk mesoscale effects: (1) energy and water budgets, (2) turbulent processes in the planetary boundary layer, (3) deep moist convection, (4) atmospheric condensation, evaporation and precipitation, and (5) long and short wave radiation under clear and cloudy sky conditions.

Recently, MASS has been used by the National Aeronautics and Space Administration (NASA) to develop a regional thunderstorm prediction system where the MASS model simulated the meteorological conditions that were conducive for thunderstorm formation.

The model demonstrated considerable skill in predicting near-surface (10 m level) hourly variations of temperature, relative humidity, winds, cloudiness, and precipitation during an evaluation period (Zack et al., 1993).

It has also been demonstrated that the spatial resolution of the MASS output can be improved with sophisticated interpolation schemes. In recent simulations it has been possible to generate weather forecast data at scales as fine as 1 km through a mesoscale model-interpolation scheme coupling. While the high resolution forecasts have yet to be thoroughly evaluated, preliminary comparisons with station observations indicate that the accuracy of the data would be acceptable for most hydrological applications. The interpolation scheme is based on elevation and location, but it has been coupled with a land-use classification and the normalized difference vegetation index (NDVI).

There are several problems involved in linking a mesoscale model to a GCM. Perhaps the largest problem is the sea surface temperature (SST) gradient. The spatial resolution of the GCM smooths the SST gradient which weakens storm intensity. Another problem is the absence of several important freshwater lakes from the GCM. Therefore, alternative methods are required to provide initial conditions for these bodies of water which include the North American Great Lakes. Finally, the resolution of the soil moisture is too large, and the representation of the water balance is inadequate for mesoscale applications.

Other approaches are used to try and estimate future probabilities or trends from the data. Bass et al. (1994) have used a grey prediction model (GPM) to extrapolate monthly averages for temperature and precipitation at a site in Northern Canada. Grey theory is a method for estimating and incorporating uncertainty when the data are too sparse for the use of standard stochastic approaches (Deng, 1984). It has proven to be an appropriate method in systems analysis in linear programming models (Huang and Moore, 1993).

Grey theory creates a model of the data from a minimum and maximum value. Grey theory can also be used in a prediction mode to estimate and extrapolate a series of maximum and minimum values from a series of observations. In this approach the series is split into two series of high and low values, and two series are generated thus yielding a dynamic grey interval.

Although a time series techniques could be used, grey theory offers a more flexible approach for nonlinear data. The GPM provides an interval or a range of values instead of a prediction. In certain applications the interval may be small enough to be considered as a point. Bass et al. (1994) demonstrate that the GPM provides a good representation of both temperature and precipitation (Figures 3.7 and 3.8). For temperature the GPM reflects the trends that have been observed for the area during the early 1990s.

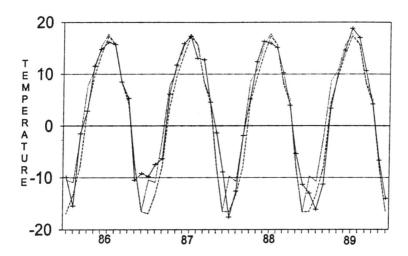

Figure 3.7. Observed and grey monthly temperature data (°C) for Cold Lake, Alberta, 1986-1989 (Bass *et al.*, 1994). Observed (—+—), grey maximum (——), grey minimum (- - - -).

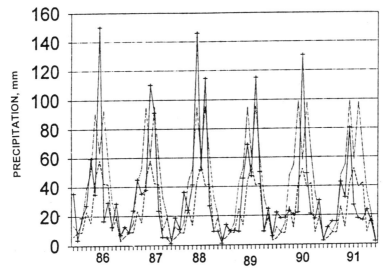

Figure 3.8. Observed and grey monthly precipitation data (mm) for Cold Lake, Alberta, 1986-1991 (Bass *et al.*, 1994). Observed (—+—), grey maximum (——), grey minimum (- - - -).

TABLE 3.1. Data requirements for ecological modelling.

Variable	Temporal Resolution[*]	Spatial Resolution	Uncertainty[**] Minimal	Desirable
Minimum set				
CO_2	monthly - annual	10 km	± 5%	± 1%
Precipitation amount and form	daily - monthly	10 km		± 5%[***]
Temperature max. and min.	daily - monthly	10 km	± 1% (systematic)	± 1%
Solar radiation				
global	⎱ daily -	⎱	± 5%	
	⎰ monthly -	⎰ 10 - 50 km		
net	annual		± 10%	± 5%
Vapour pressure deficit	daily	10 km	± 10%	
Windspeed	daily	10 km	± 25%	
Additional set				
All the above (except CO_2 and precipitation)	hourly			
Temperature wet bulb	hourly - daily	10 km	± 1° (systematic)	± 1°
Atmospheric pressure	hourly - daily	10 - 50 km	± 10%	

* Hourly data are required, for example, for leaf wetness (plant disease) modelling.

** It was either not possible or not relevant to determine both minimal and advisable uncertainty for every variable.

*** This uncertainty is very difficult to obtain for convective precipitation processes which are operative at this scale. This level of uncertainty is more realistic at a spatial scale of 100 km.

TABLE 3.2. Data requirements for hydrological modelling.

Variable	Temporal Resolution*	Spatial Resolution	Uncertainty**	
			Minimal	Desirable
Minimum set				
Precipitation amount	daily	10 km	± 20%	± 5%***
Temperature average	daily	10 km	± 1°	± 0.5°
Solar radiation global	daily	10 km	± 20%	± 10%
Specific humidity or Vapour pressure deficit	daily	10 km	± 10%	± 5%
Additional set				
Temperature max. and min.	hourly - daily	10 km	± 1°	± 0.5°
Windspeed	daily	10 km	± 20%	
Precipitation amount	hourly	10 km	± 5%	
Atmospheric pressure	hourly - daily	10 - 50 km	± 10%	

* Hourly data are required for urban and small catchment hydrology as well as the simulation of snowmelt.

** It was either not possible or not relevant to determine both minimal and advisable uncertainty for every variable.

*** This uncertainty is very difficult to obtain for convective precipitation processes which are operative at this scale. This level of uncertainty is more realistic at a spatial scale of 100 km.

3.3.4. DATA REQUIREMENTS FOR ECOLOGICAL AND HYDROLOGICAL MODELLING

The types of data and their required accuracy for ecological and hydrological modelling are listed in Tables 3.1 and 3.2. These Tables present the variables that were determined to be a minimum requirement and those that are considered advisable for modelling, by the Weather Generator working group in consultation with ecologists and hydrologists. The Weather Generator working group were informed that for most of the important research questions data are needed at a daily interval. For some issues, such as small catchments, urban hydrology, and leaf wetness a more frequent interval such as hourly

TABLE 3.3. Sensitivity of model output to meteorological inputs.

Variable	Ecology	Agriculture	Hydrology
Precipitation amount	High[*]	High (time of occurrence)	High
Precipitation form	Moderate[*]	Relatively insensitive[*]	Moderate - High (correcting obs.)
Temperature	Moderate - High	Moderate	Moderate - High (seasonality; snowmelt)
Solar radiation	Relatively insensitive - Moderate	Relatively insensitive - Moderate	Relatively insensitive
Vapour pressure deficit/ Specific humidity	Relatively insensitive	Relatively insensitive	Relatively insensitive
CO_2	Moderate	Moderate	Moderate
Windspeed	Relatively insensitive	Relatively insensitive	Relatively insensitive
Marine	Moderate		
Storm damage	Moderate		High
correcting observations			High

[*] Terms are defined thus, given ΔO is change in output and ΔI is change in input:

High	if $(\Delta O > \Delta I)$
Moderate	if $(\Delta O \cong \Delta I)$
Relatively intensive	if $(\Delta O < \Delta I)$

is necessary. On the other hand, some research interests only require data at less frequent intervals but these can always be derived from the daily data. Although the spatial resolution of 10 km may not be appropriate for all ecological and hydrological models it is an acceptable lower limit for most downscaling approaches and interpolative schemes. Nevertheless, spatial interpolative schemes have produced monthly climatologies at a resolutions of 1 km (Hutchinson, 1991).

In ecological and hydrological modelling different production processes exhibit different sensitivities to changes in environmental inputs (Table 3.3). In the current context sensitivity is defined as the normalized ratio between a change in a model's output (ΔO) for given change in input (ΔI). Where this ratio is greater than unity ($\Delta O >$ ΔI) the process is regarded as highly sensitive, while a value that is close to unity ($\Delta O \cong \Delta I$) unity implies that the process is of moderate sensitivity. A moderate sensitivity suggests that the output would vary by roughly the same magnitude as the input data. A value below unity ($\Delta O < \Delta I$) suggests that the model is relatively insensitive to the variable in question. For example, for a given change in precipitation the change in runoff would likely be magnified, whereas a large change in solar radiation would likely have little effect on runoff. In this case, the degree of uncertainty in precipitation would be of greater importance than for solar radiation (IGBP, 1993).

These uncertainties can be presented effectively as a 'numerical interval' about a simulated value. Given the various disciplines represented in BAHC and the other possible users of the weather generator, the major sources of uncertainty in input data need to be defined and quantified for the modelling efforts. The numerical interval is given as a sum of the following four contributions: (i) the bias ($X_{avg} - X_{true}$), which is the difference between a measurement or generated average value and the true value, (ii) the precision U_{prec}, which is the uncertainty due to the scatter of measured or generated values about an average value, (iii) the protocol U_{prot}, which is the uncertainty associated with the measurement practices or data-generation procedure, and (iv) the scale U_{scale}, which is related to the use of measured or generated values at different spatial and temporal scales. The contributing terms in the numerical interval can be written as a statistical addition of uncertainties:

$$\text{Numerical Interval} = (X_{avg} - X_{true}) + U_{prec} + U_{prot} + U_{scale} \qquad (3.1)$$

If the bias ($X_{avg} - X_{true}$) is assumed constant and is removed, and the remaining bias-corrected terms are distributed normally and derived statistically, a confidence level can be assigned to this truncated numerical interval. This truncated interval would be equally negative and positive on either side of the average value (X_{avg}).

An example of a complete weather generator, using a stochastic approach in a hydrological application is provided in Figure 3.9. Figure 3.10 illustrates how the MASS model is being used in an agricultural application. Both approaches can be compared in terms of uncertainty in the output, how well the output can reproduce hydrological and ecological observations, the suitability of each approach for different questions, and the ease of implementation. This evaluation is scheduled to begin in September, 1994, and the initial phase is expected to be concluded by June 1995.

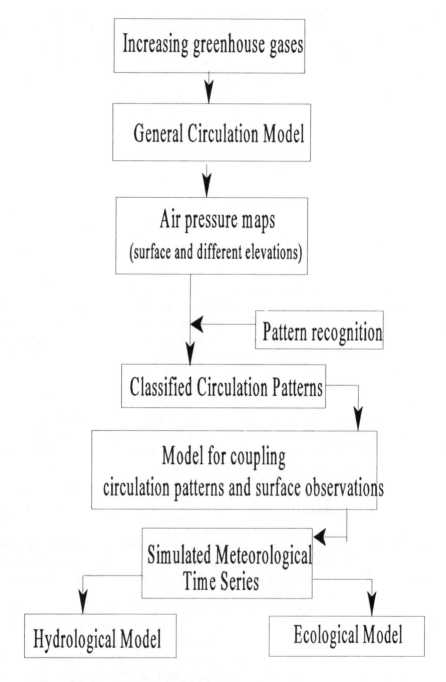

Figure 3.9. Example of a stochastic downscaling system (Bardossy and Caspary, 1990).

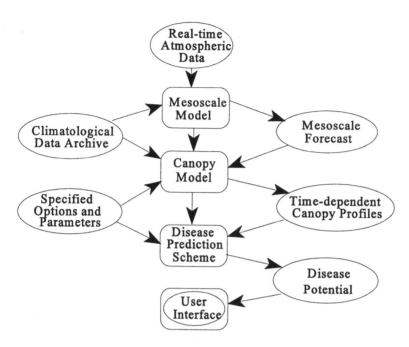

Figure 3.10. Example of a dynamic downscaling system.

3.4. Conclusions

The major goals of BAHC, to determine the biospheric controls of the hydrologic cycle and to develop the appropriate data bases are being met by two different modelling efforts. The first goal is using SVAT models at different spatial scales, beginning with the patch scale and aggregating up to the regional scales. This can only be achieved through a simplification of the parameters used in the SVAT models, but the results of the HAPEX-MOBILHY experiment suggest that simple rules of aggregation may be sufficient. An alternative approach is to use parameters that are appropriate at lower spatial resolutions and that can be derived from remotely sensed data. Regional SVAT modelling in mountainous regions is difficult due to the effects of the extreme topographical variation.

The calibration and validation of SVAT models requires weather data at higher resolutions than is currently available. The second goal is the utilization of different algorithms to move from data at the continental or similar scales down to a 10 km grid. Both stochastic and dynamic algorithms are available, but can only be implemented with certain assumptions. Spatial interpolation methods are used to extend site specific downscaled values to other sites, or to develop a high-resolution grid. This is difficult in mountainous regions due to sparseness of observations and the complexity of other factors affecting the weather at the mesoscale (BAHC, 1994). One possible solution is to use a mesoscale model to produce an initial 10 km grid which can be refined with different interpolation methods.

Acknowledgements

The authors with to acknowledge the contribution Hans-Jurgen Bolle, Ephrat Lahmer-Naim, Geoff Kite, John Porter, Mikhail Semenov and Roland Schultze for their help in editing pieces of this article. We would also like to acknowledge Dr. Alfred Becker who provided help in preparing the original presentation and Hans-Jurgen Bolle, Woo Ming-ko, Liu Changming and Paul Louie who provided generous assistance in arranging the participation of Brad Bass in the conference.

References

Akkur, N.K. (1992) *An Integrated Statistical/Knowledge Engineering Framework to predict Synoptic Ozone Risks at Brisbane, Australia*, Unpublished Ph.D Thesis, University of Queensland, Australia, November, 1992.

Akkur, N.K., Cameron, I.T., Auliciems, A., and Verrall, K. (1992) Synoptic Ozone Risks at Brisbane, Australia - Prediction using Multivariate Analysis and Rule Induction, *Proceedings of the International Clean Air Conference*, Brisbane, August, 1992.

BAHC (1993a) *Activity 4.1: Data Requirements for Ecological and Hydrological Studies and Related Management Purposes. Biospheric Aspects of the Hydrological Cycle (BAHC) Focus 4: The Weather Generator Project*, Summary Report on Prepared by the BAHC-Focus 4 working group, Bratislava, Slovakia, 16-18 September 1993, B. Bass (Chair), in BAHC Report No. 1, 1-20. (Available from BAHC Core Project Office, Freie Universität Berlin, Germany.)

BAHC (1993b) *Activity 4.2: Development of the Weather Generator. Biospheric Aspects of the Hydrological Cycle (BAHC) Focus 4: The Weather Generator Project*, Prepared by the downscaling sub-group, Toronto, Canada, 1-3 December 1993, B. Bass (Chair), in BAHC Report No. 1, 21-32. (Available from BAHC Core Project Office, Freie Universität Berlin, Germany.)

BAHC (1994) Climate-Hydrology-Ecosystems Interrelations in Mountainous Regions (CHEMSO), IGBP-BAHC/UNEP Workshop. St. Moritz, Switzerland, 2-5 December 1993. (Available from BAHC Core Project Office, Freie Universität Berlin, Germany.)

Bardossy, A., and Caspary, H.J. (1990) Detection of climate change in Europe by analyzing European Atmospheric Circulation Patterns from 1881 to 1989, *Theoretical Applied Climatology* **42**, 133-167.

Bardossy, A., and Plate, E.J. (1992) Space-time model for daily rainfall using atmospheric circulation patterns, *Water Resources Research* **28**, 1247-1259.

Bardossy, A., Duckstein, L., Bogardi, I., and Muster, H. (in press) Stochastic precipitation modeling using fuzzy classification of atmospheric circulation patterns, in K. Hipel (ed.)

Proceedings of the Unny Conference on Stochastic and Statistical Methods in Hydrology and Environmental Engineering, June 21-23, 1993, Waterloo, Canada, Volume 3, Time Series Analysis and Forecasting, Kluwer Academic Publishers, Dordecht, The Netherlands.

Bass, B., Yin, Y., and Huang, G. (1994) A grey systems approach to estimating the risk associated with general circulation models, in K. Hipel (ed.), *Proceedings of the Unny Conference on Stochastic and Statistical Methods in Hydrology and Environmental Engineering*, June 21-23, 1993, Waterloo, Canada, Volume 3. Time Series Analysis and Forecasting, Kluwer Academic Publishers, Dordecht, The Netherlands.

Bolle, H.-J. (1993) Scientific goals of the IGBP Core Project 'Biospheric Aspects of the Hydrological Cycle', in H.-J. Bolle, R.A. Feddes and J.D. Kalma (eds.), *Exchange Processes at the Land surface for a Range of Space and Time Scales*. IAHS Publication 212, 3-11.

Deng, J. (1984) *The Theory and Methods of Socio-economic Grey Systems*, Science Press, Beijing, China. (In Chinese.)

Dickenson, R. (1992) Land Surface, in K.E. Trenberth (ed.), *Climate System Modelling*, Cambridge University Press, Cambridge, UK, 149-172.

Dorman, J.L., and Sellers, P.J. (1989) A global climatology of albedo, roughness length and stomatal resistance for atmospheric general circulation models as represented by the simple biosphere model (SiB), *J. Climate and Applied Meteorology* **28**, 833-855.

Giorgi, F., and Mearns, L.O. (1991) Approaches to the simulation of regional climate change: a review, *Reviews of Geophysics* **29**, 191-216.

Hantel, M. (1987) Subsynoptic vertical heat fluxes from high-resolution synoptic budgets, *Meteorology and Atmospheric Physics* **36**, 24-44.

Huang, G.H., and Moore, R.D. (1993) Grey linear programming, its solving approach, and its application, *International J. Systems Science* **24**, 159-172.

Hutchinson, M.F. (1991) Climatic analyses in data sparse regions, in R.C. Muchow and J.A. Bellamy (eds.), *Climatic Risk in Crop Production*, CAB International, Wallingford, UK, 55-71.

International Geosphere-Biosphere Programme (1990) *The International Geosphere-Biosphere Programme. A Study of Global Change. The Initial Core Projects*, IGBP Report No. 12, Stockholm, Sweden.

International Geosphere-Biosphere Programme (1993) *Biospheric Aspects of the Hydrological Cycle: The Operational Plan*, IGBP Report No. 27, Stockholm, Sweden.

Pitman, A.J., Yang, Z.-L., Cogley, J.G., and Henderson-Sellers, A. (1992) *Description of Bare Essentials of Surface Transfer for the Bureau of Meteorology Research Centre (BMRC) AGCM*, BMRC Research Report No. 32, Melbourne, Australia.

Running, S.W. (1992) Modelling the Earth System, in E. Ojima (ed.), *Papers arising from the 1990 OIES Global Change Institute, Snowmass, Colorado, 16-27 July 1990*, NCAR/Office for Interdisciplinary Earth Studies, Boulder, Colorado, 263-280.

Running, S.W., Nemani, R.R., and Hungerford, R.D. (1987) Extrapolation of synoptic meteorological data in mountainous terrain, and its use for simulating forest evaportranspiration and photosynthesis, *Canadian J. Forest Research* **17**, 472-483.

Running, S.W., Nemani, R.R., Peterson, D.L., Band, L.E., Potts, D.F. and Pierce, L.L. (1989) Mapping regional forest evapotranspiration and photosynthesis by coupling simulation, *Ecology* **70**, 1090-1101.

Sellars, P. (1991) Modelling and observing land surface-atmosphere interactions on large scales, *Surveys in Geophysics* **12**, 85-114.

Shuttleworth, W.J. (1991) The Modellion concept, *Review of Geophysics* **29**, 585-606.

Shuttleworth, W.J. (1994) Large-scale experimental and modelling studies of hydrological processes, *Ambio* **23**, 82-86.

Zack, J.W., Waight, K.T., Young, S.H., Ferguson, M., Bousquet, M.D., and Price, P.E. (1993) Development of a Mesoscale Statistical Thunderstorm Prediction System, Final Report, Contract NAS 10-16670.

Brad Bass
Environmental Adaptation Research Group
4905 Dufferin St., Downsview, Ontario, M3H 5T4 Canada.

Naresh Akkur
Environmental and Social Systems Analysts Ltd.
1765 W 8th Avenue, Vancouver, B.C., V6L 1A8 Canada.

Joe Russo
ZedX Inc.
Boalsburg, PA, USA.

John Zack
Meso Inc,
Troy, N.Y., USA.

4 TRENDS IN HISTORICAL STREAMFLOW RECORDS

F.H.S. CHIEW and T.A. McMAHON

Abstract

This chapter investigates whether statistically significant trends or changes in means can be detected in the historical streamflow series. Five statistical tests are applied to the annual streamflow volumes and peak discharges of 142 rivers throughout the world. The analyses indicate that although statistically significant trends and changes are detected by some of the tests at several locations, they are not consistently observed throughout specific regions. As such, there is no clear evidence yet of climate change in the annual streamflow volumes and peak discharges of rivers in the world.

4.1. Introduction

Analyses of historical records can provide information about the natural climate trends and variability. For example, observed records can be used to make deductions about changes resulting from increases in greenhouse gases. However, it is difficult to detect trends in historical data because of the large variability in and relatively short period of instrumental record. Nevertheless, numerous studies have shown that the mean global surface air temperature has increased by about 0.5°C since the beginning of this century (Hansen and Lebedeff, 1987, 1988; Jones, 1988; and Houghton *et al.*, 1990, 1992).

Although temperature is commonly used as the indicator of the state of the climate system, trends in other climatic variables such as precipitation, sea level, snow, ice cover and water vapour, have also been studied. In particular, changes in precipitation are important as they directly influence society. Although there have been several large-scale analyses of precipitation changes (Bradley *et al.*, 1987; Diaz *et al.*, 1989; and Houghton *et al.*, 1990), it is difficult to draw conclusions about changes in global precipitation patterns because it varies greatly in time and space.

This chapter describes a study carried out to investigate whether statistically significant trends or changes in means can be detected in the annual streamflow volumes and peak discharges of 142 rivers throughout the world.

The study of trends in streamflow is useful for three main reasons. First, changes in precipitation are usually amplified in runoff and therefore, it should be easier to detect climate change in runoff than in precipitation. Second, streamflow data integrate some of the spatial variability within the catchment. As such, a single streamflow gauging

J.A.A. Jones et al. (eds.), Regional Hydrological Response to Climate Change, 63–68.

station can provide as much information as precipitation time series from several rainfall stations in the catchment. Third, a study of changes in runoff is important because runoff influences directly the management of land and water resources.

4.2. Streamflow Data

The annual runoff volumes and peak discharges are selected from the global streamflow dataset of McMahon *et al.* (1992). In preparing the dataset, efforts were taken to ensure that the data are for unregulated flow conditions. Only stations with at least 50 years of record and drainage areas larger than 1000 km^2 are used. Altogether, 120 annual streamflow and 83 annual peak discharge series are analysed (see Figures 4.1 and 4.2 for locations of stations). Although all six major continents are represented, most of the data come from North America, Europe and eastern Australia. The records generally extend from the beginning of this century to about the 1970s. The record lengths vary from 50 to 162 years with an average length of 68 years (a median of 61 years). The catchment areas range from 1000 to 8 060 000 km^2 with an average area of 225 000 km^2 (a median of 16 000 km^2).

4.3. Statistical Tests

Five statistical tests are used - Mann (Kendall, 1970), Cumulative Deviation (Buishand, 1982), Worsley Likelihood Ratio (Worsley, 1969), Distribution-Free CUSUM (McGilchrist and Woodyer, 1975) and Kruskal-Wallis (Sneyers, 1975) (see also Chiew and McMahon, 1993). The Mann statistic tests whether there is a statistically significant trend in the time series. The Cumulative Deviation, Worsley Likelihood Ratio and Distribution-Free CUSUM tests locate a change point and test whether the means in the two parts of the dataset are statistically different. The Kruskal-Wallis statistic tests whether flows in any period (a 10-year sub-period is used here) are significantly different from flows at other times.

The Mann, Distribution-Free CUSUM and Kruskal-Wallis are non-parametric tests, where the calculation of the test statistic is based on the ranks and not the actual values of the records. The reliability of these three statistical tests therefore does not depend on the distribution of the observations. The Cumulative Deviation and Worsley Likelihood Ratio however, assume that the observations are independent and normally distributed, although these tests can still be applied when there are slight departures from normality. The Mann, Cumulative Deviation, Worsley Likelihood Ratio and Kruskal-Wallis tests are recommended by the World Meteorological Organisation (1988) as procedures for detecting trend or change in long hydrological time series.

4.4. Analyses and Discussion

A 5% critical value for the two-sided probability is used. The statistics therefore test the null hypothesis H_0 that the observations are randomly ordered versus a trend or change in the mean, as opposed to the one-sided probability which tests against either an increase or a decrease in the data sequence. The streamflow series from each station is analysed separately because there are insufficient data to consider average runoff over specific regions. In order to make some general observations, Figures 4.1 and 4.2 illustrate locations where 'two to three' or 'four to five' of the statistical tests show significant trends or changes in the means of annual streamflow volumes and peak discharges respectively.

The analyses indicate that trends and changes in the means in the annual series are detected at the 5% level of significance by more than one statistical test at about 25% of the stations and by at least four tests at 10% of the stations.

In North America, statistically significant changes in the annual streamflow series are detected by four or all statistical tests at 20% of the stations analysed (runoff volumes at 44 stations and peak discharges at 18 stations). Negative trends (flows have been decreasing or mean of flows in the latter part of record is lower than that in the earlier part of record) in both the streamflow volumes and peak discharges are detected (usually by three or four tests) at several stations in the north-east, and positive trends in streamflow volumes are observed for some stations in the south-east.

In Europe, trends are detected by more than one statistical test (generally only two or three) in less than 10% of the streamflow volume series of 48 locations and about 25% of the peak discharge series of 36 stations. The trends, where observed, are generally negative.

In east Australia, trends in streamflow volumes and peak discharges are detected by more than one statistical test at only about 15% of the 21 stations analysed.

The limited data from the other continents show positive trends in the peak discharges of three stations in west Africa and streamflow volume of one station in north-east China. Only one of the eight annual streamflow series in South America show a statistically significant change.

4.5. Conclusions

Statistically significant trends or changes in means in the annual streamflow volumes and peak discharges are detected by some of the tests at several locations. However, except for certain parts of North America, they are not consistently observed throughout specific regions. In addition, the trends where observed, may have resulted from changes in physical catchment characteristics (although the flows are purported to be for unregulated conditions). In any case, if flow conditions in a geographical region have altered as a result of climate change, most of the statistical tests should detect changes in the streamflow records of all stations in the region.

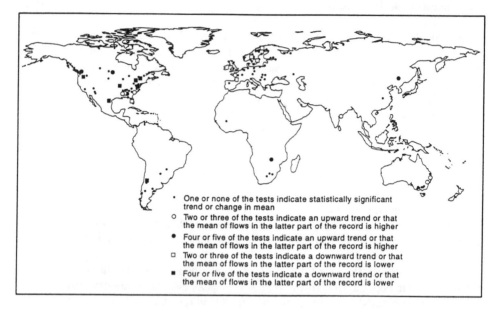

Figure 4.1. Summary of statistical analyses of annual streamflow volumes.

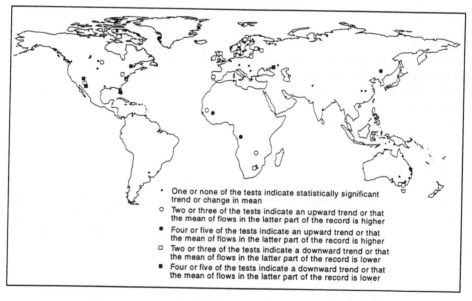

Figure 4.2. Summary of statistical analyses of annual peak discharges.

The analysis thus indicates that there is no clear evidence yet of climate change in the annual streamflow volumes and peak discharges of rivers in the world. However, it should be noted that the detection of trend or change is difficult because of the short period of instrumental record, high inter-annual variability and lag between increased concentration of greenhouse gases and the climatic variables affecting streamflow.

References

Bradley, R.S., Diaz, H.F., Eischeid, J.K., Jones, P.D., Kelly, P.M., and Goodess, C.M. (1987) Precipitation fluctuations over Northern Hemisphere land areas since the mid-19th century, *Science* **237**, 171-175.

Buishand, T.A. (1982) Some methods for testing the homogeneity of rainfall records, *J. Hydrology* **58**, 11-27.

Chiew, F.H.S., and McMahon, T.A. (1993) Detection of trend or change in annual flow in Australian rivers, *International J. Climatology* **13**, 643-653.

Diaz, H.F., Bradley, R.S. and Eischeid, J.K. (1989) Precipitation fluctuations over global land areas since the late 1800s, *J. Geophysical Research* **94**, 1195-1210.

Hansen, J., and Lebedeff, S. (1987) Global trends of measured surface air temperature, *J. Geophysical Research* **92**, 13345-13372.

Hansen, J. and Lebedeff, S. (1988) Global surface temperatures: update through 1987, *Geophysical Research Letters* **15**, 323-326.

Houghton, J.T., Jenkins, G.J., and Ephraums, J.J. (1990) *Climate Change - The IPCC Scientific Assessment*, Report prepared for IPCC by Working Group 1, Cambridge University Press, 364 pp.

Houghton, J.T., Callander, B.A., and Varney, S.K. (1992) *Climate Change 1992 - The Supplementary report to the IPCC Scientific Assessment*, Working Group 1, Cambridge University Press, 200 pp.

Jones, P.D. (1988) Hemispheric surface air temperature variations: recent trends and an update to 1987, *J. Climate* **1**, 654-660.

Kendall, M.G. (1970) *Rank Correlation Methods*, 4th Edition, London, 202 pp.

McGilchrist, C.A., and Woodyer, K.D. (1975) Note on a distribution-free CUSUM technique, *Technometrics* **17**, 321-325.

McMahon, T.A., Finlayson, B.L., Haines, A.T., and Srikanthan, R. (1992) *Global Runoff - Continental Comparisons of Annual Flows and Peak Discharges*, Catena Paperback, Cremlingen-Destedt, Germany, 166 pp.

Sneyers, R. (1975) *Sur l'analyse statistique des series d'observations*, WMO Technical Note 143, World Meteorological Organisation, Geneva, 192 pp.

World Meteorological Organisation (1988) *Analysing Long Time Series of Hydrological Data With Respect to Climate Variability*, WCAP-3, WMO/TD 224, 12 pp.

Worsley, K.J. (1969) On the likelihood ratio test for a shift in location of normal populations, *J. American Statistical Association* **74**, 365-367.

F.H.S. Chiew and T.A. McMahon
CRC for Catchment Hydrology
Department of Civil and Environmental Engineering
University of Melbourne
Parkville, Victoria 3052, Australia.

SECTION II

REGIONAL IMPLICATIONS OF GLOBAL WARMING

SUMMARY

Impacts in northern North America

■ Many GCMs show increased precipitation caused by increased storminess in the Arctic, but there is less agreement on actual amounts.

■ Annual snowfall could either increase or decrease according to equilibrium GCM scenarios. Calculations for a subarctic station based on the GISS scenario show an increase, but GFDL and OSU scenarios show higher winter snowfall being counterbalanced by higher temperatures in other seasons.

■ Warmer temperatures reduce the length of winter and increase the thawed period at lower altitudes. Lapse rates will ensure less change in winter duration at high altitudes.

■ Snowfall will probably increase at high altitudes, rainfall at lower altitudes.

■ There could be more melt events triggered by rain-on-snow events, but increased cloudiness will reduce the effectiveness of radiation melting.

■ Some small arctic glaciers that are remnants of Ice Age climate may disappear, but larger and higher ice masses may thicken with increased snowfall.

■ Ablation will increase on the lower slopes, so that icemelt runoff will increase, but in ice masses that originate in high mountains increased accumulation could steepen the glacier profile and increase the turnover or activity rate without significant change in the ice volume.

■ Where glaciers calve in tidal waters, the increased rate of glacier flow could cause more icebergs.

■ Permafrost is likely to thaw over large areas, and so increase infiltration rates and groundwater storage capacity. This will increase winter discharges supported by groundwater outflow in the subarctic.

■ It will also reduce the proportion of overland flow and decrease peak discharges during snowmelt in the subarctic.

■ Freshwater ice covers will be thinner by 0.5 -1.0 m, and the duration of open water will be weeks longer, both inland and in coastal waters.

■ Evaporation will increase due to the shorter duration of snow and ice covers and advected heat; possibly less so in the Arctic Islands due to lower vapour pressure deficits.

■ Declines in snowmelt discharge will be countered by increases in rainfall runoff.

Impacts in Europe

■ Europe lies on a climatological frontier that is highly sensitive to changes in the meridional temperature gradient and North Atlantic sea surface temperatures. Fluctuations in the planetary waves can have major impacts upon European hydrology through changing frequencies of airflow pattern types and synoptic precipitation-forming processes.

■ Global warming is likely to increase annual precipitation over a large part of Northern and Northwestern Europe, but cause decreases in the South and the Mediterranean.

■ Changes in seasonality, variability, extreme events and storminess could also be significant and cause problems, especially during the transition period.

■ Increased drought risk is expected in the South.

■ There is evidence for a more widespread increase in flood risk.

■ Many instrumental records from the past 100 years show some similarities with GCM-based predictions.

■ Palaeohydrological analogues give rather variable predictions and it is probable that many causal factors in the past were different to those of recent and predicted future climatic changes.

■ Techniques generally require considerable development before definitive predictions can be claimed, but progress is rapid in both the development of appropriate scenarios and the adaptation of methods for hydrological prediction.

Impacts in China

■ Taking the warm periods over the last 100 years as analogues of the global warming effect indicates wide regional diversity in hydrologic response.

■ Discharges were especially low in the North during the prolonged warm periods of the 1920s-1940s and the 1980s.

■ Lakes have steadily reduced in number, area and volume. Total lake area has fallen by 12%. In the West on the Qingzang/Tibetan Plateau and in the North on the Inner Mongolian Plateau this is due to warming and reduced precipitation. In the East it is due more to land reclamation.

■ Glaciers have been declining generally for most of this century. Continued reductions in glacier mass balance are predicted, along with a rise in the snow line and an eventual decrease in meltwater discharge.

■ The rivers of the plains are more sensitive to global warming because evaporation already stands at around 90% of precipitation in the North China Plain.

■ Comparing the 1950s cool period with the 1980s warm period shows:

1. General warming north of 35°N by c.1°C, compared with -0.8°C elsewhere.

2. Precipitation has decreased over most of the region, except the North and Northwest interior where there was little precipitation anyway, even in the cool 1950s.

■ GCM predictions tend to show that increased precipitation will accompany a temperature rise in the densely populated area of Northeast China, but the observational evidence of the last 100 years suggests a general decrease in precipitation.

5 HYDROLOGY OF NORTHERN NORTH AMERICA UNDER GLOBAL WARMING

MING-KO WOO

Abstract

Northern hydrology is strongly influenced by snow, ice and permafrost. Climatic warming projected by various global climate models (GCMs) for a 2 x CO_2 atmosphere suggests that while annual snowfall may or may not decrease in northern North America, the length of the thawed season will increase, at least at the low altitudes. At high elevations, thawing effect is minimized by temperature lapse rates and the ice-sheets may thicken at their accumulation zones while ablation is accelerated at their tongues. Evaporation will increase, and permafrost thaw will enlarge the groundwater storage capacity. Ice cover thickness will diminish and some northern wetlands may be altered. Rainfall generated peak flows will become more prominent as the snowmelt contribution declines. Relative magnitudes of various components of the water balance will change as the northern environment evolves in response to the changing climate.

5.1. The Northern Environment

The northern part of North America comprises the subarctic and the Arctic regions, both of which are characterized by long periods of extreme coldness during each year. The physical landscape ranges from rolling coastal lowlands in western Arctic, to rugged ranges and plateaus of the Western Cordilleras, to extensive flatlying wetlands, to glacier-clad mountains in Eastern Arctic (Graf, 1987). Sea ice covers the ocean, the straits and embayments for over nine months each year and under the land, permafrost occurs continuously in the Arctic and discontinuously in the subarctic (French and Slaymaker, 1993). The distinction between the Arctic and the subarctic is based on vegetation, with trees surviving in the subarctic while only tundra and bare ground are found in the Arctic. The distribution of permafrost and glaciers and the physiographic subdivision of the area are shown in Figure 5.1.

Northern hydrology is dominated by phase changes (melt and evaporation) and seasonal to long-term storages (snow, glaciers and ground ice). Energy considerations are highly significant in sustaining evaporation and in releasing water from snow and ice storage to generate runoff. Climatic change, involving alterations of the temperature and

73

J.A.A. Jones et al. (eds.), Regional Hydrological Response to Climate Change, 73–86.

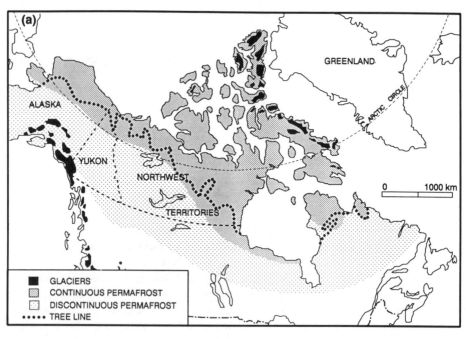

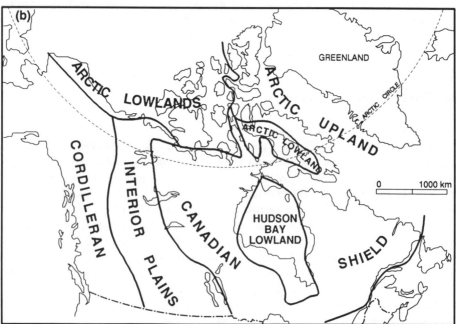

Figure 5.1. Northern North America showing (a) the boundaries of permafrost zones, treelines, glaciers and (b) major physiographic units.

precipitation patterns, will have impacts on the northern hydrological regime. The purpose of this paper is to provide an overview on the possible consequences of global warming on the hydrological environment of northern North America, using the climatic change scenarios provided by several Global Climate Models (GCMs).

5.2. Climatic Warming Scenarios

One popular approach to study climatic change impacts is to make use of scenarios produced by GCMs for a hypothetical atmosphere in equilibrium with twice the pre-industrial level of carbon dioxide concentration (or its equivalent greenhouse gases). Different GCMs produce somewhat different results. There is little agreement on how precipitation will change in northern North America under $2 \times CO_2$ modelled situation, but all scenarios point to notable temperature rises, particularly during the winter months. An example of such temperature changes (shown as temperature differences between $2 \times CO_2$ and the $1 \times CO_2$ simulated outputs) as produced by the Canadian Climate Centre GCM (Canadian Climate Program Board, 1991) is given in Figure 5.2.

Under this scenario, most parts of northern Canada will be warmer by 2 to 4°C in the summer but during the winter, temperatures of the Queen Elizabeth Islands and the Hudson Bay area will be raised by 8°C. It must be recognized that the present average winter temperature for these areas is below -20°C. An increase as proposed by the GCM scenario will still be insufficient to effect winter thaw, particularly at the latitudes where little solar radiation is received during this period. For precipitation, the Canadian Climate Centre GCM scenario suggests summer increase of over 20% for the Arctic Islands, and about 10% increase for the northern mainland. Winter precipitation may rise by about 10% over the mainland, but will change only slightly on the Islands. Despite uncertainties of the GCM scenarios, they may be used to indicate the relative magnitude of the climatic tendencies in the next century.

Computer technology and data availability limit the spatial resolution of the GCM outputs to a scale no better than several degrees of longitude and latitude. Such resolution may be suitable for large drainage basins, but is inadequate for hydrological impact studies at the scales of medium to small catchments. One possibility is to use hybrid climatic data, combining the coarse GCM outputs with historical records (Robinson and Finkelstein, 1991). This is a crude approach but will allow order-of-magnitude estimates of hydrological impacts to be inferred.

5.3. Precipitation, Snow and Glaciers

Precipitation is the major source of water input in the water balance of northern basins. Several GCMs produce scenarios of future precipitation increase due to greater storminess in the Arctic, but it must be emphasized that there is no agreement on the magnitude of change (Etkin, 1990; Maxwell, 1992). Of particularly hydrological importance is the amount of snowfall which represents the bulk of the present

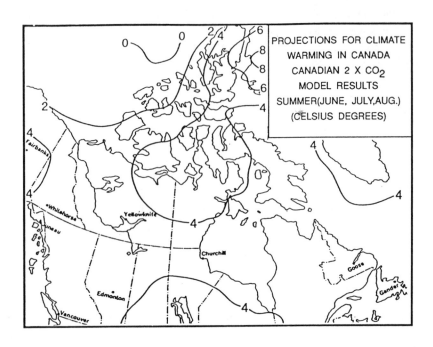

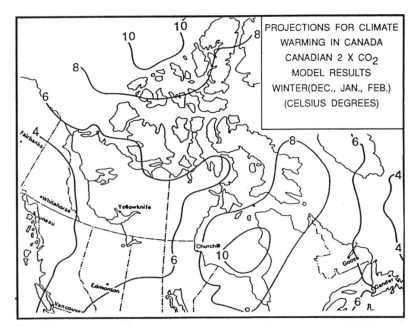

Figure 5.2 . Temperature differences between 1 x CO_2 and 2 x CO_2 for North America: (a) June to August, (b) December to February (according to the Canadian Climate Centre GCM scenario).

precipitation. Taking four GCM scenarios and applying their 2 x CO_2 temperature and precipitation changes to a subarctic station (Norman Wells, at 61°15'N, 126°30'W), the probability distributions of annual snowfall at this station were estimated. This study (Woo, 1992) made use of present and 2 x CO_2 daily values (which were stochastically simulated based on the GCM scenarios superimposed onto present-day conditions), and assumes that any precipitation falling on days with mean temperature below 0°C is snowfall. These daily amounts were cumulated over the year and the annual snowfall probabilities were then derived.

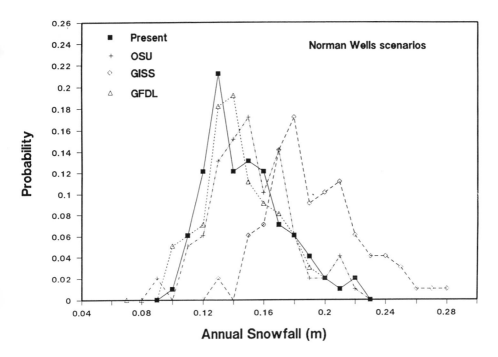

Figure 5.3. Probabilities of annual snowfall for Norman Wells, N.W.T., Canada under 2 x CO_2 scenario according to three GCMs and compared with the present conditions.

The simulation results illustrate the uncertainties regarding the future status of snowfall (Figure 5.3). According to the GISS (Goddard Institute of Space Studies) scenario, snowfall will increase, but both the GFDL (Geophysical Fluid Dynamics Laboratory) and the OSU (Oregon State University) scenarios suggest that warmer conditions in the future will be compensated by higher snowfall in winter, so that annual snowfall will not differ significantly from the present.

While mixed results are obtained for the annual snowfall, warmer conditions reduce the duration of winter, at least for the low altitudes. At higher elevations, temperature lapse rates will prevent the temperature to rise above the freezing point for a longer period. This factor, together with increased storminess and orographic influences, favours more snowfall at higher elevations while rainfall may increase at the lower zones. Rain-on-snow melt events may be more prevalent than the present, but increased cloudiness may also reduce the rate of radiation melt. A simulation was performed by altering the present-day climate of Resolute, Cornwallis Island, according to the GISS 2 x CO_2 scenario (Table 5.1), using the method described by Woo (1992a). Daily temperatures for several elevations were computed by assuming lapse rates of 0.01°C m^{-2} for a dry day and 0.006°C m^{-2} for a precipitation day. At altitudes above 1000 m, future warming may have little effect on increasing the number of thawing days which are days when mean air temperature is above 0°C (Figure 5.4). Most precipitation at these elevations will fall as snow and the exceedance probabilities of annual snowfalls will not differ from those of the present (Figure 5.4). With reduced melt due to greater storminess, more semi-permanent snowpacks at high altitudes are expected to be preserved.

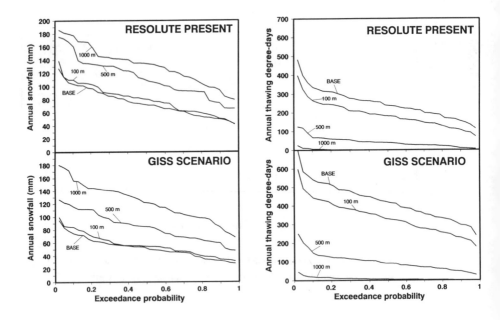

Figure 5.4. Computed exceedance probabilities of annual snowfall and annual thawing degree-days for selected elevations near Resolute, N.W.T., Canada under present climate and the 2 x CO_2 scenario according to the GISS GCM.

TABLE 5.1. Mean monthly precipitation and temperature at Resolute, N.W.T., Canada, and changes under 2 x CO₂ scenario according to GISS GCM.

	Precipitation (mm)	Change in Precipitation (%)	Temperature (°C)	Change in Temperature (°C)
January	1.0	169	-27.2	9.5
February	6.0	167	-27.8	8.9
March	4.4	157	-26.4	8.4
April	5.2	160	-19.7	3.6
May	10.3	138	-10.1	2.7
June	5.2	96	1.1	3.2
July	7.2	132	4.1	0.9
August	46.7	123	1.9	1.5
September	36.2	126	- 4.1	3.5
October	4.0	114	-12.3	11.0
November	8.6	111	-19.2	13.8
December	5.7	136	-20.3	9.9

At present, some arctic glaciers are probably remnants from the Ice Age (Bradley and Serreze, 1987). As climate warms, these small, remnant glaciers at low altitudes may disappear. Ablation will increase at the lower elevation zones of the larger glaciers, yielding higher summer runoff in the proglacial areas until the glaciers diminish in size. At high elevations, cooler conditions will reduce ablation while the possibility of precipitation increase may cause the accumulation zones of the ice sheets to thicken, as was reported to have happened during the Holocene (Miller and de Vernal, 1992). Since many ice-caps occupy a wide range of altitudes, sending glacier tongues outward from their margins, it is conceivable that there will be a steepening of the glacier profiles due to enhanced low altitude ablation and increased accumulation at the higher zones. Glacier flows may be accelerated and this will cause more calving along the margin. Calving of glaciers at the tidewater will produce more icebergs, so long as there is sufficient ice flow to sustain ice loss at the snouts.

5.4. Evaporation

Evaporation in the northern latitudes is limited by the long duration of snow and ice on the ground and the low level of energy available during the thawed season despite the long day length (Vowinckel and Orvig, 1970). A warming scenario will not only shorten the snow and ice cover duration, except for high altitudes, but will provide more heat in the summer. Other climatic factors being equal, higher summer temperatures will increase evaporation, as is demonstrated by a simulation study carried out for a central

Alaskan site (Kane *et al.*, 1992). For the Arctic Islands and coastal mainland, warmer conditions will lengthen the duration of open water conditions in the channels and inlets. Coastal fog and low clouds thus formed (Vowinckel and Orvig, 1970) will reduce the radiation. Lower energy supply and saturated air will counter evaporation enhancement due to higher temperatures. In the subarctic, unless storminess increases, evaporation is expected to increase. Should climatic change continue, the vegetation will likely be different from today, possibly at all scales from the species to the biome level. If the lichens and mosses which tend to be suppressors of evapotranspiration (Rouse *et al.*, 1977) are replaced by transpiring plants, evaporative losses will increase.

5.5. Permafrost and Groundwater

Impact models for northern regions indicate permafrost degradation under climatic warming scenarios (Kane *et al.*, 1991), though ground thaw is complicated by a host of factors besides air temperature rises (Smith and Riseborough, 1983). In the Arctic, the active layer (or the surface layer which experiences annual freeze and thaw) will thicken. In the subarctic where permafrost is discontinuous at present, some isolated patches may disappear while more thawed zones (called taliks) will form in areas with more extensive permafrost occurrence.

Permafrost usually has very low hydraulic conductivity which inhibits water movement. At present, groundwater occurs either in the taliks, or seasonally in the active layer. Active layer storage capacity is limited and is highly variable, depending on how deep thawing has reached during different times of the year (Woo, 1986). In spring when the active layer is largely frozen, meltwater will not penetrate easily and much runs off as overland flow. The intensity of overland flow lessens in the subarctic because the active layer thaws more deeply and more rapidly. There are also many taliks connecting the groundwater within or below the discontinuous permafrost. These subterranean passages allow groundwater flow to continue in winter and to maintain baseflow or seepage to slopes. Such groundwater discharges produce icings when frozen by the sub-freezing conditions at the surface (Sloan *et al.*, 1976). Winter discharge is limited in the continuous permafrost area because the active layer storage alone is insufficient to maintain much flow. There, both baseflow and icing formation are restricted.

Thickening of the active layer provides greater storage capacity for groundwater. Snowmelt may also be earlier and the intensity of meltwater release may decrease. These conditions are less conducive to overland flow in the spring. Given that more water will infiltrate and be stored, especially for rain that falls in the thawed period, there will be more groundwater available to sustain baseflow in the fall. The streamflow season will be extended in the continuous permafrost areas. There, icing may become more prevalent in the early winter. Continued streamflow in winter will enable the formation and growth of river ice until the flow ceases in the channels.

The bulk of the moisture in the permafrost occurs as ground ice which may be massive and extensive in some areas (Mackay, 1972). Degradation of permafrost will release this ice storage, adding water supply during the transient phase of climatic change. However, once the ice is melted, there will be an irreversible loss of volume, leading to land subsidence. This process of thermokarst may create many small lakes and ponds.

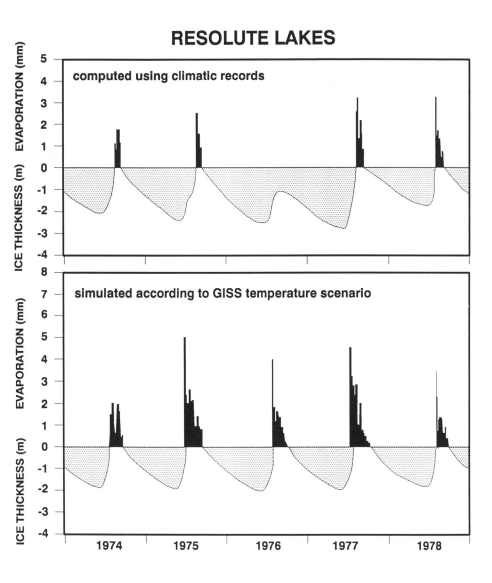

Figure 5.5 . Ice cover and evaporation during open water periods under the present and the GISS 2 x CO_2 scenario, Small Lake at Resolute, N.W.T., Canada.

5.6. Lakes and Wetlands

Arctic and subarctic lakes invariably have a seasonal ice cover which prevents open water evaporation until the ice is ablated. Taking small lakes near Resolute (74°55'N, 94°50'W) as an example, lake ice growth (using Stefan's equation) and decay (using degree-day method) for several years were calculated. For the ice-free period, lake evaporation was computed using the Priestley and Taylor (1973) model, with radiation and temperature as inputs. Assuming that radiation pattern does not change, air temperature and precipitation were increased according to values given by the GISS scenario (Table 5.1). The lake ice duration and open water evaporation were re-calculated and compared with those determined using historical data. Under future warming, lake ice thickness will decrease by 0.5 to 1 m, depending on temperature conditions of the year (Figure 5.5). Ice cover duration will shorten by weeks. A longer period with open water and higher summer temperatures will increase lake evaporation significantly (Figure 5.5). It should be noted, however, that only qualitative interpretation should be made of the results, especially because the future radiation regime which may be different was not considered in this simulation.

Deepening of the active layer or the disappearance of permafrost, coupled with higher evapotranspiration, will modify the hydrological behaviour of northern wetlands. Many small wetlands in the Arctic owe their existence to the meltwater supply from late-lying snowbanks. These isolated wetland patches may vanish if these snowbanks cannot be maintained during a warmer climate. For the extensive subarctic wetlands which occupy large tracts of the discontinuous permafrost zone, enhanced evaporation accompanying the projected climatic warming will lower the water table, followed by changes in the peat characteristics as the wetland surfaces become drier (Woo, 1992b). Whether these wetlands will be preserved depends on the water balance under the future climate, but the flat topography which retards lateral drainage will favour the preservation of these wetlands.

5.7. Streamflow

From the water balance consideration, an increase in evaporation without a compensating increase in precipitation will lead to less runoff. It is expected that under the 2 x CO_2 situation, the runoff ratio (ratio of runoff to precipitation) will decrease.

The runoff response from a permafrost basin to a given rainfall or snowmelt event is more rapid and more extreme than from a non-permafrost basin. This is due to the relatively impermeable nature of the frozen soil which hinders percolation while encouraging lateral flow; and less baseflow because the groundwater storage capacity is reduced by the frozen soils. Slaughter et al.'s (1983) study in the Caribou-Poker catchment (65°10'N) provides a modern day analogue to what the future runoff pattern may be. They found that the sub-basins with higher percentages of permafrost tend to produce larger ranges of flow conditions (yielding more high flow of large magnitude, and more low flows of lower magnitude than the basins with little permafrost).

River ice dynamics complicates the flow conditions during breakup. As noted previously, rivers in the continuous permafrost areas may produce more icing and river ice under a warmer climate. Their breakup processes will approach those experienced by the subarctic rivers of today (Gerard, 1990). In these cases, spring floods will be more severe for the arctic rivers. In the subarctic, however, warming will reduce the river ice thickness and shorten the ice-covered period (Woo, 1992a). There, the breakup dates may be advanced.

All these changes will modify the seasonal rhythm or the regime of streamflow. At present, rivers in continuous permafrost areas usually exhibit an arctic nival regime (Church, 1974). This is characterized by rapid spring melt that releases most of the snow accumulated over the long winter, generating the highest flows of the year. Afterwards, runoff recedes to baseflow, occasionally interrupted by rainfall-induced peaks. Winter is a period of no flow or, for the large rivers in the Arctic mainland, low flow conditions. As the Arctic warms, the duration of the thawed season will be extended, and warmer summers will thicken the active layer. While the capacity for suprapermafrost groundwater storage is increased, winter snow storage may be reduced. The magnitude of spring floods may decline and the baseflow may continue into winter as groundwater is discharged from the enlarged groundwater reservoir. Summer rain events will be more prevalent because the summers will be longer, and this will be accompanied by more frequent rain-induced high flows.

In the subarctic, the nival runoff regime is distinguished from the Arctic equivalent by a longer streamflow season and more prominent summer peaks generated by heavy rainfall events. The presence of discontinuous permafrost permits winter discharge which in turn enables channel icing and river ice formation. The spring snowmelt freshet is often accompanied by river ice breakup which is particularly violent for those large rivers with headwaters in warmer locations (e.g. many subarctic rivers in Canada flow northward). Warming of the discontinuous permafrost area will increase rainfall at the expense of snow accumulation. Summer floods will be more prominent while the snowmelt freshet will be less intense, especially when the river ice becomes thinner due to the shorter and warmer winters.

South of the subarctic, rainfall may constitute the bulk of annual precipitation and many high flows are generated by rain. This gives rise to the pluvial regime (Woo, 1990). Rivers with such a regime are currently located at the warmer and often drier fringes of the discontinuous permafrost area. Should warming arrive, other subarctic nival regime rivers may acquire runoff features more akin to those with a pluvial regime.

Peak flows of northern North America may be regarded to be the product of snowmelt or rainfall. Their separate probability distributions may be represented by two lines on an extremal probability paper (e.g. Gumbel). In the Arctic, the melt-related annual peaks usually exceed the rain-generated peaks, but the opposite holds for the cold temperate latitudes (Figure 5.6). The former represents nival regime rivers and the latter, pluvial regime. In between are rivers with a mixed regime, with melt being usually the producer of annual floods. Global warming will produce more heavy rainfall events and situations detrimental to intense spring melt. Perhaps with the exception of some arctic rivers where ice jams may develop in the future, the pluvial regime floods will strengthen, as is conceptually illustrated by shifts of the lines on the probability plots (Figure 5.6).

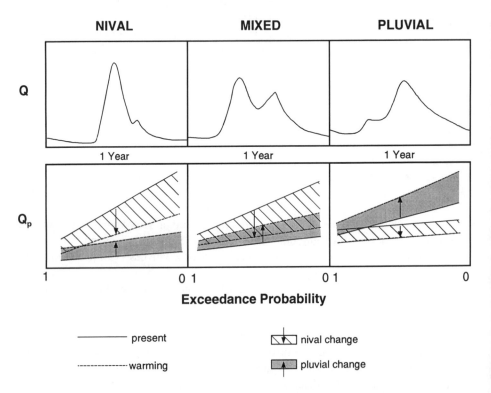

Figure 5.6 . Possible changes in the probability distributions of snowmelt-generated and rainfall-generated peak flows for nival, pluvial and mixed regime rivers.

5.8. Discussion

Climate change and its accompanying environmental changes will both affect the hydrological behaviour of northern basins. On the water supply side, rainfall will most likely increase while snowmelt contribution may or may not decline, though the snow-free season will lengthen. Ground ice melt and glacier melt will first increase, until ground ice storage is depleted, or when the glacier retreats to elevations high enough to be protected from strong ablation. On the water loss side, evaporation will increase, possibly accompanied by greater transpiration as the vegetation changes. Groundwater flow will increase at the expense of surface runoff. In terms of storage change, there may be a loss of ground ice, some gains in groundwater storage, a decrease in wetland storage or an increase in surface ponding due to thermokarst development. Glaciers at low altitudes will shrink through increased ablation and calving. This, according to some (Roots, 1989) will contribute meltwater to sea level rises.

It must be emphasized that the response of the northern environment to climatic change is unlikely to be linear (Woo *et al.*, 1992). The impacts of global warming on the hydrological system are governed by: (a) time lags (uneven storage changes causing variable response times, thresholds versus steady changes, etc.); (b) feedbacks (between the hydrological and external systems, amongst different hydrological elements); (c) random signals (climatic variability, chance occurrence of events and phenomena) and; (d) human factors (development of the North, land use modification). Our task is to deduce the hydrological tendencies under climatic change forcing constrained by uncertainties in the climate model predictions. Given also the current level of knowledge on northern hydrological and climatic processes, only generalized statements on the impacts, not detailed quantification, should warrant our confidence.

Acknowledgements

Research was supported by the Natural Sciences and Engineering Research Council of Canada. Generous logistical support was provided by the Polar Continental Shelf Project of the Canadian Department of Natural Resources to facilitate Arctic field work.

References

Bradley, R.S., and Serreze, M.C. (1987) Mass balance of two High Arctic plateau ice caps, *J. Glaciology* **33**, 123-128.

Canadian Climate Program Board (1991) Climate change and Canadian impacts: the scientific perspective, *Climate Change Digest CCD 91-01*, Environment Canada, Ottawa.

Church, M. (1974) Hydrology and permafrost with reference to northern North America, *Proceedings Workshop Seminar on Permafrost Hydrology, Canadian National Committee for IHD*, Ottawa, 7-20.

Etkin, D. (1990) Greenhouse warming: consequences for arctic climate, *J. Cold Regions Engineering* **4**, 54-66.

French, H.M., and Slaymaker, O. (1993) *Canada's Cold Environments*, McGill-Queen's University Press, Montreal and Kingston, 340 pp.

Gerard, R. (1990) Hydrology of floating ice, in T.D. Prowse and C.S.L. Ommanney (eds.) *Northern Hydrology: Canadian Perspectives*, National Hydrology Research Institute, Science Report No. 1, Canadian Ministry of Supply and Services, 103-134.

Graf, W.L.(1987) Geomorphic Systems of North America, *Geological Society of America, Centennial Special Vol. 2,* Boulder, Colorado, 643 pp.

Kane, D.L., Hinzman, L.D., and Zarling, J.P.(1991) Thermal response of the active layer to climatic warming in a permafrost environment, *Cold Regions Science and Technology* **19**, 111-122.

Kane, D.L., Hinzman, L.D., Woo, M.K., and Everett, K.R. (1992) Arctic hydrology and climate change, in F.S. Chapin *et al.* (eds.) *Arctic Ecosystems in a Changing Climate: an Ecophysiological Perspective.* Academic Press, San Diego, 35-57.

Mackay, J.R. (1972) The world of underground ice, *Annals Association of American Geographers* **62**, 1-22.

Maxwell, B.(1992) Arctic climate: potential for change under global warming, in F.S. Chapin *et al.* (eds.) *Arctic Ecosystems in a Changing Climate: an Ecophysiological Perspective,* Academic Press, San Diego, 11-34.

Miller, G.H., and de Vernal, A. (1992) Will greenhouse warming lead to Northern Hemisphere ice-sheet growth? *Nature* **335**, 244-246.

Priestley, C.H.B., and Taylor, R.J.(1972) On the assessment of surface heat flux and evaporation using large-scale parameters, *Monthly Weather Review* **100**, 81-92.

Robinson, P.J., and Finkelstein, P.L. (1991) The development of impact-oriented climate scenarios, *Bulletin American Meteorological Society* **72**, 481-490.

Roots, E.F. (1989) Climate change: high-latitude regions. *Climate Change* 15, 223-253.

Rouse, W.R., Mills, P.F., and Stewart, R.B. (1977) Evaporation in high latitudes, *Water Resources Research* **13**, 909-914.

Slaughter, C.W., Hilgert, J.W., and Culp, E.H. (1983) Summer streamflow and sediment yield from discontinuous permafrost headwater catchments, *Proceedings, Fourth International Conference on Permafrost,* Fairbanks, Alaska, 1172-1177.

Sloan, C.E., Zenone, C., and Mayo, L.R. (1976) Icings along the Trans-Alaska Pipeline Route, *U.S. Geological Survey Professional Paper* 979, 31 pp.

Smith, M.W., and Riseborough, D.W. (1983) Permafrost sensitivity to climate change, *Proceedings, Fourth International Conference on Permafrost*, Fairbanks, Alaska, 1178-1183.

Vowinckel, E., and Orvig, S. (1970) The climate of the north polar basin, in S. Orvig (ed.) *Climates of the Polar Regions. World Survey of Climatology, v. 14,* Elsevier, New York, 129-252.

Woo, M.K. (1986) Permafrost hydrology in North America, *Atmosphere-Ocean* 24, 201-234.

Woo, M.K. (1990) Consequences of climatic change for hydrology of permafrost zones, *J. Cold Regions Engineering* **4**, 15-20.

Woo, M.K. (1992a) Application of stochastic simulation to climate-change studies, *Climatic Change* **20**, 313-330.

Woo, M.K. (1992b) Impacts of climate variability and change on Canadian wetlands, *Canadian Water Resources J.* **17**, 1-7.

Woo, M.K., Lewkowicz, A.G., and Rouse, W.R. (1992) Response of the Canadian permafrost environment to climatic change, *Physical Geography* **13**, 287-317.

Ming-Ko Woo
Department of Geography
McMaster University
Hamilton, Ontario, Canada L8S 4K1

6 CURRENT EVIDENCE ON THE LIKELY IMPACT OF GLOBAL WARMING ON HYDROLOGICAL REGIMES IN EUROPE

J.A.A. JONES

Abstract

Global warming is likely to increase annual precipitation over a large part of Europe, but changes in seasonality, variability and extreme events could also be significant and cause problems, especially during the transition period. Increased drought risk is expected in the south, but this is likely to be accompanied by a more widespread increase in flood risk. Analyses based on instrumental records, palaeohydrological analogues and models are compared. Techniques generally require considerable development before definitive predictions can be claimed, but progress is rapid in both the development of appropriate scenarios and the adaptation of methods for hydrological prediction.

6.1. Introduction

The Intergovernmental Panel on Climate Change (IPCC) (1990a) report predicted a global warming of 0.3°C/decade over the next half century. The IPCC (1992) calculated an average of 0.23°C/decade and the latest update is not markedly dissimilar (IPCC, 1996). If this order of warming occurs, it will be the fastest rate of change during the historical period. The effects of such a trend on hydrological regimes and water resources in Europe are potentially extremely significant, and it was recognition of this that led to the Conference on Climate and Water convened by the WMO Regional Association VI (Europe) at Helsinki in 1989 (Academy of Finland, 1989). It also led the UK Climate Change Impacts Review Group (UKCCIRG, 1991) to state that scientific guidance on the issue is urgently needed by the water industry.

The most significant research development in Western Europe was the setting up of the Hadley Centre for Climate Prediction and Research by the UK government, combining the research programmes of the Meteorological Office and the Department of Environment. Operational since 1990, the Hadley Centre has established strong links with the 'impacts community', including liaison with the Institutes of Hydrology and Terrestrial Ecology of the Natural Environment Research Council (NERC), and has mapped out a 15 year research schedule. It has also been developing and testing a new 'Global Climate Model', based upon an enhancement of the Meteorological Office General Circulation Model (GCM) and including a high resolution European window.

J.A.A. Jones et al. (eds.), Regional Hydrological Response to Climate Change, 87–131.
© 1996 *Kluwer Academic Publishers. Printed in the Netherlands.*

The new model was fully operational in 1993, though testing and improvement is an ongoing process. Results from its High Resolution Window (*v.i.*) should provide the best available basis for predicting changes in water balances and hydrological regimes in Europe. A second European GCM has been developed at the Max Planck Institute (MPI) in Hamburg.

Meanwhile, a valuable Climate Impacts LINK scheme has been set up by the Hadley Centre, run by the Climatic Research Unit (CRU) of the University of East Anglia. The LINK scheme provides a newsletter and a series of Technical Notes for the impacts community (Viner and Hulme, 1992, 1993a and b, 1994). A major provision of LINK is the construction of standardised climate change scenarios based on GCM output from the Hadley Centre. LINK has constructed scenarios for different rates of accumulation of greenhouse gases by using a simple one-dimensional upwelling-diffusion model, STUGE (the Sea-level and Temperature Under Greenhouse Effect), to scale the output from GCMs. It is still too costly to carry out transient runs of GCMs for a full range of emissions scenarios; even the speeded-up version of the Hadley Centre model took 3 months of continuous computer time to simulate a run of 150 years. LINK has produced regional predictions for a range of emission rates by establishing ratios of regional sensitivity in temperature and precipitation per °C of GCM warming and scaling the regional predictions from the GCM by the STUGE estimates of average global warming for a particular emissions scenario (Viner and Hulme, 1994).

Other developments in the impacts community have included the establishment of a Climatic Change research group at the Institute of Hydrology (UKIH) and collaboration between the UK Water Research Centre (UKWRC) and the Climatic Research Unit (CRU) of the University of East Anglia. The UK Department of the Environment Global Atmosphere Division is also engaged in coordinating research under its Core Model Programme on biogeochemical cycling and wildlife effects, which interface with hydrological interests mainly handled by UKIH, including catchment scale data on soil hydrology and runoff, and water quality modelling (Parr and Eatherall, 1994). The new IH Core Model Demonstrator allows climate change scenarios to be linked to the Arc/Info Geographical Information System for specific catchments.

Among the significant publications arising from this impacts work are those of the UKWRC/UEA (Gregory *et al.*,1990), the UKIH, including interim guidelines for the water industry (Arnell *et al.*, 1990) and a series of reports to the UK Department of the Environment Water Directorate (Arnell and Reynard, 1993; Arnell *et al.*, 1994a) and the National Rivers Authority (NRA) (Arnell *et al.*, 1994b), the UK Institution of Civil Engineers (Binnie, 1991) and the UKCCIRG (1991). Gregory *et al.* (1990) established an homogenous daily rainfall series for regions of the UK and a probability density function model to enable perturbed series to be developed. The other reports are largely concerned to establish 'best estimates' of change as a basis for operational planning. Material of hydrological interest has also emerged from conferences on global warming organised by the British Geomorphological Research Group and the Royal Geographical Society and published in their respective journals (e.g. Gunn and Crumley, 1991; Beven, 1993). The British NERC's TIGER programme (Terrestrial Initiative in Global Environmental Research) has also progressed from focussing on greenhouse gas cycling

(TIGER I and II) to an integrated study of energy and water balances, including a project to develop continental-scale hydrological models under TIGER III, as part of the World Meteorological Organisation/International Council of Scientific Unions (WMO/ICSU) Global Energy and Water balance Experiment (GEWEX).

In essence, there are three types of approach available for prediction of climatic change: (1) extrapolation from instrumental data sets, (2) use of spatio-temporal analogues, and (3) physically-based models (GCMs). The results may then be input into stochastic hydrological models, water balance models or rainfall-runoff models, and in some cases subsequently input into water resource reservoir and systems models. All three climatic approaches have been used in Europe, although GCMs have tended to dominate.

Because of the perceived low information content of climatic predictions, most hydrological work in Europe has concentrated on sensitivity studies rather than categorical prediction of climatic change, and on using a relatively simple water balance approach. The 'cascade of uncertainty' is still widely acknowledged as a problem (Henderson-Sellers, 1993; IPCC, 1996). Nevertheless, the latest work has progressed from concentrating on single scenarios to incorporating a range of both equilibrium and transient scenarios and begun to incorporate physically-based runoff models (e.g. Arnell *et al.*, 1994a).

6.2. General Circulation Models/Global Climate Models

GCMs are based on weather forecasting models, which in their original role did not need to take much account of changing boundary conditions and external exchange processes. For medium-range forecasting the GCMs required evaluation of regional heat budget exchanges, but for long-term prediction recent experience suggests that there is almost no major process operating within the current earth- atmosphere system that may not have some influence, from ocean circulation to biogeochemical cycling, from volcanic eruptions to solar emissions and planetary alignments. The art of converting circulation models into truly *global climatic models* involves sensitivity judgements on questions of the relative importance of various processes and incorporating state-of-art submodels of those ancilliary environmental processes that are judged to be important. It also involves tradeoffs, particularly between computer capacity, spatial resolution and the scale of process that can be considered.

The current generation of computers is not capable of global modelling with a spatial resolution fine enough for hydrological purposes (Mitchell, 1991). The solution adopted in the new Hadley Centre model (UKHI) is to nest a detailed regional model for Europe within a coarser global host. Encouragingly, initial tests indicate that the regional model is behaving as a truly independent model; although this can create the problem of mismatch at the joins around the window. Whereas the host has a resolution of 2.5 x 3.75°, the European window uses 0.5 x 0.5° (around 50 km). This is a distinct improvement on the old Meteorological Office model (UKMO) in which a grid step of c.175 km meant that the UK was represented by only about four points.

The 11-level atmosphere and 50 m ocean slab of UKMO has been replaced by a 20-level atmosphere and a 17-level ocean with an ocean circulation model (OGCM). Most other GCMs that have provided predictions for Europe have also been operating with a shallow slab representation for the ocean, typically with up to 9-level atmospheres and with spatial resolutions of 4 x 5° (Princeton Fluid Dynamics Lab (GFDL), Oregon State (OSU), Goddard (GISS) fine), 4.5 x 7.5° (National Center for Atmospheric Research (NCAR)) or 8 x 10° (GISS coarse) (Bach, 1989). The resolution of the UKHI window allows much better modelling of topography and atmospheric dynamics, especially of the regional and local atmospheric convergence that is critical to European rainfall (Hadley Centre, 1992). It will also permit a number of soil and vegetation parameters to be better resolved and so pave the way to realistic prediction of local evapotranspiration and of continentally-sourced precipitable water. The realistic ocean model is also critical for Europe, much of which is dominated by advection from the North Atlantic. A recent transient run has also incorporated the effects of aerosols including sulphates, which have a significant radiative cooling effect grading from -1 to -6 W m^{-2} between coastal and interior regions of Europe (Mitchell et al., 1995; Hadley Centre, 1995).

Other important refinements being actively pursued are (1) a phytoplankton response submodel, (2) improved cloud physics, and (3) improved representation of snow and ice. The importance of assumptions about cloud physics, particularly the amount and role of ice crystals is illustrated by UKMO experiments in which simple relative humidity dependent cloud formation gave a mean global warming of 5.2°C whereas cloud formation with realistic ice/water composition and heat budget feedbacks gave just 1.9°C (Rowntree, 1990). The old UKMO model has continued in use as a vehicle for experiments to refine the cloud physics submodel, in order to economize on computer time. Eventually, UKHI should offer the prospect of resolving at least some convective rainfall at the storm level, a requirement that is likely to become more important in a 'greenhouse' Europe.

Finally, it must be remembered that most predictions for Europe are still based on equilibrium runs for double-CO_2, and that transient runs modelling a gradual buildup of greenhouse gases will show a delayed response, particularly due to thermal inertia in the ocean. Perhaps more importantly, we should expect some very different responses during the period of buildup, especially in the case of Europe.

6.3. Changes in the Global Climatic System Relevant to European Hydrology

Europe presents a peculiarly difficult set of problems for GCMs because it is an atmospheric frontier zone, remote from the driving point of the 'heat engine'. Small fluctuations in the strength of the driving force, the meridional temperature gradient, can have significant effects upon regional atmospheric convergence and baroclinicity, and thus on heat and vapour transport, eddy strength and windspeed, atmospheric stability and precipitation processes. This can affect the tracks and intensity of depressions which are linked to the latitude and wavelength of atmospheric long waves. In addition, the hydrology of European drainage basins, especially in W Europe and Scandinavia, tends to be very sensitive to slight changes in climate, so that small changes in the climatic driving force are further amplified by the hydrological system (cp. Novaky, 1985; Palutikof, 1987; Verhoog, 1987).

The strong advective component in the climate of Europe, especially in the west, makes it very dependent upon heat and vapour exchanges over the North Atlantic. There are similar problems of sensitivity with the ocean circulation that strongly influences these exchanges; moreover, the oceanic system is somewhat less well understood. This is complicated by the fact that circulation patterns depend on some factors that are only very indirectly traceable to the meridional thermal gradient. For example, the Labrador Current results from an imbalance in water budgets in the Arctic Ocean (positive) and North Atlantic (negative). In large measure, it is driven by excess runoff from the subpolar continents, plus low evaporative losses in the Arctic. Its strength affects the sea surface temperature off Newfoundland (for long used in the UK Meteorological Office monthly forecasts for Britain) and the containment of the northerly encroachment of warm water from the North Atlantic Drift. Thus the evaporative uptake in depressions leaving the eastern seaboard of Canada, and hence their instability and precipitation over western Europe, are partly controlled by precipitation and runoff in the great arctic basins of the Mackenzie, Lena, Ob and Yenisei in Canada and Russia. Cross Atlantic surface currents are also partly driven by frictional coupling with the westerly wind, and wind strengths in the boundary layer also affect the mixing depth within the ocean.

Ocean salinity patterns are important in a number of ways. Deep water formation in the northern North Atlantic and the thermohaline circulation are only just being considered by atmospheric modellers, though the effects are likely to be orders of magnitude lower than those of surface salinity. A GFDL transient run suggests that increased oceanic precipitation at high latitudes will decrease surface salinity and thus increase stability in the surface layers (Stouffer et al.,1989; Rowntree, 1990). Shallow mixing, low evaporation and high inputs of precipitation and/or freshwater runoff are important to the continued existence of sea ice in the North Atlantic and Arctic Oceans. And the response of sea ice to global warming is critical for the latitudinal thermal gradient over the ocean in winter and hence to the tracks of wintertime depressions over Europe.

A third source of complication is the location of the 'freezing line'. The latitude or altitude beyond which precipitation falls as snow is important over much of Europe for a major part of the year through its effect on heat budgets and on runoff coefficients and river regimes. The problem extends into the clouds. The dominance of the Bergeron-Findeisen process over Europe results from air temperatures fluctuating near to the freezing threshold 3 to 6 km aloft, combined with frontal and convective instability. It seems likely that increased convective instability will outweigh the increase in the altitude of the zero isotherm and so raise rainfall intensities in convective storms. Thus, summertime flood risk would increase in greenhouse Europe.

There is, however, considerable uncertainty about some aspects of the behaviour of ice crystals in clouds, particularly regarding radiative properties (Mitchell, 1991). An important complicating factor is that simple ice crystals fall out of clouds with weak updrafts faster than water droplets, thereby truncating the Bergeron process and maintaining a more liquid cloud. This tends to increase stratiform cloud cover in regions with cloud temperatures several degrees below 0°C and reduce surface air temperatures (Rowntree, 1990). Western Europe is one of the regions most affected by this complication and warming was reduced by 2°C here in experiments incorporating this process within UKMO (Rowntree, 1990).

A further problem discovered recently is the role of dimethyl sulphide released by phytoplankton as a source of condensation nuclei (Rampino and Etkins, 1990). Increased phytoplankton productivity in a warmer North Atlantic could significantly increase cloud formation over Western Europe.

Two of the more certain features of West European climates during the transition period would seem to be (1) a poleward shift in average depression tracks and (2) increased storminess. The poleward shift would be associated with a more active Hadley cell in a warmer world, the poleward spread of warmer sea surfaces and the gradual reduction in the seasonal extension of sea ice (Jones, 1990). The poleward shift would follow a northward shift in the Rossby waves. The resulting increase in angular momentum would increase the strength of the westerlies and increase the wavelength of the stationary Rossby wave over Europe, pushing the trough further into Eastern Europe (cp. Lamb, 1977). This would increase advective effects, cause maritime influences to spread deeper into Europe and reduce the frequency of incursions of arctic air, thus reducing both frost events and the duration of snowcover.

The increase in the strength of the Ferrel jet stream is likely to generate more intense depressions, which would be fed by increased evaporation over the ocean and so generate more intense precipitation. This would be intimately linked with an intensification of the oceanic thermal gradient in the northeastern North Atlantic as a North Atlantic Drift augmented by tropical warming and boundary layer drag pushes warmer water against arctic waters and towards the winter limit of sea ice. This intensification should be considerably reduced if not entirely nullified by the eventual demise of North Atlantic sea ice. For the mid-western seaboard of Europe it is also likely to be a temporary phase because the gradual poleward migration of depression tracks will eventually shift them more out of range.

Any reduction in frontal rainfall that might result will be partly offset by increases in convective rainfall, with convective storms growing more intense as warming proceeds. Most convective storms will be locally generated, and storms like the 'Great Storm' of October 1987 that devastated SE England are likely to occur more often: this storm was caused by an intense and very localized thermal low pressure cell developing over warmer than average water west of Biscay. Moreover, a parallel increase in hurricane activity (Emanuel, 1987) is likely to result in more hurricane 'rumps' getting entrained within mid-latitude depressions tracking off the eastern seaboard of North America and intensifying the mid-latitude cyclones. Some of the most severe late summer storms in Britain have been of this provenance in recent years, such as 'Hurricane Charley' in August, 1986 (Sawyer, 1987).

6.4. Precipitation

Changes in precipitation parameters are likely to be the main cause of changes in runoff patterns, and most of these parameters will be sensitive to global warming: amounts, intensities, seasonality, extreme events, phase and snowmelt dates. Four main approaches have been taken in attempts to estimate the changes: (1) analysis of instrumental data, (2) use of palaeoclimatic reconstructions, (3) temperature sensitivity models, and (4) GCMs.

6.4.1. INSTRUMENTAL ANALYSES

Analyses of past records may be used simply to discover and extrapolate trends or in a more sophisticated way to down-scale GCM predictions, which do not have sufficient spatial resolution or are regarded as not being sufficiently reliable for precipitation.

A number of analyses have concentrated on differences in precipitation receipts between the warmer period of 1934-53 and the cooler 1901-20 period, establishing empirical relationships between temperature and precipitation (e.g. Palutikof, 1987) or mapping seasonal rainfall in Europe for those periods as analogues (Chen and Parry, 1987). Work at the UEA Climatic Research Unit has sought to establish model parameters for rainfall in 9 homogenous regions of the UK and to use GCM predicted temperature change as a basis for recalculating these parameters for 2030 (Lough et al., 1983; Palutikof et al.,1984; Palutikof, 1987; Gregory et al., 1990). The methodology comprised (1) establishing homogenous regions using principal components analysis, (2) producing a standard representative daily rainfall series for each region, (3) finding a best-fit probability density function (pdf) for each record, (4) regressing the parameters of this pdf against temperature, (5) using a temperature scenario for 2030 based on an average of equilibrium predictions from 5 GCMs to perturb the pdf. An incomplete gamma distribution was selected as best fit. Unfortunately, initial work by Lough et al. (1983) indicated that correlations between temperature and precipitation vary in sign in different regions and at different seasons: generally negative in England and Wales, and positive in Scotland, N England and Northern Ireland (Palutikof, 1987). Gregory et al. (1990) found no correlation between notably wet or dry seasons in the two groups of regions. They also found negative correlations between the parameters of the gamma model for precipitation and temperature in all regions. This latter result severely affects estimates of water resources (v.i.) and is at variance with rainfall predictions from GCMs, which tend to show overall increases in precipitation with rising temperatures. It is a situation that is only theoretically possible for a small region.

Wilby (1993) has taken a different approach to down-scaling British rainfall by using Lamb's (1972) classification of synoptic weather types. Wilby correlated the observed daily rainfall totals with the daily Lamb Weather Type (available from 1861), which gave him precipitation event probabilities and magnitudes for 8 dominant Weather Types. He proceeded to test the sensitivity of the discharge and hydrochemistry of a Midlands catchment to a range of synoptic scenarios using this stochastic rainfall model, a simple one-store conceptual catchment model and an empirical pH rating curve model. He concluded that the hydrological response is principally determined by the A/C index, the frequency ratio of Anticyclonic to Cyclonic Weather Types. Wilby used this as an exercise in sensitivity analysis, but to apply it to down-scaling GCM output would require rather more detailed output statistics than normally on general release.

Other analyses of instrumental records in Europe have looked simply for trends in recent records. Analyses by Bradley *et al.* (1987) and Groisman (1991) indicate increased annual precipitation this century in Scandinavia, rising at 13%/century, accompanied by a decrease of 5%/century in Western Europe (37-55°N). Both trends are significant at the 5% level (Groisman, 1991). National scale analyses show (1) marked reduction in rainfall in Italy over the last 30 years (Conte *et al.*, 1989), (2) reduced snowfall in the Austrian Alps during the warm periods of the 1930s and 1975+ (Mohnl, 1991), (3) increases in central Finland but reductions in the south (Heino, 1989), and (4) drier summers and wetter autumn/winters in Britain 1974-85 (both being the most extreme decadal figures this century) causing increased seasonality (Morris and Marsh, 1985). These shifts are generally consistent with the theoretical shifts outlined above. The UKCCIRG (1990) report and Arnell (1992) conclude that the rate of change in mean seasonal precipitation in the UK 1970-90 provides the best estimate of global warming trends.

Inspite of the discovery of numerous trends and temperature relationships in the recent instrumental record (Table 6.1), however, caution is required in using them as a basis for prediction. In the first place, they are for the most part evidence of short term shifts within this climatic 'frontier' zone which are not supported by sustained system-wide change of the type expected next century. A sustained shift in sea surface temperatures, for example, is likely to produce rather different responses. Secondly, it seems likely that the forcing factors responsible for warmings earlier this century were different and it is therefore likely that system responses will not be the same in shifts forced by greenhouse gases. Budyko (1989) has suggested (arguably) that only the 1970+ warming can be regarded as due to greenhouse forcing, and (even more arguably) has suggested that projections should be based only on this very short recent period.

6.4.2. PALAEOCLIMATIC ANALOGUES

Budyko has been an exponent of palaeo-analogues and has reconstructed rainfall patterns for the Holocene Hypsithermal (5-6 ka BP) and the Pliocene Optimum (3-4 Ma BP) (Budyko and Izrael, 1987). Budyko argues that, since these periods represent full scale natural 'experiments' containing all the natural systematic feedbacks and interactions, they are superior to what can be achieved in GCMs (Budyko, 1989). Unfortunately, reconstruction is fraught with difficulties, especially for the preglacial period, and Budyko's argument is significantly weakened by the lack of explanation of methods or evaluation of errors, at least in his presentation in English (Budyko, 1989).

TABLE 6.1. Some trends in hydrological components identified from recent records in Europe.

Precipitation	Evaporation, soil moisture & groundwater	Streamflow

United Kingdom
-20% summer, wetter autumn/winter
  1974-85 (Morris and Marsh, 1985)
England and Wales:
  More in winter 1900-1985
      (Thomsen, 1989)
  More in spring 1960-1990
  Less in summer 1960-1990
      (Gregory *et al.*, 1990)
	United Kingdom
	Higher soil moisture deficits 1927-
	  77 (Wigley & Atkinson, 1977)
	Marked reduction in forest ET in
	  Wales 1972-88
	    (Hudson & Gilman, 1993)	

Scandinavia
+13%/century 55 - 70°N
    (Bradley *et al.*, 1987)
More in winter 1950-1985
  in Denmark (Thomsen, 1989)
More in C Finland, less in S
    (Heino, 1989)
	Scandinavia
	More evapotranspiration 1945-87
	  in Sweden/Denmark
	    (Jutman *et al.*, 1989)
	Lower soil moisture 1890+
	  in Sweden (Andersson, 1989)
	Higher water tables 1964-83 in
	  Denmark (Thomsen, 1989)
		Scandinavia
		Increased 1980s in Finland
		  (Hyvarinen & Leppajarvi,
		1989)
		More river ice breakup
		  floods 1980s in
		  Sweden/Finland
		  (Zachrisson, 1989)

Western Europe
-5%/century 37 - 55°N
    (Bradley *et al.*, 1987)
		Western Europe
		Increased in Rhine 1970+
		  (Schadler, 1989)

Southern & Central Europe
Less 1960+ in Italy
Desertification in Sicily 1980+
    (Conte *et al.*, 1989)
Less snowfall in warmer 1930-40 &
  1970-90 in Austria
    (Mohnl, 1991)
	Southern & Central Europe
	Droughts increased 1980s in
	  Hungary (Zoltan, 1989)
		Southern & Central Europe
		Water temperatures
		  increased 0.5-1.25°C
		  in Austria this century
		  (Webb & Nobilis, 1994,
		  1995)

NO TRENDS
Seasonal/annual (1931-88) in
  Scotland & N Ireland
    (Gregory *et al.*, 1990)
	Groundwater 1962-89 in Finland
	  (Soveri & Ahlberg, 1989)
		River ice breakup dates -
		  last 150 years in Finland
		  (Kajander, 1989)
		65% of European rivers
		  show no trends in
		  discharge, but % with
		  trends is 10% above
		  global average
		  (Mitosek, 1995)

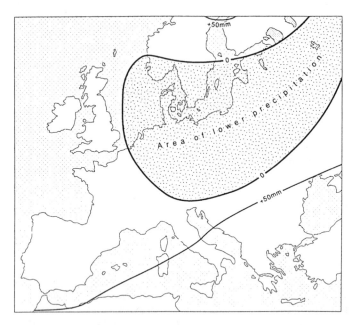

Figure 6.1. Difference between annual precipitation (mm) in the Hypsithermal and the present, as reconstructed by Budyko and Izrael (1987). Estimated temperature difference +1-1.5°C.

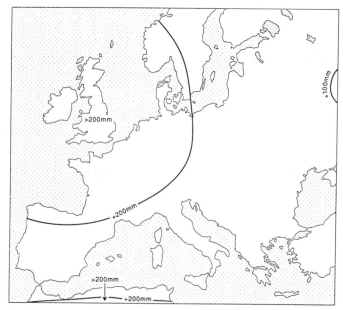

Figure 6.2. Difference between annual precipitation (mm) in the Pliocene Optimum and the present, as reconstructed by Budyko and Izrael (1987). Estimated temperature difference +3-4°C.

Figures 6.1 and 6.2 show Budyko's comparisons between the reconstructions and the present-day annual rainfall, and they reveal a marked discrepancy between the two periods in both the patterns and the amounts of change. Whilst the Hypsithermal pattern shows a large area of reduced receipts extending from the Low Countries to Moscow, the Pliocene shows a higher and universal increase with maximum rates in the northwest and minimum in the Volga basin. This need not imply disagreement. It may quite logically imply that differing storm patterns are associated with differing degrees of warming, and that a universal increase in precipitation across Europe may only occur in a global warming scenario which is at the upper extreme of the range of predictions for next century. Interestingly, Palutikof's (1987) analysis of recent data does not fit in too badly with the Hypsithermal trend over Britain, and some GCMs concur with reductions over central parts of Europe, e.g. GFDL. There has been some conflict within UKHI between the high resolution window and the global host as to the existence of a 'drought' area in E Europe.

6.4.3. TEMPERATURE SENSITIVITY MODELS

On the basis of the Clausius-Clapeyron relationship, the water vapour capacity of the air should increase at 6-7%/°C. However, not all this will be converted into increased precipitation and the UKMO model suggests a global increase of just 2-3%/°C, because of heat balance considerations (Mitchell *et al.*, 1987; Rowntree, 1990; Mitchell, 1991). On the other hand, maximum precipitation rates do seem to be controlled more directly by the water content, so that the most intense rainfall events are likely to be intensified (Rowntree, 1990).

Although these arguments may be applied at a global scale, the prime focus of Groisman's (1991) temperature sensitivity studies has been to relate trends in *regional* precipitation to changes in *global* temperature. This immediately shifts the argument out of the safe bounds of physical laws into the realm of probabilities. The logical underpinning for this approach is that global temperature change is more easily and consistently modelled by GCMs than is regional precipitation. It is suggested that, until GCMs can produce better predictions for precipitation, correlating observed global temperatures with regional records of precipitation should provide an empirical method for prediction which inherently subsumes the characteristic behaviour of the regional precipitation-creating processes. Unfortunately, this approach assumes that the instrumental record includes the full range of atmospheric states that may be encountered in a double-CO_2 world, even though it accepts that the frequencies of those states are expected to change. In reality, we may expect changes, for example in North Atlantic sea surface temperatures, that represent states beyond the range of the instrumental record.

Thus, Groisman (1991) attempted to estimate the sensitivity of regional precipitation in Europe to global mean temperature by comparing his own instrumental analysis, Budyko's palaeoclimatic results and 2 GCMs. Table 6.2 is based on his results, with some additional interpolation by the present author based on Groisman's source material

and Bach (1989). Despite the crudity of the comparison, some useful conclusions seem
possible: (1) the estimates from the GCMs seem conservative compared with the
empirical estimates, and (2) all the relationships are positive with the exception of W
Europe according to Groisman's own analysis of instrumental records (a point he omits
to comment on).

TABLE 6.2. Sensitivity of regional annual precipitation in Europe to global mean temperature.
Based on Groisman (1991) with additions. Percentage change per degree Celsius of warming.

	Last 100 years (1)[*]	Hypsithermal (2)	GISS (3)	GFDL (4)	OSU (5)
Scandinavia	26%	5%	6%	--	--
W Europe	-10%	3%	2%	--	--
European CIS	8%	8%	2%	--	--
Europe excluding CIS	2%	5%	1%	2%	5%
CIS	18%	12%	--	3%	4%

[*]Sources: (1) Groisman (1991), (2) Budyko (1989) and Budyko and Izrael (1987), (3) Bach (1989),
(4) Manabe and Wetherald (1987), (5) Schlesinger and Zhao (1989).

6.4.4. GCM PREDICTIONS

The main disagreement between GCMs seems to be the trend in summer rainfall,
especially in continental interiors. Overall, GFDL, UKMO and OSU indicate generally
reduced summer rainfall in northern mid-latitude continental areas, whilst GISS and
NCAR show increases. The problem is partly due to (1) the difficulties of simulating
relatively small scale convection at the resolution of the GCMs currently operational, (2)
the relatively unsophisticated modelling of ground hydrology, i.e. the surface and
subsurface drainage processes which affect the rate of drying of the soil surface, and (3)
the importance of the amount and timing of snowmelt to summer soil moisture levels
and hence atmospheric moisture in many continental areas. Thus, deviations in
simulated winter precipitation are compounded in the summer rainfall estimates.
Because summer convective storms tend to depend more on local evapotranspiration
than upon atmospheric convergence of moisture, modelling the soil moisture content
correctly is important for good results. Consequently, the development of improved
models of ground hydrology has been given priority in recent years. This includes
improving the representation of vegetation cover. The intensity of summer drying is also
affected by windspeeds, about which models disagree and where improvements are
actively being sought.

    Hence, Rind (1988) concluded that winter precipitation was better modelled than summer in both the coarse and fine versions of GISS. In fact, although the fine version resolved topography and the sea level pressure field better, control runs showed twice as much rainfall in the US prairies than observed: a worse fit than the coarse version. Rind notes that, despite current efforts, the problems of matching spatial resolution to the processes which need to be considered, and determining the cutoff point at which parameterization can replace explicit process modelling, are 'formidable' and will probably take 10 years to solve.

Whilst these improvements are being implemented, therefore, the best that can be done is to judge GCM precipitation output in the light of changes expected on the basis of past evidence and theoretical deductions.

Warrick *et al.* (1990) compiled 'consensus' maps for winter and summer precipitation over Europe based on averages of 4 GCMs for equilibrium scenarios. These indicate increases in winter receipts everywhere north of Sicily, reaching a peak of 0.6 mm/day centred on the Alps, and a large area of reduced rainfall in summer extending over SW Europe to a line from S Ireland to N England, SW Germany and down to Italy.

Bach (1989) published maps from an equilibrium run on fine GISS which showed reduced rainfall in parts of W, SW and SE Europe virtually all year (Figure 6.3). The summer and winter patterns were very similar to the GCM averages published by Warrick *et al.* (1990).

The present author has compiled maps of annual changes (Figures 6.4 and 6.5) based on Bach's maps and using Bach's 4 x 5° graticule for interpolation. The map of annual change (Figure 6.4) shows a tendency for greatest increases in NE Europe, with a ridge from Scandinavia to the Alps, and negative changes restricted to Biscay and some Greek islands. The overall pattern seems physically plausible. Figure 6.5 converts this into percentage change based on the current pattern as given in the UNESCO (1978) atlas. It shows a moderate 5-10% increase in W Europe, which is in the same range as most other estimates, but extreme increases of 25-50% in NE Europe, which may be rather high. The general pattern seems believable, but it is very different from either the Pliocene or the Hypsithermal reconstructions. Again, this could reflect differing degrees of warming, or the effects of different forcing factors, rather than indicating that either the model or the reconstructions are wrong.

Figure 6.6 plots the season of maximum change implied in the GISS maps. The greatest increases in W Europe are in winter, gradually shifting into spring to the north and east, presumably due to increased cyclonic activity. In NE Europe, summer is the season of maximum change, possibly due to a northerly restriction in depression tracks and more local convection. In the southern Mediterranean, autumn is the main season of change, probably because of a buildup in convective activity over warmer seas.

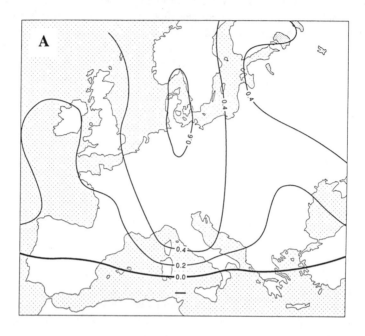

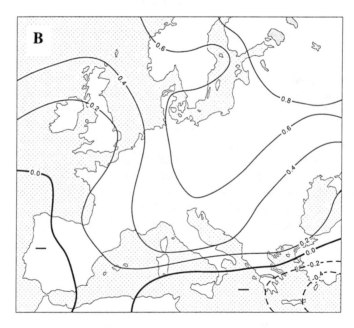

Figure 6.3. Seasonal rainfall differences (mm d^{-1}) in a double-CO$_2$ equilibrium run of GISS, based on maps from Bach (1989). A) Winter, B) Spring.

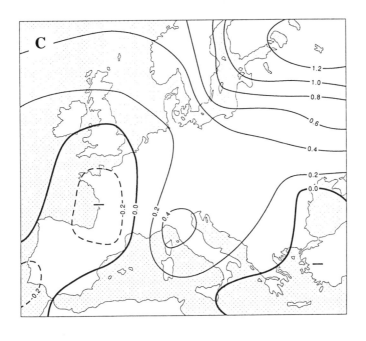

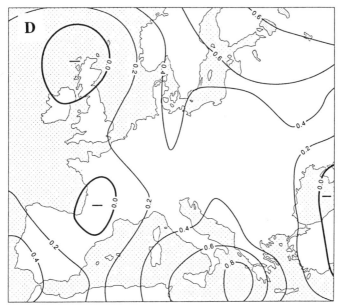

Figure 6.3. Seasonal rainfall differences (mm d^{-1}) in a double-CO$_2$ equilibrium run of GISS.
C) Summer, D) Autumn.

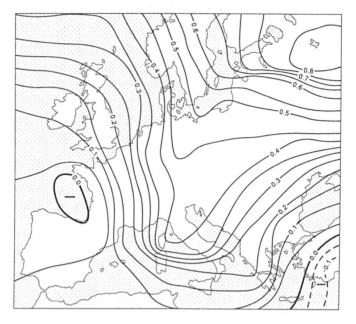

Figure 6.4. Changes in mean annual precipitation (mm d^{-1}) derived from Figure 6.3.

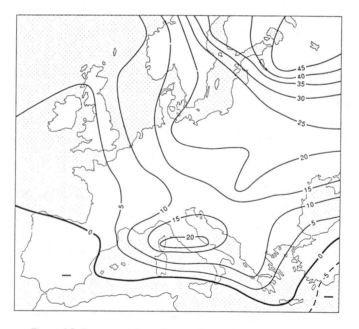

Figure 6.5. Percentage change in annual precipitation in Figure 6.4.

Figure 6.6. Seasons of maximum change in precipitation, derived from Figure 6.3.

Figures 6.7 and 6.8 are derived from trial runs of UKHI, again using Bach's 4 x 5° graticule for interpolation: this is coarser than the UKHI output, but it seems justifiable for this preliminary exploration. The maps reveal a winter pattern similar to GISS, but the amount of change is rather larger than the general average. The summer pattern again contains substantial areas of lower rainfall in the mid-west of Europe and in the southeast. The marked winter increases seem to be associated with a notable enhancement of westerly winds and increased advection over W Europe in the new model.

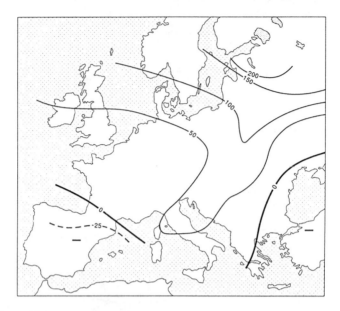

Figure 6.7. Percentage change in winter precipitation in a double-CO_2 equilibrium scenario run of UKHI, based on output from the Hadley Centre, with permission.

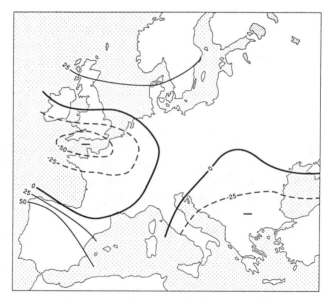

Figure 6.8. Percentage change in summer precipitation in a double-CO_2 equilibrium scenario run of UKHI, based on output from the Hadley Centre, with permission.

6.4.5. EXTREME EVENTS

Hansen *et al.* (1991) predict that both ends of the events spectrum will be enhanced in mid-latitudes, with more frequent and severe droughts and more extreme wet periods, and that the changes at the extremes will offset any regional advantages accorded by shifts in the mean. Warrick and Farmer (1990) have explained this in terms of increased variance normally associated with an increase in the mean. Rind (1991) suggests that increased variance is potentially a severe problem for the 21st century. Physically, drought risk is expected to increase in the south as depression tracks migrate north and more generally as soil moisture deficits reduce local sources of atmospheric moisture. Flood risk would increase in response to heavier winter rainfall and to more severe convective cloudbursts. There are numerous references to these potential problems for the UK (e.g. UKCCIRG, 1991; Binnie, 1991; Jones, 1990; Newson and Lewin, 1990). Hansen *et al.* (1991) predict pessimistically that drought events that occurred 5% of the time in 1958 will have a 25% risk by 2020.

Although GCMs cannot resolve at convective storm level, Hansen *et al.* (1991) identify certain aspects of GCM output that are diagnostic of greater convective storminess: (1) increased depth of moist convection, (2) increased mass flux due to this and to a decrease in shallow convection in mid-latitudes, and (3) increased precipitation accompanied by little change in stratiform clouds, resulting from lower synoptic scale wave activity as the meridional gradient declines. With the expansion of the area of North Atlantic with sea surface temperatures exceeding the critical 26°C for hurricane development, and a reduction in the minimum sustainable surface pressure from 880 to 800 hPa, there is likely to be increased risk of hurricane-sourced storms reaching Europe (Emanuel, 1987).

6.5. Evapotranspiration and Soil Moisture

Changes in evapotranspiration are inherently more difficult to predict than changes in precipitation, because of the wider range of influencing factors, and a recent analysis by Lockwood (1995) emphasises doubts about the response of botanical processes (*v.i.*).

Increases of 14-19% in evaporation losses have been estimated for lakes and reservoirs in Poland by Jurak (1989). Jurak used GCM estimates of temperature and irradiance in his own 'combined model' evaporation formula, but he had to make assumptions regarding relative humidity (RH). In fact, Rowntree (1990) concluded that difficulties in predicting RH and windspeed with GCMs make it impossible at present to estimate even potential evapotranspiration accurately. Arnell (1992) has attempted to do this by assuming no change in RH, windspeed and net radiation, although in fact all are likely to change in varying degrees. Using Penman's equation and the UKCCIRG (1991) scenario of a 2.1°C warming in the UK by 2050, Arnell calculates a 7% rise in potential evapotranspiration (PET), but he notes that in the sunnier summer of 1989 PET was 15% higher in SE England though temperatures were only 1.3°C above normal. Atmospheric transmissivity can have a significant effect. Rowntree (1990) calculated that increased water vapour and CO_2 would reduce clear sky radiation over S England by about 1 W m^{-2} per °C of warming, but noted that this could be literally overshadowed by an increase in cloud amount and/or optical thickness, for example, if soil moisture deficits reduce the efficacy of rainmaking processes. GCM predictions on windspeed are particularly varied, and in any case transient effects are likely to dominate over Europe for some decades yet and few GCM runs attempt to cover this phase.

Predicting actual evapotranspiration (AET) is even more difficult because it depends on many of the worst modelled features in current GCMs, such as rainfall intensity and frequency, cloudiness, soil moisture content and vegetational response to CO_2 fertilization. Some of these are both controls and products, so that modelling errors can be compounded. Uncertainties about total summer rainfall are compounded by doubts about rainfall intensities and the length of dry periods, which affect infiltration and thence soil moisture. The overlays in Figure 6.9 show very little agreement between GISS, UKMO and NCAR on projected Mediterranean soil moisture in Bach's (1989) maps.

The main alternative has been to use empirical correlations with temperature for a given site or area. Rather misleadingly termed the 'geographical' method, this has been used by Gregory *et al.* (1990) and Cole *et al.* (1991) to estimate actual evaporation. Palaeoclimatic analogues do not seem to be of much use for Europe where the sources of evidence are dominated by rainfall or runoff.

A major concern at present is to improve the modelling of vegetation responses, which are complex and not fully understood. Plant response will be affected by changes in irradiation, temperature, soil moisture, windspeeds, RH and CO_2. Initially, most changes will affect growth - photosynthesis, length of growing season and biomass production. There is likely to be a significant increase in the area of NW Europe experiencing an all-year-round growing season (Parry *et al.*, 1989). Warming will extend the growing season more in maritime climates because of the lower gradients in the annual temperature curves in spring and autumn compared with continental climates. Parry *et al.* (1989) estimate increases of 6 weeks or more in parts of Britain and point to a shift in agricultural potential towards NW Europe and away from the south and east.

Increased biomass production resulting from CO_2-fertilization, higher temperatures, and in some areas increased moisture, should increase transpiration and evaporative losses from interception. However, greater leaf area could be offset by greater aerodynamic resistance from the canopy. By fitting the Penman-Monteith model to field data and perturbing it according to GCM predictions, Martin *et al.* (1989) showed that ET is only increased unequivocally if all meteorological variables are favourable: if some are against, then plants may moderate the change. Whilst stomatal resistance will tend to reduce with warmer temperatures, increasing CO_2 will stimulate stomatal closure and may in the long run encourage fewer stomata. Evidence presented by Woodward (1987) indicates a 40% reduction in stomatal densities on specimens from the English Midlands between 1750 and the 1980s.

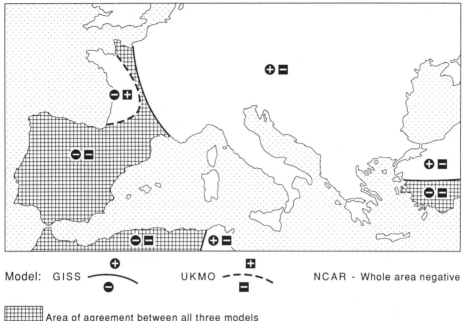

Model: GISS ⌒ UKMO ⊞/⊟ (dashed) NCAR - Whole area negative

▦ Area of agreement between all three models

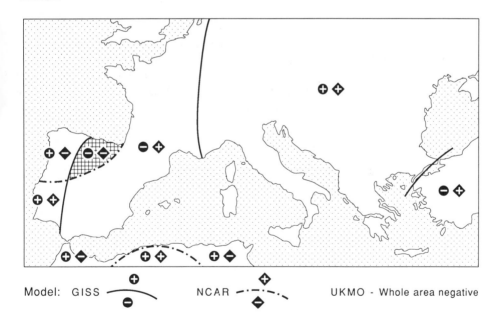

Model: GISS ⌒ NCAR ◈ (dashed) UKMO - Whole area negative

Figure 6.9. Comparison of predicted soil moisture changes in S Europe in double-CO_2 runs of GISS, UKMO and NCAR, based on overlay of maps published by Bach (1989). The shaded areas mark 'total agreement' on the sign of change. A) Winter, B) Summer.

Moisture tension may also restrict transpiration, and different species respond differently. Increased moisture stress could fortify the physiological effect of CO_2 concentrations to reduce transpiration, especially in S Europe. Plants of the C3 photosynthetic group (like wheat) have been shown to develop 10-20% lower transpiration rates in double CO_2 atmospheres, whereas C4 plants (like maize) show a 25% reduction (Rabbinge et al., 1993). Conversely, growth rates are likely to increase more in C3 plants in double CO_2 than in C4 plants, which might increase interception and direct evaporation. The exact balance between physiological responses to the wide range of environmental changes envisaged is difficult to predict, and many responses are poorly understood and the subject of active research (Wilson et al., 1993).

Lockwood (1995) believes it is not clear whether ET will increase or decrease in NW Europe. In addition to the effects of CO_2 on stomatal resistance, most temperature increases will occur at night, reducing the diurnal range, but ET response is essentially a daytime phenomenon. One could add that the same applies to likely seasonal changes in temperature: most warming is expected in winter, when photosynthesis is at a minimum. Lockwood's calculations suggest that over the period from 1951 to 1990 a 0.28°C rise in maximum temperatures in Britain would have increased transpiration by 2.3% in C3 forests and 0.6% in grassland, but that this was more than counterbalanced by CO_2-induced reductions, creating net reductions of 0.2-2.7% in forests and a larger 1.9-4.4% reduction from grassland. The effects would be greater in C4 vegetation, with a net reduction of up to 5.4%. But note that this refers to transpiration not ET, and it is difficult to obtain field verification of recent trends in transpiration on its own or to predict future trends in ET.

Recent evidence from one of the most comprehensive and long-term hydrological experiments in W Europe suggests a dramatic reduction in ET is already underway. Hudson and Gilman (1993) found a drop of approximately 40% in ET in the UKIH Plynlimon research catchments from 1972 to 1988. The change was about the same in both the moorland and the forested catchment. Clearly, some of this fall was caused by lower insolation, air temperatures and saturation deficits, which gave a 14% reduction in PET, but there must be other factors. Evapotranspiration may be reduced by acid rain damage or ageing of the trees, or differing frequency/intensity patterns in rainfall that affect interception, which has been shown to be a major factor in ET rates at least in the forested catchment. Whatever the cause, therefore, global warming does not presently seem to be an important factor in these changes. But it does indicate the precarious nature of predictions based solely on meteorological factors.

Lockwood (1995) suggests that CO_2-suppression of ET will have a knockon effect in reduced rainfall. He believes this will be most marked in C4 grass-covered continental interiors and least in maritime climates, although it could aggravate the summer rainfall shortage in SE England.

One long term response could be major shifts in natural vegetation belts in Europe. The International Institute for Applied Systems Analysis (IIASA) in Austria has run a computer simulation which indicates Mediterranean vegetation types spreading to the UK and the demise of boreal forest in Scandinavia (Warrick et al., 1990).

To date the main hydrological studies in Europe have all ignored the potential effects of plant physiology on ET (Bultot *et al.*, 1988; Arnell and Reynard, 1989; Berndtsson *et al.*, 1989; Arnell, 1992). The evidence of Martin *et al.* (1989) suggests that the consequences may not have such a dire effect on water surplus as was estimated by Gleick (1987), but we cannot be sure until the plant physiologists have reported.

In general, however, the expectation of drier land in summer in northern mid-latitudes implied by UKMO, GFDL and OSU seems reasonable. Mitchell (1991) explains this in terms of lower rainfall, less cloud, higher ET and earlier snowmelt. Mitchell and Warrilow (1987) found that in some regions summer dryness in UKMO was sensitive to the treatment of runoff from frozen ground. Bultot *et al.* (1988) put an assumed increase of 7% actual ET (AET) into a conceptual hydrological model for three Belgian basins, and found a marked reduction in the frequency of soil saturation and an increase in the days of deficit. Episode durations and total number of days of saturation were reduced by 15-18% in a lowland clay basin, against only 4-7% in the upland and aquifer basins. The number of days of less than 60% capacity increased by 37% in the lowland and 47% in the upland, and durations by 20% and 16% respectively.

6.6. Runoff

Despite the current limitations of GCM predictions for the water budget, it is important that likely scenarios are tested in hydrological models in order to study process sensitivities, to advise the water industry and to improve hydrological models so that when firmer climatic predictions become available hydrologists will be ready to convert them into hydrological responses. Ultimately, physically-based hydrological simulation models are needed with sets of criteria or submodels that will permit recalibration to changed environmental parameters, such as vegetation and soil macropore properties. In the meantime, it is reasonable that nearly all studies in Europe have used simpler water balance or conceptual models, or even temperature sensitivity models.

As with precipitation, a number of analyses have been devoted simply to looking for trends in the instrumental record (Table 6.1). Mitosek (1995) has recently tested world discharge records for stationarity. This work included 46 time series of discharge from Europe, of which 35% proved non-stationary. Globally, only a quarter of the records were non-stationary. The difference seems to support the premise that European hydrology is more sensitive or susceptible to climate change than the global average.

Others have taken a temperature sensitivity approach. Gunn and Crumley (1991) developed a temperature sensitivity model to predict river stage in the Mechet Creek, Burgundy, based on an 18-year record, that was rather more elaborate than Groisman's (1991) differential ratios for predicting precipitation. They used a polynomial regression relationship between mean monthly stage during the growing season and average annual temperature in the Northern Hemisphere, as supplied by NOAA. This provided a 'transfer function' in order to hindcast riverflows using average Northern Hemisphere temperatures reconstructed from an astronomical model. Despite the two stages of reconstruction, they found good agreement between the hindcasted flows and secondary evidence from diatoms, pollen and sedimentary sequences. Using GCM predictions of future energy budgets as input to the model, they concluded that the Burgundian basin would suffer from progressive summer drying early next century and that winters will become progressively wetter, which appears to support Flohn's expectations of a warmer and drier south-central Europe (cp. Flohn and Fantechi, 1984).

Even so, it must be emphasized that statistically-based temperature sensitivity models are essentially crude and 'black box' affairs, and that there is no guarantee that the same relationships can be extended outside the period of the records used in the calibration, for example, because of changing circulation patterns; a point clearly made by the WMO (Klemes, 1985).

A step more sophisticated than a temperature sensitivity model is represented by Novaky's (1991) modification of Budyko's equation, in which runoff is an exponential function of potential evaporation (PE) and precipitation. He concluded that a 0.5°C increase in temperature and an accompanying reduction in rainfall rates by 0.08 mm d^{-1} could lead to a 25-30% reduction in runoff in small basins in Hungary.

Bultot *et al.* (1988) used a daily conceptual water balance model in three representative basins in Belgium: a sandy-clay lowland, a sandy-loam basin on a deep aquifer and a rock-floored upland site. Although their present climates are very similar and Bultot *et al.* applied a uniform change in precipitation (+54 mm p.a. or 5-7%) based on GFDL, the responses of the individual basins were very different and depended very much on soils and geology. Where infiltration capacity was high and an aquifer present, increased groundwater storage gave increased baseflow and annual runoff despite higher ET and less summer rain. Where surface runoff dominated, there was an increase in flood frequency in winter and lower summer discharges. Bultot *et al.* noted that changes affect the extremes more than mean water balances.

Their model also predicted the responses of differing drainage processes and Figure 6.10 is based on their tables. It shows: (1) higher annual flows in all pathways, except for throughflow (interflow) and groundwater flow in the upland, (2) general reduction in all summer flows, except for groundwater and streamflow in the aquifer basin, and (3) increases in all flows in winter, except subsurface flows in the upland. The latter seems more due to a 23% increase in surface runoff (+6.3 mm) from wetter soils than to higher winter AET (+1.5 mm) and lower snow water equivalents (-2.7 mm).

Gellens (1991) used the same model to extend Bultot's analyses to cover changes in the variability of flows and in the frequency of high and low flows. He concluded that variance will increase in winter along with total discharge and that there will be an increased probability of high flow events as a result. Overall, all three basins would suffer from more flooding, with longer flood episodes and higher peak discharges. In summer, there was a general tendency for more low flow events, except in the aquifer basin. The aquifer basin was the only one to show a joint increase in mean and variance during the summer. Thus, summer low flow events would be fewer and shorter in the aquifer basin, the exact opposite of the situation in the uplands. In the lowland catchment without aquifer support, there would be more low flow episodes but they would be shorter and less severe.

In Holland, Jongman and Souer (1991) also used a water balance approach. They predicted increasing water deficiencies in summer and reductions in groundwater storage. In cases where increased winter surplus drains away before the summer, the mean annual water surplus could be almost halved.

Berndtsson *et al.* (1989) used a monthly water balance approach to study sensitivities in S Sweden. With an input of +20% precipitation and +2°C, they obtained a 50% increase in annual streamflow. Their scenario fits the GISS projection in Figure 6.5 and seems to suggest even more marked increases in Scandinavian streamflow and hydropower resources than the already high increases in precipitation.

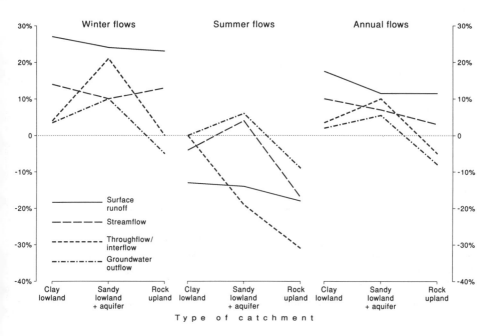

Figure 6.10. Percentage changes in seasonal and annual flows via different hydrological pathways in three Belgian basins, derived from hydrological modelling by Bultot *et al.* (1988) for a double-CO_2 scenario.

The amplifying effect of runoff generating processes and the marked differences between geological areas are confirmed in work at UKIH. Arnell (1989, 1992), Arnell and Reynard (1989) and Arnell *et al.* (1990) used a multiple regression equation relating runoff to rainfall and PE in combination with a simple water balance model in different hydrological regions of the UK. Initial work concentrated on sensitivity, using purely notional changes in precipitation and ET: (1) +20% precipitation throughout the year or (2) +20% in winter and -15% in summer, with a fixed +15% PET. This showed greater seasonality in discharges. Like Bultot *et al.* (1988), Arnell (1989) found higher summer flows in groundwater-fed catchments even when summer rainfall was reduced.

Arnell and Reynard (1989) showed that in SE England runoff amplified rainfall change threefold for increases and 2-3 times in decreases. In the southwest, amplification reduced to twofold, and to 1.5 and 1.2 times in the northwest and Scotland respectively. Thom and Ledger (1976) reached similar conclusions. This general trend accentuates the NW-SE gradient in British water resources.

One of the more remarkable conclusions from Arnell's latest work is that the annual runoff map of Britain is extremely sensitive to relatively small changes in PET. Arnell (1992) produced a map for the UKCCIRG's best estimate rainfall scenario of +8% winter rain and zero summer change, with +15% PET from Rowntree (1990), applying each percentage change to current mean monthly figures derived from the Meteorological Office Rainfall and Evaporation Calculation System (MORECS) (Meteorological Office, 1981) and using water balance parameters for an 'average' basin. This gave *reduced* annual streamflow over most of the country, with increases confined to the extreme west and northwest and the south coast,and mostly less than +4%. In contrast, reducing PET to +7% (near Bultot *et al.*, 1988) gave *increased* yield over the whole country, though again less than +5%.

Adding the extreme values from the UKCCIRG (1991) range of 'confidence', ±8% winter rain and ±16% in summer, gave even starker contrasts. Combining these with +7% or +15% PET, Arnell (1992) obtained a 'worst possible dry' scenario in which annual streamflow was reduced by 10-30% across the country, and a 'worst possible wet' case in which runoff increased by 17-30%. This seems an unnecessarily wide range of scenarios, given the unlikelihood of like-signed extremes occurring in both winter and summer of the same year. But he does observe that the extremes for individual seasons are comparable to current extremes in decadal variability.

On the other hand, given his illustration of the importance of basin characteristics, it seems likely that his maps for 'average' basins underestimate the variability within each scenario. In fact, this is clear from his application of the 'best' scenario to three characteristic basins: one from the north with a runoff coefficient of 0.65, one from the south with a coefficient of 0.29, and a chalk basin. These showed that the difference between +7% and +15% PET on the 'best estimate' rainfall is critical for the chalk aquifer basin (River Lambourn), with +4% or -8% annual runoff. This is the more critical since the Lambourn is one of the rivers that has suffered most from the extreme drought and probable over-abstraction of groundwater in recent years. It is also notable that in the 'best' scenario, summer flows in the upland catchment are much more sensitive to the higher rate of PET than in the lowland, with up to -15% runoff against -7% in the lowland. Conversely, for annual flows the sensitivities are reversed, with totals ranging across the scenarios from + to - 30% in the lowland and + to - 15% in the upland.

Finally, Arnell (1992) considers low flow frequencies. His main conclusion is that 'hydrological drought' will be less common, despite increased PE and less summer rain. This is because in the UK such droughts tend to be due to dry autumn-spring periods rather than to hot summers. However, it should be realized that his analyses were only for monthly flows, and that in the 'best estimate rainfall minus +7% PE' scenario he only observed a maximum improvement in the frequency of exceedance from the present 95% monthly runoff up to 95.8%, i.e. just 0.8%! In contrast, the extreme UKCCIRG scenarios produced a 2-4% improvement for the 'wettest estimate minus +7% PE' and a massive 2.6-9.2% deterioration in the 'driest rainfall estimate minus +15% PE' case. His conclusion may therefore be a little premature.

Arnell and Reynard (1993) and Arnell et al. (1994a) adopted a new methodology, based on work on down-scaling GCM output carried out at the CRU (Viner and Hulme, 1993b). In this the scenarios are based upon output from the UKHI 2.5 x 3.75° model, which gives 15 gridcells around Britain. Since this is still a relatively low spatial resolution for applying their conceptual daily rainfall-runoff model to the 25 chosen catchments scattered around the UK, they used the CRU interpolations of output from the UKHI equilibrium run onto the 40 x 40 km MORECS gridcells. Arnell and Reynard (1993) proceeded to compare the hydrological results using this down-scaled equilibrium scenario with those from an artificially constructed transient scenario.

The transient scenario was created from a transient run of the simple STUGE model scaled according to the temperature sensitivity of the relevant gridcell as indicated by the down-scaled results from the UKHI equilibrium run. In other words, the ratio of the change in local precipitation to the change in mean global temperature, derived from UKHI, was multiplied by the change in mean global temperature according to STUGE for each 40 x 40 km gridcell. Arnell and Reynard (1993) generated additional scenarios by improving upon the PET predictions in the UKCCIRG scenarios. For this, they recalculated PET using the Penman-Monteith formula according to three different assumptions about changes in temperature, net radiation, RH, windspeed, stomatal conductance and leaf area index.

The results indicate large differences in estimated change, with a general increase in discharge throughout the country under the wettest scenario and up to 30% reduction in SE England under the driest. The flow exceeded 95% of the time (which is used in issuing abstraction licences and wastewater discharge consents) would be significantly reduced in all but the wettest scenario. In most scenarios, the seasonality of flows changed more than mean annual discharges, and the southern and eastern catchments showed the greatest sensitivity. Snowfall would generally show significant reductions, although the number of raindays did not change much. However, the net changes generally remained within current interannual variability may not be more extreme than experienced during the 1980s.

Holt and Jones (1993; this volume; in press) also compared transient and equilibrium scenarios for catchments in Wales, for which they used the UKCCIRG equilibrium scenarios and direct output from the UKHI transient run (or 'UKTR'), empirically scaled for local catchments.

Elsewhere in Europe, Olejnik and Kedziora (1991) developed a joint model comprising both the heat and water balances, which also considers the phenological characteristics of crops, and applied it in Poland. They concluded that the general increase in discharges (commensurate with the expected shift in European rainfall patterns discussed above) would mean that water resources would be under less strain. However, if deforestation continues at the current rate this could be translated into a catastrophic 250% increase in flood discharges. They therefore identify the main factor determining whether the changes are beneficial or detrimental as landuse change, both through the direct effects of forest clearance on interception and evapotranspiration and through its indirect effects via the alteration of local heat budgets. Interestingly, Jongman and Souer (1991) also point out the major complications introduced by changing agricultural landuse, and they add the extra complicating effect of acid rain which is affecting 87% of German forests and is thereby likely to reduce ET losses. Beven (1993) notes the potential difficulty in distinguishing changes in the hydrological regime induced by climate change from those induced by landuse changes.

Confirmation of the relative importance of landuse changes comes from Kwadijk's (1991) sensitivity study on the Rhine. Kwadijk concluded that large scale landuse changes in the lowland areas of France and Germany resulting from the EC agricultural set-aside policy could eventually have an effect equal to the reduction in snow-covered area in the Alps associated with a 4°C rise in winter temperatures. The Dutch Rijkswaterstaat is currently planning for a 10% reduction in the summer discharge of the Rhine as it crosses the German border, but a 50% reduction in the area under agriculture, the higher ET from the new forests and increases in PE and the length of the growing season could dramatically exacerbate the summer drought problem.

In 1989, the International Commission for the Hydrology of the Rhine Basin (CHR) initiated a collaborative research programme to develop a basinwide hydrological model to simulated these effects. Parmet and Mann (1993) describe a methology which is being tested to reduce the number of landscape units (which could approach 100 000 for the basin as a whole) to a tractable amount for a physically-based model to handle.

6.6.1. MELTWATER RUNOFF

Interesting analyses of alpine snow and ice regimes indicate the complexity of predicting overall trends in meltwater runoff. Mohnl's (1991) analysis of recent snow records in Austria has demonstrated the complex relationship of snowcover and snowfall variations with elevation and climatic region, with some adjacent areas showing conflicting trends in precipitation receipts and accumulation rates. Martin et al. (1994) found spatial differences in snowcover response to temperature and insolation in the French Alps and classified them in terms of altitude. At 1500 m a.s.l. snowcover was sensitive to temperatures 50% of the time, whereas at 3 km it was sensitive for barely a third of the time. Above 2400 m they found a homogeneous response throughout the French Alps, but below this the southern Alps were more sensitive to temperature rises.

Bultot et al. (1992; 1994) have extended the approach they took in Belgium to two prealpine basins in Switzerland. In the first simulation, the Murg basin showed a shorter snowcover period and an 11% increase in winter discharge balanced by a 14% reduction in summer runoff. The second, in the Broye basin, indicated a decrease in meltwater floods, but overall flood risk was increased by higher rainfall.

A common observation is that snowmelt will tend to occur earlier in the spring. In some regions, this may be accompanied by an intensification of melt rates. In Norway, Saelthun et al. (1990) suggest that the intensity of the spring flood will increase in most areas as a result of more intense melting. Despite increased annual precipitation, they predicted that glaciers will decrease in volume and that the period of snowcover will reduce by between 1 and 3 months. Much of the increased precipitation will be in the form of rain and this could increase flood risk in autumn and early winter. Similarly, earlier and more intense melting in the spring will contribute to reduced summer runoff.

Vehvilainen and Lohvansuu (1991) obtained similar results for Finland using the conceptual water balance model HBV with a GISS scenario that indicated annual temperature increases of 2-6°C (5-6°C in winter) and precipitation increases of 10-30 mm per month. Evaporation losses were estimated to increase by 5-30 mm per month. Less snowfall would reduce the period of snowcover by 2-3 months and reduce the annual peak discharge in spring. But for the rest of the year discharges would increase by 20-50% due to higher rainfall, or in winter due to more mid-season snowmelt.

Earlier and more intense melting could also increase flood risk on rivers like the Rhine issuing from the Alps, despite some reduction in snowfall and snowcover in France and Switzerland (Kwadijk, 1991, 1993). Higher temperatures could also maintain spring flood levels on the Danube despite less snowcover, but the floods will occur slightly earlier according to Gauzer (1993).

The degree of glacierization in a basin is likely to have a strong influence on meltwater response. Collins (1989) compared glacier response between basins with high and low glacierization in the upper Rhone, and concluded that highly glacierized basins are more sensitive to temperature changes, whereas basins with low glacierization are more responsive to changes in total snow receipts. A major factor in basins with lower levels of glacierization is likely to be the changing extent of snowcover under global warming. The findings of Chen and Ohmura (1990) suggest that smaller seasonal snowcovers in a warmer climate will cause increased icemelt by lowering the albedo. Eventually, however, alpine glaciers are expected to continue their current retreat under global warming, so that equilibrium lines will increase in altitude and annual discharges will gradually decrease over the decades (Ohmura et al., 1992). This will be partly due to air temperatures, partly reduced snowfall at least in the west, and partly due to attendant changes in the heat budget components.

6.6.2. TEMPORAL AND SPATIAL ANALOGUES

Most effort has been devoted to projecting changes in runoff via hydrological modelling. However, some would argue, as with rainfall, that spatial or temporal analogues could be used. Spatial analogues do not appear to have been used yet (Arnell and Beran, 1989; Arnell and Reynard, 1989). The WMO World Climate Programme is, however, encouraging further research into palaeohydrologic analogues and the IAHS International Commission for Surface Waters has been collating source material on hydrological proxy data to aid reconstructions (Liebscher, 1987). Schadler (1989) has reconstructed a 450-year sequence for the Rhine using regression, based on modern rainfall and temperature, applied to a proxy set of weather data. From this and other 20th century discharge records from Switzerland, he concluded that the most recent period has been the warmest and wettest, and that this is the only viable analogue period for global warming. Runoff in Finland has also been found to have increased by as much as 20% this century, especially in the 1980s, parallelling the warming curve (Hyvarinen and Leppajarvi, 1989). Supporting the latest GCM predictions for a wetter NW Europe, Thom and Ledger (1976) reconstructed water balances for each major climatic period of the Holocene and concluded that England and Wales had 11-22% more runoff during the warm Atlantic period.

6.6.3. EXTREME EVENTS

One of the most crucial and difficult questions is how global warming will affect flood and low flow frequency curves. Nobilis (1989) commented that no methods used to date seem to be capable of giving the necessary guidelines. GCM predictions for shifts in rainfall patterns (section 6.4) may give a qualitative indication, but it is too costly to carry out sufficient reruns of GCM simulations to derive statistical estimates of the probabilities.

Although Beven (1993) concurs with the general view that Britain will be more prone to flooding, he concludes that the uncertainties as to potential changes in flood frequency remain extremely high and he suggests that new models are required.

Newson and Lewin (1990) argue for studying the Holocene flood record, mainly from geomorphological evidence, and they point to the flood-prone period around the demise of the Little Ice Age as a possible analogue for future developments in Europe. However, there are still some serious practical and methodological problems, both in creating and using such analogues, despite considerable refinements in recent years in palaeohydrological techniques for reconstructing flood histories (Starkel et al., 1990). Perhaps the most important of these is the difficulty of finding a close enough analogue to the state of the regional atmospheric circulation system. The climatic system is in a very different state now from that 150-200 years ago, when the stationary Rossby trough was beginning to migrate east from W Europe, the Ferrel jetstream was beginning to shift north from a markedly lower latitude, and sea surface temperatures to the west of Europe were decidedly lower (Lamb, 1977). Secondly, there are the practical problems of reconstruction: (1) the depositional record is often only partial, as later floods have erased earlier evidence, (2) techniques for estimating palaeovelocity and palaeodischarge from deposits still require considerable improvement (Maizels, 1983; Thornes and Gregory, 1990), and (3) the fluvial system contains many complex interactions and whilst climate may be the 'drive' there is a considerable 'cultural blur' in the history of European rivers (Newson and Lewin, 1990). Both Thornes and Gregory (1990) and Newson and Lewin (1990) point out that considerable research is still needed before quantitative estimates can be made from palaeohydrology.

Holt and Jones (this volume, Chapter 18; in press) have taken a different route, using Wigley's (1989) assumption that extreme events follow a Gaussian distribution. This predicts probabilities for a perturbed climate in terms of the predicted shift in means and the present-day variance. Though this represents a first step towards quantative prediction, the likely errors remain high.

6.7. Water Quality

Most research so far has been concerned with quantitative aspects of water regimes, and yet global warming has important implications for water quality, both directly through the water budget and indirectly via the effects of temperature or cultural changes upon chemical and biological processes and upon the supply of pollutants. The IPCC (1990b) report acknowledges that this is an area about which very little is known. Birley (1991) underlines the current problems in predicting the effect of changing hydrological regimes upon public health. And Beran (1989) outlines the major gaps where research is needed, especially as regards the water quality effects of extreme events.

There are basically three main types of problem associated with changing regimes: (1) the Dilution factor, by which low flows tend to increase concentrations, (2) the Flushing factor, in which high flows entrain more sediment and pollutants from both channels and hillslopes, and (3) the Biogeochemical factor, whereby higher temperatures tend to speed up chemical and biochemical reactions. Hence, for example, higher water temperatures can be expected to increase eutrophication, algal blooms and microbial populations, and lead to a decrease in dissolved oxygen. Webb and Nobilis (1994, 1995) have demonstrated that water temperatures in Austrian rivers have been steadily rising throughout the 20th century by about 0.5-1.25°C, due to a variety of causes, especially milder winters and lower autumn discharges. The latter have been aggravated by increasing river regulation. From a series of multiple regression equations, they predict that mean monthly temperatures will continue to rise a further 1°C by 2030 and 2°C in autumn low flows. Jurak (1989) estimated increases of 0.9-1.5°C in mean monthly water temperatures in Poland with double CO_2. Webb (1992) and Arnell et al. (1994b) conclude that by 2050 the mean temperature of most UK rivers could be over 10°C, with over 15°C in some rivers in the Midlands and S England. This would represent an increase of 1.0-2.5°C in most of Britain and 4°C in N Scotland. As with atmospheric temperatures, minimum temperatures are likely to rise more than maxima. This will reduce the average temperature range and the probability of subzero temperatures will reduce threefold in SW England. But this is matched by an increase of similar order in the risk of water temperatures exceeding 25°C. This could increase the occurrence of low dissolved oxygen (DO) content in summer and have damaging ecological effects (Jenkins et al., 1993). Higher temperatures are also likely to increase the biological oxygen demand (BOD) in organically polluted waters, through stimulated bacteria activity and algal blooms (Arnell et al., 1994b). Contamination of public water supply reservoirs has caused a number of alarms during recent hot summers in England: toxic algal blooms at Rutland Water in 1989, and Cryptosporidium contamination at Farmoor Reservoir, Oxfordshire, in 1989 and again in Devon in 1995.

Some basins are likely to undergo a change in the ratio of surface to subsurface runoff, as demonstrated in Bultot's Belgian catchments, which could have important implications for water quality. Gellens (1991) concluded that his Belgian aquifer basin would suffer less from lack of dilution because of the increased summer discharges, but this is unlikely to be the case in SE England with its over-exploited aquifers.

Changes in water quality could result from higher rainfall intensities, less snowmelt, lower soil moisture contents and summer desiccation - especially cracking in histic soils - or from persistent saturation in winter. The most common effects are likely to be more rapid runoff, higher coefficients of runoff and shorter residence times as surface and near-surface flows become more dominant. This would reduce the buffering of acid rain and increase the chance of acid flushes with toxic levels of monomeric aluminium and so increase fish-kills (Jones, 1990). Greater acidity would increase the solution of heavy metals in general and aggravate the dangers for fisheries in mineralized fold mountain environments. Mountain areas with peaty soils and low base status rocks, as in the west of Britain, are particularly vulnerable, the more so since increases in annual rainfall will tend to increase the frequency of high flows. Similarly, the resulting increase in stream power, perhaps assisted by drier soils in summer, should increase sediment yields, destroying breeding habitats for fish and increasing their ingestion of solid pollutants. Arnell et al. (1994b) suggest that nitrate pollution could increase as higher temperatures speed up the rate of mineralization of organic N in the soil and rapid runoff following summer droughts flushes out the NO_3.

Boardman and Favis-Mortlock (1993) estimate that soil erosion, and therefore suspended sediment concentrations, will increase in Britain as a result of higher winter rainfall, increased summer storm frequency, increased irrigation and the introduction of erosion-susceptible crops such as maize. Kwaad (1991) reached a similar conclusion, extrapolating the results from experimental soil erosion plots and considering the likely seasonal increases in rainfall amounts and intensities in The Netherlands. One unique report from Shiklomanov et al. (1995) on the Dnipro River basin in the Ukraine expresses the fear that increased discharges, which are predicted from all the GCMs investigated (GFDL, UKHI and MPI) and have already been rising since the 1970s, could entrain sediments that are heavily polluted by radioactive deposits from the Chernobyl nuclear meltdown. Six reservoirs are at risk from reworking of bottom deposits.

As with ET and runoff (e.g. Lockwood, 1995; Kwadijk, 1991), however, it is worth noting that Arnell et al. (1994b) believe that changes in landuse and effluent disposal are likely to have a greater effect than the envisaged degree of climate change upon future water quality. Nevertheless, this should certainly not be taken to imply that global warming effects are unimportant. Indeed, there could be additional subtle feedbacks between climate change, pollution and landuse. For example, the reduced space heating requirements in NW Europe demonstrated by Warrick et al. (1990) might help the acid rain problem. Changes in the frequency of synoptic weather types could also affect acid deposition, as shown by Davies et al. (1986). And landuse change will be sensitive in part to the changing distribution of agricultural potential stimulated by climate change (cp. Warrick et al., 1990; Parry et al., 1989).

6.8. Sea Level

The consensus view is that sea level will rise 200-300 mm by 2050 (IPCC, 1990a, 1990b; Warrick and Farmer, 1990; Warrick et al., 1990; UKCCIRG, 1991), as much due to thermal expansion as to icemelt. The predicted rate of sea level rise is greater than any long-term average over the last 7000 years (Shennan, 1993). Greater vertical mixing in the ocean could cause more (Wigley and Raper, 1987).

This has a number of potential implications for terrestrial hydrology which are begging for more research. In general, there is likely to be an increase in landward penetration of saltwater and an increase in flood risk in the coastal and estuarine zone. The UK National Rivers Authority recognizes a 'blue belt' below 5 m a.s.l. that is currently vulnerable to marine flooding (Whittle, 1989) and this will require extension (Pugh, 1990; UKCCIRG, 1991), not simply because of a rise in mean sea level, but possibly more because of probable increases in storminess and lower atmospheric pressure in depressions. These could increase the risk of storm surges, with waves and water mass piling on top of high tides, as in the 1953 North Sea disaster that spawned the Rhine Delta Project or the North Wales floods of February 1990 in the UK. Greater coastal water depths combined with increased storminess are also likely to enhance the erosivity of the waves. Jones (1990) has mapped the vulnerable coasts in Wales, and many of them are currently designated Sites of Special Scientific Interest (SSSIs) or of outstanding scenic value.

Extension of saltwater intrusions, both above and below ground, could threaten some coastal water supplies using groundwater and some surface intakes currently near the tidal limit (UKCCIRG, 1991). However, Arnell et al. (1994b) conclude that reductions in freshwater discharges are likely to be more important than sea level rise for saline penetration in estuaries. The rise in the base-level of erosion might also cause problems with extra sedimentation in estuaries, especially if stream competence is increased upstream by more frequent high discharges. Indeed, increased freshwater flooding in estuaries could aggravate marine floods and be triggered by the same intense depressions that create coastal storm surges.

6.9. Demand and Water Resource Management

Despite some professional reluctance amongst water managers to consider implications whilst predictive errors remain so large, Arnell et al. (1994b) note that the impacts of global warming may be very large and be felt within the time horizons of current planning.

Predicting the implications for water management is doubly difficult because it requires modelling not only reservoir inflows but also the pattern of consumer demand. Holt and Jones make an attempt to tackle the first of these for equilibrium (this volume, Chapter 18) and transient (in press) scenarios for direct supply and river regulation reservoirs in Wales by using site specific equations based on the MORECS archive and Welsh Water's mass balance equations for the regulation reservoirs. Law (1989) and Cole et al. (1991) have also modelled reservoir response in England (v.i.). Otherwise, most water resource modelling in the UK has been undertaken by the UKIH team of Arnell and colleagues in terms of riverflows (section 6.6). As with the other riverflow predictions discussed in section 6.6, this is relevant both to reservoir inflows and to direct abstraction of water supplies from 'unsupported' river sources. Numerically, unsupported river sources actually dominate in Wales (over 50%), although they only account for about one-eighth of the total public water supplies.

Predicting trends in demand is rather more difficult and there is very little work in the field. Cole *et al.* (1991) point out that, although some of the adjustments in reservoir storage that may be necessary to compensate for changing inflows are quite substantial (*v.i.*), it is possible that they will prove to be minor compared with attendant increases in consumer demand. The 125% increase in demand in SW England during July 1995, the third hottest on record, is a clear indication of the likely problem.

The NRA (1994) specifically excluded any global warming effects from their predicted trends for England and Wales. However, economists at the University of Leicester have undertaken a study for the UK Department of Environment (Herrington and Hoschatt, 1993). The study suggests that whereas demand is projected to rise by 21% in S England between 1991 and 2021 from non-climatic causes, global warming could add a further 5%. Arnell *et al.* (1994a) calculate a 4% increase in annual public water supply demand in S and E England by 2021, but point out that summertime peak demands are likely to be the main problem.

Herrington and Hoschatt (1993) also suggest that demand from spray irrigation will increase by 28% with a $1.1°C$ warming by 2050, but they point out that overall demand from agriculture is also very sensitive to prices and production potential. The calculation in Arnell *et al.* (1994a) is much more extreme: an overall increase in demands from spray irrigation in S and E England between 1991 and 2021 from 313 to 704 Ml d^{-1}, of which 156 Ml d^{-1} (50% of current demand) would be due to climate change. A separate study in Poland indicates a 12% increase in irrigation demand with a $1°C$ warming (Jurak, 1992).

It is likely that the combination of a northward shift in warming degree days within Europe, as indicated by Warrick *et al.* (1990), and increased water stress in the south will shift the centre of gravity of agricultural potential within the EU towards the NW. This is therefore likely to be an important driving factor which will make irrigation more economically viable in NW Europe.

Da Cunha (1989) undertook a comprehensive study of water resources in the then EEC based on early GCM projections, which pointed to potential management problems arising from (1) increased drought in Iberia, S Italy and Greece due to reduced annual rainfall and increased rainfall variability especially in summer, and (2) increased rainfall all the year round elsewhere especially in N Europe, which could create more flood hazard. Da Cunha discussed a wide range of water resource impacts under the headings of social, economic and environmental in qualitative terms. These included the probable spread of desertification in the south and emigration from S Europe, plus temperature-generated increase in demand in the Mediterranean tourist areas. Mimikou and Kouvopoulos (1991) and Mimikou *et al.* (1991a and b) observe that the reductions in discharges indicated by their conceptual water balance model could significantly reduce hydropower production in C Greece.

Studies in both the Alps and Scandinavia identify changes in snowmelt patterns as a potential management problem. For hydropower in Norway, Saelthun *et al.* (1990)

suggest that the increased seasonality of flows will be offset by the higher annual precipitation, giving a net increase of 2-3% in power generating potential. Lake ice covers will reduce or even disappear, with local effects on transport (both positive and negative). In Hungary, Novaky (1992a and b) forecasts more summer water shortages as a result of less snowfall and reduced spring flood levels, and although Gauzer (1993) predicts that the level of snowmelt flooding on the Danube will be maintained by meltwaters from the upper basin, the floods will occur earlier.

In Britain, the main concern amongst water managers has been maintaining supplies through the summer (e.g. Parry and Read, 1988; Binnie, 1990; UKCCIRG, 1991). The UKCCIRG point out that reservoir reliability is likely to be significantly reduced in areas where summer rainfall is approximately equal to ET and where runoff is more rapid, and that the long lead times for implementing most system developments make the need for guidance urgent. The risk of events like the 1976 drought is estimated to increase a hundred-fold to 10% by 2030 (UKCCIRG, 1991).

The increased vulnerability of resources in the southern UK identified in Arnell's analyses (section 6.6) is likely to be aggravated by increasing demand in the same area, through increased domestic and agricultural requirements. This could further increase the frequencies of low flows in rivers, with many knockon effects for water management. Many French nuclear reactors were forced to shut down partially or completely during the 1991 drought for lack of riverwater coolant. Many rivers in SE England appear to be drying up through overexploitation of groundwater. Many more rivers in the UK have their flows significantly reduced by direct abstraction. More than 20% are covered by river regulation schemes and therefore could be materially affected by any changes in operating rules or abstraction rates.

The Thames basin has been one of the few areas of Britain with a sustained increase in demand over the last decade. Following the most severe multi-year drought this century in SE England, the NRA identified 40 endangered rivers in 1992 and many others were under review; some like the R. Darent in Kent were already dry. Farmers were fined for unlicensed irrigation and plans were mooted for long distance interbasin transfers to the SE from the N and W, perhaps from the Kielder dam in Northumbria, Britain's largest, or from Wales, with a long term possibility of reviving a plan shelved more than a decade ago to construct the largest reservoir in Europe at Craig Goch in mid-Wales (NRA, 1992, 1994). Alternative proposals to develop new reservoirs in S England have met with strong environmental opposition.

The UKICE has identified the need for 'demand management', perhaps by metering and tariff structuring, and for a sustained programme to reduce system leakage which is currently estimated at 25% or more over much of Britain (Binnie, 1991). A metering experiment covering the Isle of Wight has succeeded in reducing demand by 20%. In Germany, where metering is normal, losses are typically only 5% and demand has tended to remain static for the past decade. Unfortunately, there seems to be very little data to indicate likely demand stresses in a global warming scenario. Certainly, in Britain any data tend to relate to situations in which operational measures such as drought orders have been used to regulate demand. But summer demand in S England can be 25% higher than normal winter levels (UKCCIRG, 1991), and holiday areas are likely to face increased stress.

Serious attempts are being made in Britain to determine effects on reservoir yield and to improve methods of deriving yield-storage curves, for example, by UKIH (Law, 1989). Jack and Lambert (1992) highlight deficiencies in traditional methods of calculating yield shown up by the major droughts since 1976, and propose the use of 'operational yield' with control rules being included: in many cases this produces a significantly lower yield. Holt and Jones (this volume; in press) have adapted this method to look at likely reductions in operational yields under global warming.

One of the most significant research projects has been the collaboration between WRC, the CRU and the Hadley Centre in the UK. Building upon the work of Gregory *et al.* (1990), Cole *et al.* (1991) have used synthetic daily rainfall series, produced by perturbing the incomplete gamma model according to the projected temperature change, together with regression estimates of seasonal AET, as input to water balance calculations for each homogeneous region. The resultant 50-year sequences of daily runoff were then used to simulate the fluctuating contents of various direct supply reservoirs, and the reservoir contents searched for periods of deficit. They concluded that runoff will be 4% lower in the NW and 8% lower in SE England by 2030, which would result in a 4-25% loss of yield in the NW and 8-15% in the SE. This would require either a commensurate reduction in drawoff or else increased reservoir storage ranging from 7 to 33% in the NW and between 10 and 21% in the SE, assuming that the probable duration of deficit continues to be contained at the 1% risk level. The larger percentages would apply to the smaller reservoirs. Interestingly, the proportional adjustments needed in storage capacity are somewhat larger than the reductions in runoff and in this regard the NW is no better off than the SE. Indeed, the NW UK may not benefit much from wetter winters, since reservoirs tend to be full now at the end of winter and extra runoff would mean more spillage. However, although Cole and his colleagues do not compare their results with Law (1989), it is notable that Law's calculations for a reservoir in the British Pennines seem to show much lower sensitivities.

6.10. Conclusions

Any conclusions at this stage must be tentative. It is, nevertheless, extremely important for both practical and scientific reasons to test hydrological sensitivities and to assess the order of magnitude of likely impacts. The main conclusions seem to be:

(1) Most of Europe should receive higher annual precipitation, more so in the north, but there will also be increased variance, greater storminess and more frequent extreme events at both ends of the spectrum. A transition period is likely to involve more intense depressional rainfall, until the meridional temperature gradient in mid-latitudes is reduced and depressions track further north: more intense convective storms should replace them.

(2) Evapotranspiration will increase and so will the risk of summer drought, especially in the south, not solely because of higher ET. As noted in IPCC (1990b), the

Mediterranean, where water supply often already poses problems, is likely to be particularly at risk.

(3) Greater extremes in runoff are expected, but net changes will also depend on the physical characteristics of the drainage basins and be more varied than precipitation and ET.

(4) Water management problems are likely to increase, even in areas of net positive water balance, because of a) increased climatic variability, b) increased demand, especially for irrigation and domestic use, c) the common double amplification within the water catchment system, whereby runoff tends to amplify changes in rainfall and reservoir schemes will tend to amplify changes in mean annual runoff in the projected scenarios, d) increased winter precipitation may not be of much use without increasing storage, because it will tend simply to increase spillage losses.

(5) Increased risks of pollution incidents and of coastal floods seem assured.

(6) Methodological debates will continue, but in reality all approaches are to be welcomed for what they may offer at present. In the long run, true Global Climate Models in combination with physically- based hydrological simulation models will be the best route forward. In the meantime, despite uncertainties in reconstruction and despite differences in forcing factors and circulation systems, historical analogues can provide useful yardsticks. Similarly, extrapolations from recent trends are needed, but if the projected rate of global warming is correct then their usefulness will be strictly limited.

Acknowledgements

I am indebted to the Hadley Centre for permission to base Figures 6.7 and 6.8 on UKHI output, and particularly to Drs David Bennetts and Richard Jones. I am most grateful to Nigel Arnell, Logan Jack, Hans Mohnl, Jim Hudson, Bruce Webb and B W Parmet, for offprints and manuscripts, and to Chris Holt and Charlie Pilling for keeping me supplied with papers.

References

Academy of Finland (1989) *Conference on Climate and Water*, Academy of Finland, Helsinki, 2 volumes.

Andersson, L. (1989) Soil moisture dynamics in South-Central Sweden in a 100 years perspective, in *Conference on Climate and Water*, Academy of Finland, Helsinki, vol. 1, 252-261.

Arnell, N.W. (1989) The potential effects of climatic change on water resources management in the UK, in *International Seminar on Climatic Fluctuations and Water Management*, WMO, Cairo, Paper II.11.

Arnell, N.W. (1992) Impacts of climatic change on river flow regimes in the UK, *J. Institution of Water and Environmental Management* **6**, 4, 432-442.

Arnell, N.W., and Beran, M.A. (1989) The use of regional databases for the assessment of the hydrological impact of climate change and variability, European Conference on Landscape Ecological Impact of Climate Change, Lunteren, Netherlands.

Arnell, N.W., Brown, R.P.C., and Reynard, N.S. (1990) *Impact of Climatic Variability and Change on River Flow Regimes in the UK*, Wallingford, UK, Institute of Hydrology Report 107.

Arnell, N.W., A. Jenkins, P. Herrington, B. Webb, and M. Dearnaley (1994a) *Impacts of Climate Change on Water Resources in the United Kingdom: Summary of Project Results*, Report to the Department of Environment Water Directorate, Institute of Hydrology, NERC, Wallingford, UK, 50 pp.

Arnell N. W., A. Jenkins, D. G. George (1994b) *The Implications of Climate Change for the National Rivers Authority*, National Rivers Authority, Bristol, UK, R & D Report 12, 94 pp.

Arnell, N.W., and Reynard, N.S. (1989) Estimating the impacts of climate change on river flows) some examples from Britain, in *Conference on Climate and Water*, Academy of Finland, Helsinki, vol. 1, 426-436.

Arnell N. W., and N.S. Reynard (1993) *Impact of Climate Change on River Flow Regimes in the United Kingdom*, Institute of Hydrology, Wallingford, UK, 129 pp.

Bach, W. (1989) Projected climatic changes and impacts in Europe due to increased CO_2, in *Conference on Climate and Water*, Academy of Finland, Helsinki, vol. 1, 31-50.

Behr, O. (1989) Long-term variability of precipitation in Austria, in *Conference on Climate and Water*, Academy of Finland, Helsinki, vol. 1, 93-102.

Beran, M. (1989) The impact of climatic change on the aquatic environment, in *Conference on Climate and Water*, Academy of Finland, Helsinki, vol. 2, 7-27.

Berndtsson, R., Larson, M., Lindh, G., Malm, J., Niemczynowicz J., Zhang, T. (1989) Climate-induced effects on the water balance - preliminary results from studies in the Varpinge experimental research basin, in *Conference on Climate and Water*, Academy of Finland, Helsinki, vol. 1, 437-449.

Beven, K. (1993) Riverine flooding in a warmer Britain, *The Geographical J.* **159**, 2, 157-161.

Binnie, C.J.A. (ed.) (1991) Policy paper on: Securing adequate water supplies in the United Kingdom in the 1990s and beyond, Institution of Civil Engineers, London.

Birley, M.H. (1991) Methods of forecasting the vector-borne disease implications in the development of a water resource project, in R. Wooldridge (ed.) *Techniques for Environmentally Sound Water Resources Development*, Pentech Press, London, 50-63.

Boardman, J., and Favis-Mortlock, D.T. (1993) Climate change and soil erosion in Britain, *The Geographical J.* **159**, 2, 179-183.

Bradley, R.S., Diaz, H.F., Eischeid, J.K., Jones, P.D., Kelly, P.M., and Goodess, C.M. (1987) Precipitation fluctuations over Northern Hemisphere land areas since the mid-19th century, *Science* **237**, 171-175.

Budyko, M.I. (1989) Climatic conditions of the future, in *Conference on Climate and Water*, Academy of Finland, Helsinki, vol. 1, 9-30.

Budyko, M.I., and Izrael, Yu. A. (eds.) (1987) *Anthropogenic Climate Changes*, Gidrometeoizdat, Leningrad, 406 pp.

Bultot, F., Coppens, A., Dupriez, G.L., Gellens, D., and Meulenberghs, F. (1988) Repercussions of a CO_2-doubling on the water cycle and on the water balance: a case study from Belgium,

J. Hydrology **99**, 319- 347.

Bultot, F., Gellens, D., Spreafisco, B., and Schedler, B. (1992) Repercussions of a CO_2 doubling on the water balance - a case study in Switzerland, *J. Hydrology* **137**, 199-208.

Bultot, F., Gellens, D., Spreafisco, B., and Schedler, B. (1994) Effects of climate change on snow accumulation and melting in the Broye catchment (Switzerland), *Climatic Change* **28**, 339-363.

Chen, R.S., and Parry, M.L. (1987) *Policy-orientated Impact Assessment of Climatic Variations*, IIASA, Laxenburg, Austria, RR87/7.

Chen, J., and Ohmura, A. (1990) On the influence of Alpine glaciers on runoff, *International Association of Hydrological Sciences Publication* No. 193, 117-125.

Cole, J.A., Slade, S., Jones, P.D., Gregory, J.M. (1991) Reliable yield of reservoirs and possible effects of climate change, *Hydrological Sciences J.* **36**, 6, 579-598.

Collins, D.N. (1989) Influence of glacierisation on the response of runoff from Alpine basins to climate variability, in *Conference on Climate and Water*, Academy of Finland, Helsinki, vol. 1, 319-328.

Conte, M., Giuffrida, A., and Tedesco, S. (1989) The Mediterranean oscillation: impact on precipitation and hydrology in Italy, in *Conference on Climate and Water*, Academy of Finland, Helsinki, vol. 1, 121-137.

Davies, T.D., Kelly, P.M., Brimblecombe, P., Farmer, G., and Barthelmie, R.J. (1986) Acidity of scottish rainfall influenced by climatic change, *Nature* **232**, 359-361.

Da Cunha, L. (1989) Water resources situation and management in the EEC, *Hydrogeology* **2**, 57-69.

Dunkel, Z. (1989) Monitoring of vegetation period course of soil moisture based on measurements and evaluation in Hungary, in *Conference on Climate and Water*, Academy of Finland, Helsinki, vol. 1, 262-271.

Emanuel, K.A. (1987) The dependence of hurricane intensity on climate, *Nature* **326**, 483-485.

Flohn, H., and Fantechi, R. (1984) *The Climate of Europe: Past, Present and Future*, Reidel, Dordrecht, Netherlands, 356 pp.

Gauzer, B. (1993) Effect of air temperature change on the Danube flow regime, VITUKI, Budapest. (In Hungarian.)

Gellens, D. (1991) Impacts of a CO_2-induced climatic change on river flow variability in three rivers in Belgium, *Earth Surface Processes and Landforms* **16**, 619-625.

Gleick, P.H. (1987) Regional hydrologic consequences of increases in atmospheric CO_2 and other trace gases, *Climatic Change* **10**,137-161.

Gregory, J.M., Jones, P.D., and Wigley, T.M.L. (1990) *Climatic change and its potential effect on UK water resources*, Report to the Water Research Centre, Medmenham, UK, Parts 1 and 2.

Groisman, P.Ya. (1991) Data on present-day precipitation changes in the extratropical part of the Northern Hemisphere, in M.E. Schlesinger (ed.) *Greenhouse-gas-induced Climatic Change: a Critical Appraisal of Simulations and Observations*, Elsevier, Amsterdam, 297-310.

Gunn, J., and Crumley, C.L. (1991) Global energy balance and regional hydrology: a Burgundian case study, *Earth Surface Processes and Landforms* **16**, 579-592.

Hadley Centre (1992) *The Hadley Centre transient climate change experiment*, Meteorological Office, Bracknell, UK, 20 pp.

Hadley Centre (1995) *Modelling climate change 1860-2050*, Meteorological Office, Bracknell, UK, 13 pp.

Hansen, J., Rind, D., Delgenio, A., Lacis, A., Lebedeff, S., Prather, M., and Ruedy, R. (1991) Regional greenhouse climate effects, in M.E. Schlesinger (ed.) *Greenhouse-gas-induced Climatic Change: a Critical Appraisal of Simulations and Observations*, Elsevier, Amsterdam, 211-229.

Heino, R. (1989) Changes of precipitation in Finland, in *Conference on Climate and Water*, Academy of Finland, Helsinki, vol. 1, 111-120.

Henderson-Sellers, A. (1993) An antipodean climate of uncertainty, *Climatic Change* **25**, 203-224.

Herrington, P., and Hoschatt, M. (1993) *Climate change and the demand for water*, University of Leicester, Department of Economics, Report to Department of the Environment.

Holt, C.P., and Jones, J.A.A. (1993) Potential impact of global warming on water resource management in upland Wales, Cardiff, Proceedings of the British Hydrological Society 4th National Hydrology Symposium, 1.17-1.22.

Holt, C.P., and Jones, J.A.A., (in press) Implications of the equilibrium and transient global warming scenarios for water resources in Wales - a comparison, *Water Resources Bulletin*.

Hudson, J.A., and Gilman, K. (1993) Long-term variability in the water balances of the Plynlimon catchments, *J. Hydrology* **143**, 355-380.

Hyvarinen, V., and Leppajarvi, R. (1989) Long-term trends in river flow in Finland, in *Conference on Climate and Water*, Academy of Finland, Helsinki , vol. 1, 450-461.

Intergovernmental Panel on Climate Change (IPCC) (1990a) *Climate Change. The IPCC Scientific Assessment*, J.T Houghton, G.J Jenkins, and J.J. Ephraums (eds.) for WMO/UNEP, Cambridge University Press, Cambridge, UK.

Intergovernmental Panel on Climate Change (IPCC) (1990b) *Climate Change. The IPCC Impacts Assessment*, W.J.McG. Tegart, G.W. Sheldon, and D.C. Griffiths (eds.) for WMO/UNEP, Commonwealth of Australia Government Publishing Service, Canberra.

Intergovernmental Panel on Climate Change (IPCC) (1992) *Climate Change 1992, The Supplementary Report to the IPCC Assessment*, J.T. Houghton, B.A. Callander, and S.K. Varney (eds.), Cambridge University Press, Cambridge, UK.

Intergovernmental Panel on Climate Change (IPCC) (1996) *Climate Change 1995*, J.T. Houghton, B.A. Callander, and S.K. Varney (eds.), Cambridge University Press, Cambridge, UK.

Jack, L., and Lambert, A.O. (1992) Operational yield, in M.N. Parr, J.A. Charles and S. Walker (eds.) *Water Resources and Reservoir Engineering*, Thomas Telford, London, 65-72.

Jenkins, A., McCartney, M., and Sefton, C. (1993) *Impacts of Climate Change on Water Quality in the United Kingdom*, Wallingford, UK, Institute of Hydrology Report to Department of the Environment, 39 pp.

Jones, J.A.A. (1990) Greenhouse Wales, *Planet* **82**, 48-62.

Jongman, R.H.G., and Souer, M.A. (1991) Landscape ecological and spatial impacts of climatic change in two areas in The Netherlands, *Earth Surface Processes and Landforms* **16**, 639-652.

Jurak, D. (1989) Effect of climate change on evaporation and water temperature, in *Conference on Climate and Water*, Academy of Finland, Helsinki, vol. 1, 138-148.

Kajander, J.M. (1989) 150 years since C.G. Hallstrom's studies on ice break-up dates as climatic indicators, in *Conference on Climate and Water*, Academy of Finland, Helsinki, vol. 1, 329-338.

Klemes, V. (1985) Sensitivity of water-resource systems to climate variations, *World Climate Program Report No. 98*, World Meteorological Organization, Geneva.

Kwaad, F.J.P.M. (1991) Summer and winter regimes of runoff generation and soil erosion on cultivated loess soils (The Netherlands), *Earth Surface Processes and Landforms* **16**, 653-662.

Kwadijk, J.C.J. (1991) Sensitivity of the River Rhine discharge to environmental change, a first tentative assessment, *Earth Surface Processes and Landforms* **16**, 627-637.

Kwadijk, J.C.J. (1993) The impact of climate change on the discharge of the River Rhine, *Netherlands Geographical Studies* No. 171, 299 pp.

Lamb, H.H. (1972) British Isles weather types and a register of the daily sequence of circulation patterns, 1861-1971, *Geophysical Memorandum* 116, HMSO, London.

Lamb, H.H. (1977) *Climate: Present, Past and Future*, vol. 2, Methuen, London, 835 pp.

Law, F.M. (1989) Identifying the climate-sensitive segment of British reservoir yield, in *Conference on Climate and Water*, Academy of Finland, Helsinki, vol. 2, 177-190.

Liebscher, H.J. (1987) Paleohydrologic studies using proxy data and observations, in *The Influence of Climate Change and Climatic Variability on the Hydrologic Regime and Water Resources*, International Association of Hydrological Sciences Publication No. 168, 111-121.

Lockwood, J.G. (1995) The suppression of evapotranspiration by rising levels of atmospheric CO_2, *Weather* **50**, 9, 304-308.

Lough, J.M., Wigley, T.M.L., and Palutikof, J.P. (1983) Climate and climate impact scenarios for Europe in a warmer world, *J. Climatology and Applied Meteorology* **22**, 1673-1684.

Maizels, J.K. (1983) Palaeovelocity and palaeodischarge determination for coarse gravel deposits, in K.J. Gregory (ed.) *Background to Palaeohydrology*, Wiley, Chichester, UK, 101-139.

Manabe, S., and Wetherald, R.T. (1987) Large scale changes of soil wetness induced by an increase in atmospheric carbon dioxide, *J. Atmospheric Science* **44**, 1212-1235.

Martin, P., Rosenberg, N.J., and McKenney, M.S. (1989) Sensitivity of evapotranspiration in a wheat field, a forest, and a grassland to changes in climate and direct effects of carbon dioxide, *Climate Change* **14**,117-151.

Meteorological Office (1981) *The Meteorological Office Rainfall and Evaporation Calculation System*, Hydrological Memorandum No. 45, Bracknell, UK.

Mimikou, M., and Kouvopoulos, Y. (1991) Regional climate change impacts: I Impacts on water resources, *Hydrological Sciences J.* **36**, 247-258.

Mimikou, M., Hadjisavva, P.S., Kouvopoulos, Y., and Atrateos, H. (1991a) Regional climate change impacts: II Impacts on water management works, *Hydrological Sciences J.* **36**, 259-270.

Mimikou, M., Kouvopoulos, Y., Cavadias, G., and Vayiannos, N. (1991b) Regional hydrological effects of climate change, *J. Hydrology* **123**, 119-146.

Mitchell, J.F.B. (1991) The equilibrium response to doubling atmospheric CO_2, in M.E.Schlesinger (ed.) *Greenhouse-gas-induced Climatic Change: a Critical Appraisal of Simulations and Observations*, Elsevier, Amsterdam, 49-61.

Mitchell, J.F.B., and Warrilow, D.A. (1987) Summer dryness in northern mid-latitudes due to increased CO_2, *Nature* **330**, 238-240.

Mitchell, J.F.B., Wilson, C.A., and Cunnington, W.M. (1987) On CO_2 climate sensitivity and model dependence of results, *Quarterly J. Royal Meteorological Society* **113**, 293-322.

Mitchell, J.F.B., Johns, T.C., Gregory, J.M., and Tett, S.F.B. (1995) Climate response to increasing levels of greenhouse gases and sulphate aerosols, *Nature* **376**, 501-504.

Mitosek, H.T. (1995) Climate variability and change within the discharge time series: a statistical a pp.roach, *Climatic Change* **29**, 101-116.

Mohnl, H. (1991) Fluctuations of snow parameters in the mountainous region of Austria within the last 90 years. *Extended abstracts, International Association of Hydrological Sciences Symposium H5, XX General Assembly of IUGG*, Vienna, 264-268.

Morris, S.E., and Marsh, T.J. (1985) United Kingdom rainfall 1975-1984: evidence of climatic instability, *J. Meteorolology* **10**, 103, 324- 332.

National Rivers Authority (NRA) (1992) *Water Resources Development Strategy: a Discussion Document*, National Rivers Authority, Bristol, UK, 12 pp.

National Rivers Authority (NRA) (1994) *Water - nature's precious resource. An environmentally sustainable water resources development strategy for England and Wales*, HMSO, London, 93 pp.

Newson, M.D., and Lewin, J. (1990) Climatic change, river flow extremes and fluvial erosion - scenarios for England and Wales, *Progress in Physical Geography*, **15**, 1, 1-17.

Nobilis, F. (1989) Flood potential, an uncertain estimate resulting from climatic variability and change, in *Conference on Climate and Water*, Academy of Finland, Helsinki, vol. 2, 348-356.

Novaky, B. (1985) Water resources, in R. Kates (ed.) *Climate Impact Assessment*, Wiley, Chichester, UK, 187-214.

Novaky, B. (1991) Climatic effects on the runoff conditions in Hungary, *Earth Surface Processes and Landforms* **16**, 593-599.

Novaky, B. (1992a) Effect of climate change on excess waters of winter/spring and minimum flows of summer/autumn, VITUKI, Budapest. (In Hungarian.)

Novaky, B. (1992b) Effect of climate change on maximum flows of winter/spring, VITUKI, Budapest. (In Hungarian.)

Ohmura, A., Kasser, P., and Funk, M. (1992) Climate at the equilibrium line, *J. Glaciology* **38**, 130, 397-411.

Olejnik, J., and Kedziora, A. (1991) Model for heat and water balance estimation and its application to land use and climate variation, *Earth Surface Processes and Landforms* **16**, 601-617.

Palutikof, J.P. (1987) Some possible impacts of greenhouse gas induced climatic change on water resources in England and Wales, in S.I. Solomon, M. Beran, and W. Hogg (eds.) *The Influence of Climate Change and Climatic Variability on the Hydrologic Regime and Water Resources*, International Association of Hydrological Sciences Publication No. 168, 585-596.

Palutikof, J.P., Wigley, T.M.L., and Lough, J.M. (1984) *Seasonal Climate Scenarios for Europe and North America in a High-CO_2 Warmer World*. DOE/EV/10098-5, US Dept. of Energy, Washington, D.C.

Parmet, B.W.A.H., and Mann, M.A.M. (1993) Influence of climate change on the discharge of the River Rhine - a model for the lowland area, In *Exchange Processes at the Land Surface for a Range of Space and Time Scales*, International Association of Hydrological Sciences, Publication No. 212, 469-477.

Parr, T., and Eartherall, A. (1994) *Demonstrating Climate Change Impacts in the UK: the DoE Core Model Programme*, Institute of Terrestrial Ecology, NERC, Huntingdon, 18 pp.

Parry, M.L., Carter, T.R., and Porter, J.H. (1989) The greenhouse effect and the future of UK agriculture, *J. Royal Agricultural Society of England*, 120-131.

Parry, M.L., and Read, N.J. (eds.) (1988) *The Impact of Climatic Variability on UK Industry*, AIR Report No. 1, Atmospheric Impacts Research Group, University of Birmingham, Birmingham, UK, 71 pp.

Pugh, D.T. (1990) Is there a sea-level problem? *Proceedings Institution of Civil Engineers*, Part 1, **88**, 347-366.

Rabbinge, R., Van Latesteijn, H.C., and Goudrian, J. (1993) Assessing the greenhouse effect in agriculture, in Lake, J.V., Bock, G.R., and Ackrill, K. (eds.) *Environmental Change and Human Health*, Ciba Foundation Symposium, Wiley, Chichester, UK, 62-79.

Rampino, M.R., and Etkins, R. (1990) The greenhouse effect, stratospheric ozone, marine productivity, and global hydrology: feedbacks in the global climate system, in R. Paepe (ed.) *Greenhouse Effect, Sea Level and Drought*, Kluwer, Dordrecht, Netherlands, 3-20.

Rind, D. (1988) The doubled CO_2 climate and the sensitivity of the modeled hydrologic cycle, *J. Geophysical Research* **93**, D5, 5385-5412.

Rind, D. (1991) Climate variability and climate change, in M.E. Schlesinger, (ed.) *Greenhouse-gas-induced Climate Change: a Critical Appraisal of Simulations and Observations*, Elsevier, Amsterdam, 69-78.

Rowntree, P.R. (1990) Estimates of future climatic changes over Britain. Part 2: Results, *Weather* **45**, 3, 79-89.

Saelthun, N.R., *et al.* (1990) *Climate Change Impact on Norwegian Water Resources*, Norwegian Water Resources and Energy Administration Publication NR V42.

Sawyer, M.S. (1987) The rainfall of 22-26 August 1986, *Weather* **42**, 4, 114-117.

Schadler, B. (1989) Water balance investigations in Swiss Alpine basins- tool for the improved understanding of impacts of climatic change on water resources, in *Conference on Climate and Water*, Academy of Finland, Helsinki, vol. 1, 462-475.

Schlesinger, M.E., and Zhao, Z.C. (1989) Seasonal climatic changes by doubled CO_2 as simulated by the OSU atmospheric GCM/mixed layer ocean model, *J. Climatology* **2**, 459-495.

Shennan, I. (1993) Sea-level changes and the threat of coastal inundation, *The Geographical J.*, **159**, 2, 148-156.

Shiklomanov, I., Georgeosky, V., and Shereshevsky, A. (1995) An assessment of the influence of climate uncertainty on water management in the Dnipro River Basin, Technical Report, Russian State Hydrological Institute. (In Russian.)

Soveri, J., and Ahlberg,T. (1989) Multiannual variations of the groundwater level in Finland during the years 1962-89, in *Conference on Climate and Water*, Academy of Finland, Helsinki, vol. 1, 501-510.

Starkel, L., Gregory, K.J., and Thornes, J.B. (eds.) (1990) *Temperate Palaeohydrology*, Wiley, Chichester, UK, 548 pp.

Stouffer, R.J., Manabe, S., and Bryan, K. (1989) Interhemispheric asymmetry in climate response to a gradual increase of atmospheric CO_2, *Nature* **342**, 660-662.

Thom, A.S., and Ledger, D.C. (1976) Rainfall, runoff and climate change, *Proceedings Institution of Civil Engineers* **61**, 633-652.

Thornes, J.B., and Gregory, K.J. (1990) Unfinished business - a continuing agenda, in L. Starkel, K.J. Gregory and J.B. Thornes (eds.) *Temperate Palaeohydrology*, Wiley, Chichester, UK, 521-536.

UK Climate Change Impacts Review Group (1991) *The Potential Effects of Climate Change in the United Kingdom*, First Report, HMSO, London, 124 pp.

UNESCO (1978) *World Water Balance and Water Resources of the Earth*, UNESCO, Paris, 663 pp.

Vehvilainen, B., and Lohvansuu, J. (1991) The effects of climate change on discharges and snow cover in Finland, *Hydrological Sciences J.* **36**, 2, 109-121.

Verhoog, F.H. (1987) Impact of climate change on the morphology of river basins, in S.I. Solomon, M. Beran, and W. Hogg (eds.) *The Influence of Climate Change and Climatic Variability on the Hydrologic Regime and Water Resources*, International Association of Hydrological Sciences Publication No. 168, 315-327.

Viner, D., and Hulme, M. (1992) *Climate Change Scenarios for Impact Studies in the UK: General Circulation Models, Scenario Construction Methods and Applications for Impact Assessment*, Report to Department of the Environment, University of East Anglia, Climatic Research Unit, Norwich, UK, 70 pp.

Viner, D., and Hulme, M. (1993a) *The UK Met Office High Resolution GCM Equilibrium Experiment (UKHI)*, University of East Anglia, Climatic Research Unit, Norwich, UK, Technical Note No. 1, 16 pp.

Viner, D., and Hulme, M. (1993b) *Construction of climate change scenarios by linking GCM and STUGE output*, University of East Anglia, Climatic Research Unit, Norwich, UK, Technical Note No. 2, 19 pp.

Viner, D., and Hulme, M. (1994) *The climate impacts LINK project*, Report to Department of the Environment, University of East Anglia, Climatic Research Unit, Norwich, UK, 24 pp.

Warrick, R.A., Barrow, E.M., Wigley, T.M.L. (1990) *The Greenhouse Effect and its Implications for the European Community*, Commission of the European Communities, Brussels, EUR 12707 EN.

Warrick, R.A., and Farmer, G. (1990) The greenhouse effect, climatic change and rising sea level: implications for development, *Transactions Institute of British Geographers* **15**,1, 5-20.

Webb, B.W. (1992) *Climate change and the thermal regime of rivers*, Report to Department of the Environment, University of Exeter, Department of Geography, Exeter, UK, 79 pp.

Webb, B.W. (1995) Trends in stream and river temperature, *Hydrological Processes* **10**,

Webb, B.W., and Nobilis, F. (1994) Water temperature behaviour in the River Danube during the twentieth century, *Hydrobiologia* **291**, 105-113.

Webb, B.W., and Nobilis, F. (1995) Long term water temperature trends in Austrian rivers, *Hydrological Sciences J.* **40**, 1, 83-96.

Whittle, I.R. (1989) The greenhouse effect - lands at risk: an assessment, Conference of River and Coastal Engineers, Loughborough, UK, session no. 8, paper 6.5, Ministry of Agriculture Fisheries and Food, London, 6 pp.

Wigley, T.M.L. 1989. The effect of changing climate on the frequency of absolute extreme events, *Climate Monitor* **17**, 44-55.

Wigley, T.M.L., and Atkinson, T.C. (1977) Dry years in SE England since 1698, *Nature* **265**, 5593, 431-434.

Wigley, T.M.L., and Raper,S.C.B. (1987) Thermal expansion of sea water associated with global warming, *Nature* **330**, 127-131.

Wilby, R.L. (1993) The influence of variable weather patterns on river water quantity and quality regimes, *International J. Climatology* **13**, 447-459.

Wilson, D., Thomas, H., and Pithan, K. (eds.) (1993) *Crop Adaptation to Cool, Wet Climates, Particularly Temperature, Light and Water Variations and Limitations*, Commission of the European Communities, Brussels, COST 814.

Woodward, F.I. (1987) Stomatal numbers are sensitive to increases in CO_2 from pre-industrial levels, *Nature* **327**, 617-618.

Zachrisson, G. (1989) Climate variation and ice conditions in the River Tornealven, in *Conference on Climate and Water*, Academy of Finland, Helsinki, vol. 1, 353-364.

J.A.A. Jones
Institute of Earth Studies
University of Wales, Aberystwyth, SY23 3DB, UK.

7 THE IMPACT OF CLIMATIC WARMING ON HYDROLOGICAL REGIMES IN CHINA: AN OVERVIEW

LIU CHANGMING and FU GUOBIN

Abstract

Analyses of records of temperatures, riverflow, lake levels, lake areas and glacier response over the past century have been used to determine the sensitivity of hydrological regimes to global warming in China. The results show general trends which are already consistent with the predicted effects of global warming. They also reveal marked inter-regional differences in response. The factors responsible for these different patterns of response are discussed.

7.1. Introduction

At present, the world pays close attention to global warming resulting from the increase of carbon dioxide and other greenhouse gases released by the overuse of fossil fuels, forest fires, volcanic activity and urbanization. A detailed analysis of the impact of global warming on hydrological regimes in various regions of China is urgently required, especially the effects upon the temporal and spatial distribution of precipitation, evaporation, stream runoff, glacier mass balances, snowline altitudes, groundwater and lake storage. This is essential for the management, control and reasonable utilization of water resources, large water projects and regional economic development.

Changes to rivers, lakes and glaciers over the past 100 years have been used to analyze the climatic sensitivity of hydrological regimes in China.

7.2. Climatic Background to Hydrological Regimes in China

Hydrological regimes are generated by complicated interactions with various physical geographical factors, the climate being the principal factor. There are three basic climatic types in China: 1) the monsoon climate, with summer and winter precipitation related to the advance and retreat of the monsoon; 2) a continental climate, with a larger seasonal range than in coastal areas at the same latitude; 3) a climate with mixed features, combining elements of 1 and 2.

J.A.A. Jones et al. (eds.), Regional Hydrological Response to Climate Change, 133–151.
© 1996 *Kluwer Academic Publishers. Printed in the Netherlands.*

The characteristics of climate provide the basic controls for the hydrological regimes in China. Areas strongly affected by the monsoon have large variations in the seasonal distribution of water balance components, and large differences between years. Western China, where continental conditions prevail, has an arid or semiarid climate. There, the hydrological regime shows high evaporation, and rather low, seasonally distributed precipitation, with a low runoff coefficient and large variations between years. Mixed climates create an immense variety of hydrological regimes in China.

The area of exterior drainage basins covers 64% of the total land area of China and stream runoff from these basins accounts for 95.65% of the total stream runoff of China. On the other hand, the area of closed drainage basins is 36% of the total area of China, and its stream runoff represents only 4.35% of the total runoff in the country.

Among the elements of water balance in the exterior drainage areas of China, precipitation is similar to the average for exterior drainage areas in continents across the world, whereas stream runoff is comparatively high, and evaporation low. The values of water balance elements in the hinterland of China are lower than the average for continental interiors, so the hinterland is unusually dry. Over all of the country, precipitation and evaporation are both below average for continents. However, the runoff is relatively high, and the runoff coefficient ($R/P = 0.43$) averages about 20% more than that of the continents in general ($R/P = 0.36$). This is related to the climatic and geomorphological environment of China. Under the monsoon climate, precipitation is concentrated, and floods occur frequently during the rainy season. In the mountainous regions, runoff is readily generated. The average precipitation of China is 4% lower than that of the average continent, due to the large arid and semiarid areas in the western part of China and the monsoon climate of the Southeast. Liu (1986) developed a relationship between runoff, R, evapotranspiration, ET and precipitation, P, based on the boundary conditions of three water balance elements.

$$R = P - C[1-\exp(-P/C)] \tag{7.1}$$

where the evaporation parameter, C, is the maximum annual average evaporation in the drainage basin. An analysis of data from about 100 middle size and small size drainage basins by Liu (1986), indicated that in the mountainous drainage basins of China C is 800 mm, thus,

$$R = P - 800[1-\exp(-P/800)] \tag{7.2}$$

or

$$ET = 800[1-\exp(-P/800)] \tag{7.3}$$

Although the value of C varies greatly in the plains, it usually falls between 1000 and 1200 mm. This indicates evident differences in water balance elements under different precipitation regimes (Figure 7.1).

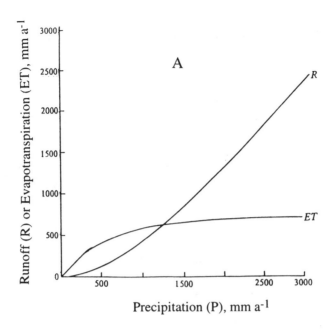

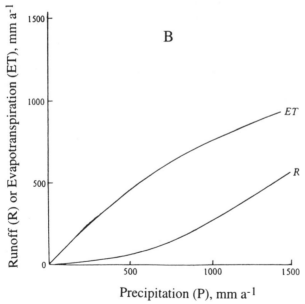

Figure 7.1. Precipitation - Runoff - ET relationships for rivers in China:
A) Mountainous rivers; B) Plains rivers. From Liu (1986).

7.3. Impact of Global Warming on Regional Hydrological Regimes

Although the relationship between global warming and the increase in carbon dioxide concentration, as well as other greenhouse gases in the atmosphere, cannot be proved from a strictly mathematical point of view, global warming and greenhouse gas concentrations follow a similar trend. The average global temperature has risen 0.3-0.7°C, while the concentration of carbon dioxide has risen 25%, from 275 ppm to 343 ppm, over the past 100 years. Global warming will accelerate if human beings continue to release carbon dioxide and other greenhouse gases as rapidly as they are now doing. There have been some well-known experiments with atmospheric General Circulation Models and mixed-layer oceanic models to predict the effects of global warming and changes in precipitation when the concentration of carbon dioxide is doubled. The models indicate that it is possible for the average global temperature to rise 1 to 2°C in the coming 30-50 years. Several factors should be taken into account in the analysis of future hydrological regimes when the global temperature rises. A considerable variety of modelling predictions is now available which indicate the range of predictions for changes in temperature and precipitation.

7.3.1. REGIONAL DIVERSITY OF GLOBAL WARMING IMPACT ON HYDROLOGICAL REGIMES IN CHINA

Changes in the distribution of precipitation are of paramount importance to the average river relying on rainfall to generate flow. In addition, changes in regional evaporation caused by the rise in temperature will have an impact. As for rivers relying on melted snow, the increase in snowmelt due to the rise in temperature should result in increased stream runoff, and a rise in snowline altitude. In addition to this, a negative mass balance is likely for glaciers, with glacier extent decreasing, thus causing stream runoff to be reduced after a few years. For rivers which rely more on groundwater, the situation is more complicated, and depends very much upon the geology and geomorphology of individual basins.

   Obvious differences in water balance between mountain and plain rivers are seen in Figure 7.1. In the mountainous area, runoff increases notably as precipitation increases. When annual precipitation is over 1200 mm, the change in evaporation relative to precipitation is close to zero and annual runoff exceeds evaporation. When precipitation is below 200 mm, and runoff is near to zero and most precipitation is evaporated. In the plains, evaporation is much greater than runoff. The amount of annual precipitation in the North China Plain is 90% of annual precipitation. Therefore, plain rivers are much more sensitive to global warming than mountain rivers.

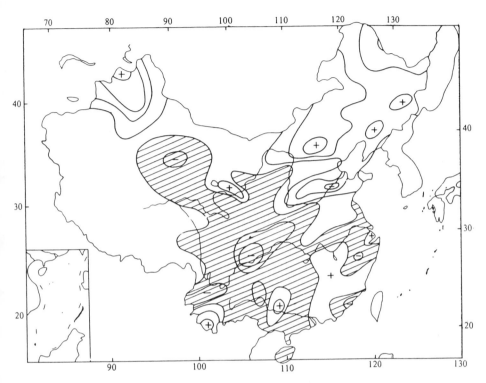

Figure 7.2. Difference in mean annual temperatures between the 1980s and the 1950s.
Regions shaded were cooler in the 1980s. The isoline interval is 0.4°C.
From Bureau of Social Development, CST (1990).

7.3.2. REGIONALITY OF GLOBAL WARMING EFFECTS

The 1980s were the hottest years this century. Using data on average annual temperature and average annual precipitation from 160 stations, differences in temperature and precipitation between the 1980s and the 1950s give some idea of what it would be like in the future when the temperature rises (Figure 7.2).

The area to the North of 35° North latitude and between 113-117° East longitude is warming; the rest is cooling. There were several warming centres in the warming area, mainly found in the North-east with temperature increments 0.83-1.14°C, in the North with temperature rises 1.31-1.55°C and in the Northwest with temperature increments 0.41-1.75°C. In addition, there is a clear metropolitan heat island effect, since these centres are mostly concentrated around large cities. This regional distribution of rises in temperature coincides with the patterns predicted by numerical models when the concentration of carbon dioxide is doubled. The result indicates that the higher the latitude, the greater the rise in temperature. Some areas were cooling, such as the area to the south of Mount Qinling and the Huihe River, and areas east of the Qingzang Plateau, especially the Sichuan and Hanzhong Basins, the northern part of Yunnan Province, all of Guizhou Province, and the coastal area in the Southeast. These areas have opposing temperature changes because: 1) there are many factors in climatic change, such as natural oscillations in the climatic system, solar activities, and volcanic eruptions; 2) regional physical geographical conditions cause diversity in mesoclimate and microclimate.

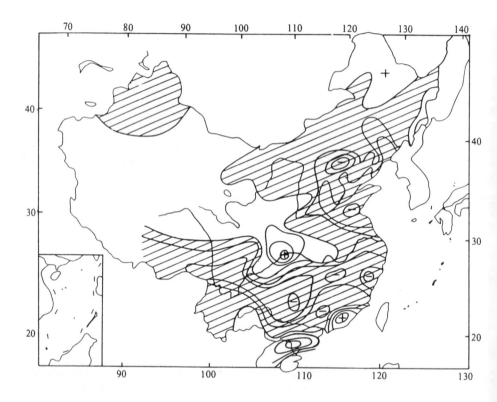

Figure 7.3. Difference in mean annual precipitation between the 1980s and the 1950s.
The area with reduced precipitation in the 1980s is shaded. The isoline interval is 60 mm.
From Bureau of Social Development, CST (1990).

In Figure 7.3, the distribution of precipitation differences between the 1980s and the 1950s shows that precipitation has been reduced over most of the country, except for some Provinces in the northwestern and northern parts, and a few Provinces in coastal regions of the southeastern parts. There were some negative centres scattered in the North and Southeast. The situation in which Northwest China and North China are getting warmer and drier is consistent not only with numerical model predictions for doubled carbon dioxide concentration, but also with the conclusion that the middle latitudes would become warmer and drier when planetary wind systems move Northward. The models indicate that when the concentration of carbon dioxide is doubled, Northeast China should become warmer and more humid. Actually, Northeast China might be warmer, and a little drier.

The comparisons between the model predictions and measured change from the 1950s to the 1980s, shows that the effects of future climatic change could be quite different among various regions.

7.3.3. NONLINEAR RELATIONSHIP BETWEEN CLIMATIC CHANGE AND THE WATER RESOURCES SYSTEM

A nonlinear relationship exists between climatic change and the water resources system, such that a rather small change in a climatic parameter would cause a great variation in hydrological regime.

A 10% reduction in regional precipitation means 15-25% reduction in stream runoff, and a 10% reduction in precipitation combined with a 2°C rise in temperature, means a 25-35% reduction in regional runoff, according to some research conducted in China and abroad in recent years (Fu and Liu, 1991; Stockton, 1982, 1983; Flaschka, 1985). In addition, some physical geographical parameters which influence runoff have changed markedly under a different climatic regime in recent years (Fu and Liu, 1991; Stockton, 1982, 1983; Irgmard Flaschka, 1985).

7.3.4. HYPOTHETICAL SCENARIOS

It is not correct to use GCMs output to predict regional effects directly. The numerical model mesh is very crude, with mesh points typically at intervals of 4.5 to 5.0° latitude and 5.0 to 5.5° longitude. The prediction based on the mesh is reliable from the global point of view, but not for specific regions. To simulate the effects in specific regions, we analyze the responses of hydrological regimes to a few hypothetical scenarios.

There are m possible changes in regional precipitation in the future, P_1, P_2,....P_m and n possible temperature changes T_1, T_2,....T_n. Thus each of the possible precipitation and temperature changes create scenarios of regional climate in the future. The response of hydrological regimes in China to global warming could be predicted by analyzing characteristic value assemblages (C_{ij}) of hydrological elements, such as perennial average runoff (including variance and mean), soil storage capacity, snowline altitude, glacier melt, lake level, lake capacity, and groundwater levels: i.e., we can formulate a scenarios matrix.

7.3.5. SPATIAL SCALE OF GLOBAL WARMING IMPACT ON HYDROLOGICAL REGIMES

Table 7.1 illustrates various time and spatial scales within the hydrological cycle from micro-scale to macro-scale. Hydrological phenomena at different temporal and spatial scales respond differently to global warming. On a macro-scale, for example, a rise in temperature would speed up the hydrological cycle, thus increasing average annual global precipitation. However in mid-latitudes, due to Northward movement of the planetary wind system, precipitation would decrease. The internal structure and state of the hydrological system would change notably with the shift of scale. For example, the phenomenon of flow, which is isotropic at the micro-scale, shows considerable spatial variation at the macro-scale.

TABLE 7.1. Hydrological time-space scale (Dooge, 1986).

Type	Spatial scale			Time scale	
	System	Length represented (m)	Area represented (km^2)	Type	Magnitude (years)
Macro	Planet	10^7	10^8	Planet evolution	10^8
	Continent	10^6	10^6	Eccentricity period of Earth	10^5
	Large basin	10^5	10^4	Sunspot period	10
	Small basin	10^4	100	Sun revolution period	1
Meso	Catchment	10^3	1	Moon revolution period	1 month
	Catchment block	10^2	10^{-2}	Earth rotation period	1 day
	Represented unit	10^{-2}	10^{-10}	Experimental area	1 second
Micro	Continuous point	10^{-5}	10^{-16}	Continuous point	10^{-6} seconds
	Water molecule set	10^{-8}	10^{-20}	Water molecule set	10^{-12} seconds

7.3.6. MODELS OF THE IMPACT OF GLOBAL WARMING ON HYDROLOGICAL REGIMES IN CHINA

Models of hydrological response to global warming should satisfy the following requirements (Fu and Liu, 1992); a) the models are applicable to different areas; b) the models are strictly based on physical law; c) inputs of the models are easily obtainable; d) inputs to the models are independent; e) outputs of the models, such as runoff, agree with measurements and they are independent; f) the application scale of models is clearly stated; g) because of uncertainty as to whether model parameters will be suitable under global warming in the future, as few parameters as possible are used.

At present, there are several kinds of models covering a variety of spatial scales:

1) GCMs model on the global scale.

General Circulation Models (GCMs) include simple radiation circulation models, energy balance models (EBMs), atmospheric general circulation models and oceanic general circulation models and combined atmosphere-ocean models. The United States Environmental Protection Agency (USEPA, 1984), analyzed the potential response of American hydrological regimes by applying GCMs. Manabe and Wetherald (1986) analyzed the response of global soil water content when the concentration of carbon dioxide was doubled by using GCMs. Zhao Zhongci (1990) analyzed the potential impact of the greenhouse effect on Chinese climate through the application of various GCMs.

Because the grid meshes in GCMs are crude and the results are mainly for climatic variables only, correction is necessary. The correction is, on the one hand, to improve the presentation of hydrological processes in GCMs, and on the other hand, the establishment of transformation models for different scales. Combination of GCMs with regional hydrological models will be a useful method of analyzing responses of hydrological regimes. Bultot (1988) analyzed the impact of atmospheric CO_2 increment on Belgian hydrological processes by applying outputs of GCMs and EBMs.

2) Models of hydrological regime response to global warming on the scale of a drainage basin or region include statistical models, analytical models and numerical models.

Statistical models are designed to establish correlations between hydrological elements and climatic parameters. Revelle and Waggoner (1989) analyzed hydrological regimes in Colorado drainage basins, where

$$R = b_0 + b_1 P + b_2 T \qquad (7.4)$$

Accordingly, Fu (1991) analyzed regional water resources in North China, obtaining

$$R = kPT \qquad (7.5)$$

and

$$R = f(X, b) \qquad (7.6)$$

where the bs are nonlinear parameters, $X = \{P, T...\}$ is variable, and the coefficients, k, are parameters specific to individual regions.

Other statistical models could reveal the connection between the hydrological regime and climatic change by analyzing historical evidence, which is usually based on tree ring analysis, ice cores, pollen analysis, stratigraphic deposits and historical literature. Kellogg (1981) analyzed global vegetation and soil moisture during the Megathermal period 4500 - 8000 years ago, and obtained a map of global soil moisture distribution.

Analytical models are based on physical mechanisms of mass and momentum balance. Liu and Fu (1991) analyzed the responses of regional water resources in the Wanquanhe drainage basin, Hainan Province, to global warming by applying a water balance model. In the parameterization of complicated hydrological processes in analytical models, however, some uncertainty remains as to whether parameters are suitable and how the parameters would respond to global warming.

Numerical models are based on strict physical mechanisms. They accurately represent physical processes in differential equations, arriving at numerical solutions through finite element and finite difference methods. The application of numerical models is limited by three technological problems. In the first place, accuracy depends on the understanding of hydroclimatological processes as they relate to soil, landform, vegetation and geologic setting. Secondly, they are limited by the accuracy of input parameters. Most input parameters depend on estimation, because direct observation is impossible. Finally, because observational data are presently collected only at discrete points, they cannot be used in continuous partial differential equations, or as estimates of model parameters under different climatic conditions.

3) Micro-scale analyses.

This research work consists of local experiments at hydrological stations, covering the dynamic state of soil water and soil water balance, and correlations between crops and moisture contents. In particular, this involves the distribution of root systems and their resistance, boundary layer properties, stomatal and aerodynamic resistances of the canopy and soil, and elemental parameters of soil water movement. The research work has been conducted using continuous flow equations, Darcy's law, and the SPAC (Soil-Plant-Atmosphere Continuum) model. Research on the microscale response of the hydrological regime is essential to the whole research effort, because not only does it provide a foundation for research on macroscale hydrological response, it also provides parameters for models of global and regional hydrological processes.

7.4. Changes within Water Bodies in China over the Last 100 Years and Evidence for Change due to Global Warming

There has been a clear tendency toward global warming over the past 100 years. During the slow process of global warming, the impacts of warming on various water bodies were added to their intrinsic change.

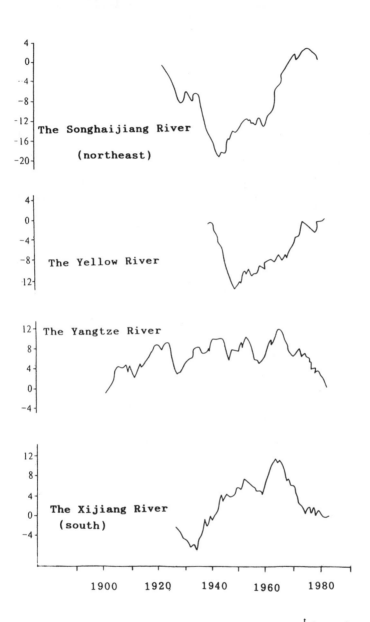

Figure 7.4. Differential mass curves for China's major rivers. Ordinates are: $\dfrac{\sum\limits_{1}^{t}(K-1)}{C_v}$ where K is

the modulus coefficient (ratio of variable to its mean, $K = \dfrac{X_i}{\overline{X}}$); C_v is the coefficient of variation.

7.4.1. RIVERFLOW CHANGES DURING THE LAST 100 YEARS

Figure 7.4 shows changes of annual runoff of some major rivers in China during the period. There are some regularities in the patterns:

1) The alternation of periods of low flow with periods of high flow. Each period lasted for several years.

2) A more obvious alternation of low flow with high flow is found in the North than in the South. The range is also wider, and the low flow periods lasted longer in the North. Thus, drought and water shortage caused by prolonged low flow has had a greater impact on northern industry and agriculture.

3) An apparently greater variability in flow during periods of high flow. Although the alternation of low flow with high flow was not as obvious in the southern rivers as in northern rivers, floods and waterlogging caused by excessive rain during the period of high flow did considerable damage in the South.

4) An apparent phase lag of about a half cycle in the alternation of low flow with high flow between the South and North. This phase lag between southern and northern rivers is a principal cause of floods in the South combining with floods in the North.

5) During the period of global warming in the 1920s and 1940s, low flow lasted for about 30 years in the North, with high flow persisting in the South.

During the last 40 years, drought was particularly frequent during the periods 1957-63, 1971-78, 1980-81 and 1985-89 (Figure 7.5).

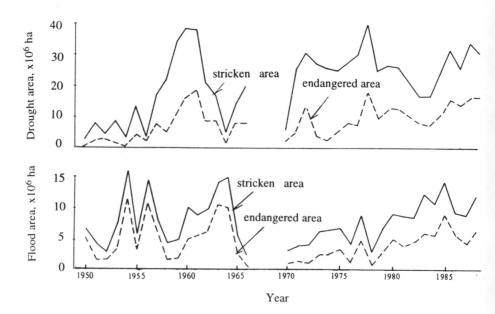

Figure 7.5. Annual changes in the area of China stricken by drought or floods.
From Bureau of Social Development, CST (1990).

Certain artificial factors lay behind drought in the 1960s. Drought was more severe following the 1970s, and drought in the 1980s may be related to global warming. From 1980 to 1989, drought occurred in most parts of North China. Because of lack of rain (average precipitation during this period was reduced by 20-50%), reservoir storage decreased, and groundwater tables fell continuously. Water shortage in North China restrained social and economic development in these regions. Though these were no detailed, accurate data to extend Figure 7.4 to the 1990s, it is clear that the 1990s are once again a period of low flow in the North.

Flooding, or waterlogging, is another primary natural hazard in China. Under global warming, flood control in the South calls for attention. Flooding and waterlogging often occur in regions where industry and agriculture are concentrated. It must be pointed out that in China the big floods usually result from typhoon storms, while the magnitude and frequency of the typhoon storms is closely related to surface temperature. The higher the temperature is, the greater the frequency of typhoon occurrence. According to results obtained by others (Wang Jian *et al.*, 1991), we obtained a regression relationship between typhoon occurrence in China (N) and global surface temperature change (ΔT), which predicts an increase of 0.8°C in the global surface temperature will increase typhoon frequency by about 60%.

7.4.2. CHANGING LAKE LEVELS IN CHINA

Global warming, silting and reclamation of land from lakes in China have been the principal causes of sharp reductions in the area and volume of lakes. From the early 1950s to the middle 1980s, the total area of lakes with an individual area of over 1 km^2 has reduced from 2800 to 2300 km^2, and the whole area of China's lakes has been reduced from 80 600 km^2 to 70 988 km^2.

An increasing warm-dry climate was the principal cause of reduced lake area on the Qingzang Plateau, in Northwest China, and the Inner Mongolian Plateau, and on the North China Plain.

The presence of many lakes on the Qingzang Plateau today demonstrates that there was far more rain when the lakes began to form. The continuous tectonic rise of the plateau has restrained water vapour from entering this area. Thus, precipitation on the plateau has been reduced sharply since the turn of the century, especially during the last 20 years. It became progressively drier on the plateau and the aridity increased lake evaporation.

Qinghai Lake is the largest inland lake in China. Due to increasing aridity in the area, the water balance of the lake is now negative (Figure 7.6). Its area decreased from 4568 km^2 in 1957 to 4304 km^2 in 1986. Water level in the lake dropped 3.36 m, averaging 108 mm annually. If climate changes and artificial disturbance in the future follow the current trend, the water balance of the lake will attain equilibrium and the lake will stop shrinking once the lake level has dropped another 1.8 m and the area has decreased to 4080 km^2. This process will take about 25 to 40 years. Qinghai Lake will enlarge only if the circulation patterns change and the warm, wet climate of the early and middle

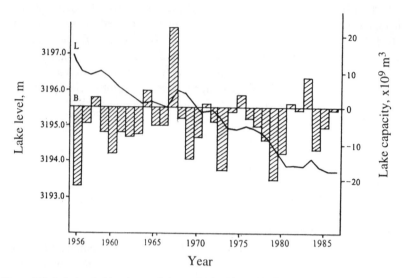

Figure 7.6. Lake level (L) and water balance (B) for Qinghaihu Lake between 1956 and 1985.
From Qin Buoqiang (1990).

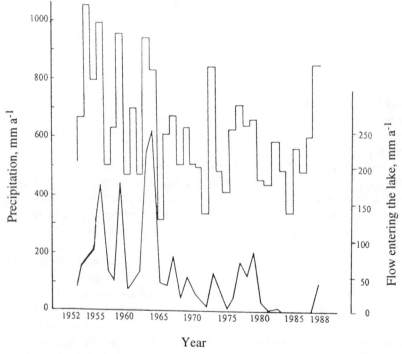

Figure 7.7. Annual precipitation in the mountains and flow entering the lake in the Baiyangding
drainage basin. Precipitation and inflows both in mm per annum. From Wang Huangbang (1990).

Holocene reappears (Qin Buoqiang, 1990).

Although lakes in these regions have shrunk, they will enlarge in years of high flow. Baiyangdian Lake is well known as the "pearl of North China". Since the 1970s, the weather has become much drier and flows entering the lake have decreased. The lake dried up four times during the 1970s and almost every year during the 1980s. However, Baiyangdian Lake was restored during heavy rains in 1988 (Figure 7.7).

Land reclamation has been a principal cause of reduction in lake area along the middle and lower reaches of the Yangtze River. Dongtinghu Lake was the biggest freshwater lake in China in 1949, when it had an area of 4350 km². The area decreased to 2691 km² in 1983 due to land reclamation, and Dongtinghu Lake became the second largest freshwater lake (Table 7.2).

Table 7.2. Changing size of Dongtinghu Lake.
After Wang Hongdao (1989).

Year	1896	1949	1954	1958	1971	1977	1983
Area (km²)	5400	4350	3915	3141	2820	2740	2691
Capacity (km³)	293	268	--	210	--	178	174

The lakes of Jianghan have decreased in number from 1066 to 326 since 1949, and their area was reduced by 6000 km². As a result, Hubei Province did not live up to the reputation of "a province with over 1000 lakes", and the current lake area is less than one-third of the area in 1949 (Wang Hongdao, 1989).

Reduction in lake area will result in an increase in flows from the lakes, because of a reduced ability to store flood water. Table 7.3 shows measurements of floodwater regulations after land reclamation around Poyanghu lake.

TABLE 7.3. Water stage and flows at different stages
of land reclamation around Poyanghu Lake.

Reclamation area	Increase in lake stage (cm)	Increase in lake discharge (m³ s⁻¹)	Increase in river stage (cm)	Increase in river discharge (m³ s⁻¹)
200	2-3	400	3-4	200
600	6-7	800	9-10	500
1000	8-9	1100	13-14	700

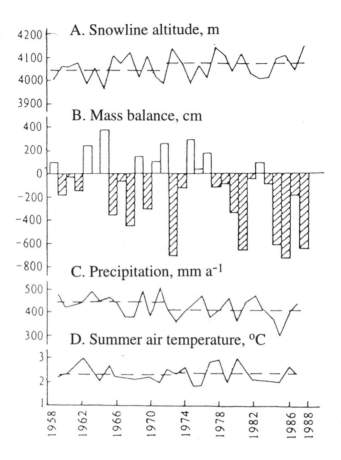

Figure 7.8. Snowline altitude, mass balance, annual preciptitation and summer temperature (May to September) on No. 1 Glacier, Urumchi River, Tian Shan Mountains, over the last 30 years. From Shi Yafeng (1990).

The rise in water stage and the longer duration of high flows in the wet season on trunk streams in the Yangtze River basin in 1980 was partly related to this reduced capacity for flood water regulation and storage in lakes in the middle and lower reaches of the Yangtze.

7.4.3. CHANGE IN GLACIER ADVANCE AND RETREAT IN CHINA

Changes in glacier mass balance are an inevitable consequence of climatic fluctuation. Precipitation and temperature profoundly influence the growth and decline of glaciers. Data are inadequate, but from 1900 to the 1930s glaciers generally shrank. During the

1960s, the shrinking slowed down and some glaciers became stable, while some even began to advance. During the middle 1970s, the number of advancing glaciers increased, but in the 1980s the numbers decreased, and ice melt increased under global warming. Overall, the western glaciers of China showed a tendency toward retreat during this century. According to Shi Yafen (1990), of the 227 glaciers documented in China and the former Soviet Union between 1950 and 1980, 73% retreated, 12% were stable, or in a state of equivocal advance and retreat, and 15% advanced.

Figure 7.8 shows the snowline altitude, mass balance, precipitation and summer temperature (May-September) of No.1 Glacier, in the Tian Shan Mountains, from 1958 to 1988. During that period, a negative mass balance occurred for the first 20 years. However, a weak positive regime lasted for 3 years following 1973, despite global warming and reduced precipitation. The annual mass balance was -28 mm from 1958 to 1972, and -169 mm from 1973 to 1988. Snowline altitude rose sharply after 1973.

In short, the hydrological regimes of various water bodies in China have fluctuated in response to climatic change over the past century. The tendency in northern rivers has been for stream runoff to decrease year by year, while in southern rivers there has been little overall change. The numbers and volumes of lakes are also being reduced, but the causes of reduction in lake area are different in the middle and western areas of China. The decline of glaciers has been prominent.

These tendencies became increasingly obvious with global warming during the 1980s. They seem likely to continue, if the rise of carbon dioxide and other greenhouse gases increases global warming.

7.5. Conclusions

The impact of global warming on hydrological regimes is a new problem in the development of hydrology. The analyses in this paper are based on the climatic background of hydrological regimes in China, and they yield the following conclusions:

1) The characteristics of the monsoon climate in China determine much of the hydrological background, as does the geomorphic environment.
2) The range and scale of global warming in China demonstrates marked regionality, and the impact on hydrological regimes in China are regionally diversified.
3) The relationship between global warming and hydrological regimes in China is nonlinear.
4) Hypothetical scenarios can be useful to analyze the responses of hydrological regimes to global warming.
5) The impacts of global warming on hydrological regimes are different at different spatial scales, as are the appropriate models.
6) Changes in river flows in China during the last 100 years have basically involved alternating periods of low flow and prolonged high flow. The period of prolonged low flow in the North was relatively long, especially during the hot years from the 1920s to the 1940s, and during the 1980s.

7) The number, area and volume of lakes have been reducing. However, different causes apply to the eastern and the western areas of China.
8) Glaciers in China have tended to decline over this period.

Acknowledgements

Many thanks to the reviewers for their comments, and the kind assistance of Dr. Li Suxia and Mr. Liu Tao in the typing of this paper is acknowledged.

References

Bultot, F., Coppens, A., Dupriez, G.L., Gellens, D., and Meulenberghs, F. (1988) Repercussions of a CO_2 Doubling on the Water cycle and on the Water Balance: A Case Study from Belgium, *J. Hydrology* **99**, 319-347

Bureau of Social Development of National Sciences and Technology Committee (1990) *Global Climatic Change and the Policy*, 68 pp.

Editorial Committee of Physical Geography in China, Chinese Academy of Sciences (1985) *Physical Geography-General Introduction*, Science Press, 20 pp.

Fu Guobin (1990) Global Warming and Global Water Resources, *Water Resources Research* **11**, 3, 1-6.

Fu Guobin, and Liu Changming (1991) Estimation of Regional Water Resources Responses to Global Warming - A Chinese Academy of Sciences Study of Wanwuan Basin, *Acta Geographica Sinica* **46**, 3, 277-288.

Fu Guobin, and Liu Changming (1991) Evaluation of the Model of Impact of Global Warming on Water Resources, *Collection of Theses of the Fifth Hydrological Session of the Geography Society of China*, Science Press.

Fu Guobin (1991) Preliminary Analysis on Impact of Global Warming on Water Resources in North China, *Geography and Territory* **7**, 4, 22-26.

Irmgard Flaschka (1984) Climate Variation and Surface Water Resources in the Great Basin Region, *Water Resources Bulletin* **23**, 1.

Kellogg, W.W., and Schware, R. (1981) *Climate Change and Society*, Westview Press.

Lanzhou Institute of Glaciology and Geocryology, Chinese Academy of Sciences (1988) *An Introduction to Glaciers in China*, Science Press, 184 pp.

Liu Changming (1986) Analysis on Water Balance and Research of Water Resources in China, *Collection of Theses in the Third Hydrological Session of the Geography Society of China*, Science Press, 113-118.

Liu Changming, and Fu Guobin (1989) *Change of Runoff in Main Rivers in China during a Period of Near 100 years*, Institute of Geography, Chinese Academy of Sciences, Beijing.

Liu Chuizhen, and Tian Yuying (1991) Research on Impact of Climatic Change on Runoff with Improved Nonlinear Model of Water Balance, *Advances in Water Sciences* **2**, 2, 120-126.

Manabe, S., and Wetherald, R.T. (1986) Reduction in Summer Soil Wetness Induced by an Increase in Atmospheric Carbon Dioxide, *Science* 232, 626-628.

Qin Buoqiao (1990) Hydrological Characteristics and Water Balance of Qinghaihu Drainage Basin, in Shi Yafen *et al.* (eds.) *Climate change in China and Change of Sea Level*, Ocean Press, 113-115.

Revelle, R.R., and Waggoner, P.E. (1989) Effects of a Carbon Dioxide induced Climate change on Water Supply in the Western United States, in *Changing Climate*, National Academy of Sciences, National Academy Press, Washington, D.C.Shi Yafen (1990) Climatic Tendency toward Dryness and Warming in the Middle of Asia where Mountainous Glaciers and Lakes Shrank and a Prospect, *Acta Geographica Sinica*, **44**, 1.

Stockton, C.W. (1982) Climatic Variability and Hydrologic Processes : An Assessment for the Southwestern United States, *International Symposium on Hydrometeorology*, June, 30-45.

Stockton, C.W. (1983) Climate Change and Surface Water Availability in the Upper Rio Grande Basin, in *Regional and State Water Resources Planning and Management*, October, 23-34.

United States Environmental Protection Agency (1984) *Potential Climatic Impacts of Increasing Atmospheric CO_2 with Emphasis on Water Availability and Hydrology in the United States*, Strategic Studies Staff, Office of Policy Analysis, Office of Policy, Planning and Evaluation, EPA. 230-04-84-006 USEPA, Washington, D.C., 96 pp.

Wang Gongdao (1989) *Lake Resources in China*, Science Press, 28-35.

Wang Huanbang, and Lu Keyan (1990) Impact of Climatic Change on Volume of Baiyangdian Lake, in Shi Yafen *et al.* (eds.) *Climate Change in China and Change of Sea Level*, Ocean Press, 144-146.

Wang Jian, and Liu Zechun (1991) The Possible Response of Hurricane Frequency in the Northwest Pacific to Global Warming, *Quaternary Sciences* **24**, 1, 173-177.

Zhao Zongci (1990) Simulation of Impact of Greenhouse Warming Effects on Climatic change in China, *Meteorology* **15**, 3.

Liu Changming and Fu Guobin
Institute of Geography and United Research Centre for Water Problems
Academia Sinica, Beijing 100101, China.

SECTION III

PRECIPITATION CHANGE AND VARIABILITY

SUMMARY

Linking precipitation to General Circulation patterns in the USA
■ Relationships between the daily occurrence/non-occurrence of local precipitation and General Circulation patterns were studied as a method of downscaling GCM scenarios for small basins.
■ Precipitation occurrence probabilities are seen as functions of location, season and topography. The method therefore separates the data according to 4 criteria - summer, winter, mountain and plain - and relates the conditional probabilities of precipitation occurrence at pairs of stations to their interstation distances. Finally, it links these probabilities to the Pacific-North America (PNA) Index of atmospheric circulation.
■ Probability analyses in North Carolina and Colorado indicate:
    1. Precipitation events are less frequent in Colorado than in North Carolina.
    2. The Colorado mountains have higher probabilities than the plains in all seasons.
    3. North Carolina was more complex: the summertime plains showed the highest probabilities and steepest probability/distance gradient, indicating frequent localised convective storms; in winter the plains showed higher probabilities but lower gradient, indicating more widespread frontal precipitation; the mountains had less convective activity and fewer summer events.
■ A clear relationship is demonstrated between the PNA and precipitation probabilities, but the relationship is highly variable and other factors also play a significant role.
■ It should be possible to extend this approach to precipitation amount, duration, area and watershed volumes.

Neural networks for modelling temporal precipitation patterns
■ Tests on data from Memphis, Tennessee, show that artificial neural networks can be applied to predicting temporal patterns in precipitation.
■ Far less data are required than in conventional time series or stochastic modelling.
■ Reliability is, however, very dependent upon proper selection of parametric values.
■ It is argued that the technique might be adaptable for predicting monthly patterns under climate change.

Downscaling using spatial statistics in Poland.
■ Both GISS and GFDL equilibrium scenarios indicate positive shifts in temperature and precipitation.
■ 1° x 1° gridcells based on output nodes from the GCMs were downscaled to 6 x 6 km, using kriging to interpolate temperature and a distance-decay function for precipitation.
■ Total changes for Poland were calculated from the resulting isotherm and isohyet maps by conventional methods.
■ This shows that the average annual precipitation in Poland would rise from 639 mm to 749 mm (+17%) under the GISS scenario and to 655 (+3%) in the GFDL scenario.
■ The spatial median mean air temperature would rise from 7.7°C to 10.5°C and 11.9°C respectively.

■ The interpolated patterns could be the starting point for applying catchment-scale hydrological models.

Trends and variability on the Tibetan Plateau
■ Records show rising temperatures and reduced precipitation, especially in summer and mid-winter, since 1951.
■ Seasonal variations bear a close relationship to El Niño and La Niña events, as follows:

1. Lower temperatures and precipitation occur in El Niño summers and La Niña winters.

2. Higher temperatures and precipitation occur in El Niño winters and La Niña summers.

3. Lower temperatures and higher precipitation occurs in La Niña winters.

■ Patterns at Lhasa in the west of the plateau are rather different from those at the other 3 stations which are 5-9° to the east.
■ The patterns can be related to changes in monsoonal activity.

Precipitation variability in Ireland
■ Winter precipitation showed no linear trend over the period 1957-1987.
■ Moderate to strong associations were, however, found with each of the hypothesised sources of variation: the North Atlantic Oscillation (NAO), the El Niño/Southern Oscillation, and sunspot numbers modulated by the phase of the stratospheric Quasi-Biennial Oscillation (QBO).
■ Precipitation is above average with NAO positive, i.e. strong westerlies, especially in the Southwest and Northeast.
■ Precipitation is generally below average with ENSO positive, i.e. during a South Pacific 'cold' event when the upper westerly circulation tends to display a strong ridge over the Northeast Atlantic, steering depressions north and west of Ireland.
■ Precipitation is generally above average during sunspot maxima, especially when the QBO is in its West Phase. This is best demonstrated in the Northern half of the island.
■ These relationships interact. ENSO appears to be the primary control, but the Northwest is dominated by the sunspot/QBO relationship and the Southeast responds weakly to all three effects.
■ Long-range weather forecasting might be possible if these relationships were supplemented by Sea Surface Temperatures (SSTs) for the evaporative source areas.

8 THE INFLUENCE OF TOPOGRAPHY, SEASON AND CIRCULATION ON SPATIAL PATTERNS OF DAILY PRECIPITATION

P.J. ROBINSON

Abstract

Estimates of future precipitation regimes on small watersheds and daily time scales are frequently required in hydrologic analyses. Such estimates must be based on projections from coarse resolution global climate models combined with analyses of current observational data. Possible combination methods are reviewed. One approach is to link historical precipitation records to observed and modelled atmospheric circulation patterns. This requires knowledge of the spatial scale of daily precipitation. A method of establishing this scale as a function of location, season, and topography is developed and linked to circulation patterns. The method relates the conditional probabilities of precipitation occurrence at station pairs to their inter-station distance. It was tested for mid-latitude continental interior and coastal conditions, Colorado and North Carolina, USA respectively. Separate analyses were undertaken for summer and winter and for mountainous and flat regions. A logarithmic relationship between inter-station correlation and distance was established. Storms in Colorado were less frequent and smaller than in North Carolina, while for the latter state winter storms were significantly larger than summer ones. Mountainous regions tended to have more widespread storms than did flat areas. The spatial scale varied little with circulation in Colorado but meridional flow over North Carolina led to more frequent and localized storms than did zonal flow, especially in winter. Thus this simple method allows specification of the impact of circulation changes on the scale of precipitation, and hence some indications of possible consequences of climate change.

8.1. Introduction

Most of the quantitative information concerning potential climatic change comes from the results of general circulation model (GCM) experiments. These explicitly model the physical processes creating climate, and climatic changes are commonly determined by comparing results from model runs for the present atmospheric composition with those for a postulated future one. Usually the conditions resulting from a doubling of atmospheric carbon dioxide are considered (Houghton *et al.*, 1990). The information

J.A.A. Jones et al. (eds.), Regional Hydrological Response to Climate Change, 157–172.

from these models, however, is primarily restricted to temporally and spatially averaged values, where the averaging is in the order of a month over a 4° latitude x 5° longitude grid cell. Further, many modellers would suggest that even the use of individual grid cell results is somewhat suspect. In contrast, many practical problems, including many related to hydrology, are likely to be associated with much smaller temporal and spatial scales. Concern for the frequency and extent of flash floods, for example, cannot be addressed directly by GCM outputs. The need to undertake appropriate scale matching and incorporate the associated uncertainties when assessing the potential consequences of climate changes has led to the development of 'scenarios', suites of possible future climatic conditions, rather than to definitive predictions. The many potential scenario development techniques have been reviewed by Robinson and Finkelstein (1991), but they can be summarized as those directly using models developed on the required scales and those requiring observational data for the current climate.

One promising line of approach for scenario development, linked directly to the GCM outputs, is to use nested models (Dickinson et al., 1989). Such models, notably the regional ones with a resolution of a few tens of kilometers, offer probably the best long-term solution and are beginning to provide realistic simulations of regional climates (Giorgi et al., 1992). For the foreseeable future, however, they have several severe drawbacks for precipitation studies. First, they must be developed for a specific area, and must explicitly incorporate the influence of local topographic features on airflow and the resultant precipitation. This presents a major challenge in complex, mountainous, terrain. Further, the regional models produce results for individual precipitation episodes. Since these represent typical episodes, not actual ones, it is difficult to compare the model results with actual observations and thus to test the model itself. A series of model runs can be used to simulate the climatological regime, but the models are computationally intensive, and it remains difficult to test their results against reality. While all of these problems will undoubtedly be overcome in time, at present it is impractical to use regional models to assess precipitation regime changes in areas of complex terrain.

The major alternative scenario development approach is to use the observational information for the current climate in a way that can be translated into estimates of future consequences. In general, this involves development of a set of transfer functions relating the GCM output for the present atmospheric composition to the observations of the present climate. Those climatic elements of particular interest can be emphasized. The transfer functions are then applied to the GCM results for the future atmospheric composition to estimate the new conditions. Transfer function development has become known as the 'climate inversion problem', and several methods of solution have been proposed. The best established takes temporally averaged grid cell output and uses statistical methods to develop short-period small-area transfer functions (e.g. Kim et al., 1984; Chen and Robinson, 1991). Since GCM outputs implicitly assume that the output parameter is continuous in space and time, these methods are most appropriate for continuous variables such as temperature. Precipitation is discontinuous and it is unclear what the GCM precipitation values actually represent (Robinson et al., 1993). Certainly it is inappropriate to use the same inversion methods to estimate both temperature and precipitation. In addition, the GCM precipitation estimates are commonly regarded as much less reliable than the temperature outputs (Houghton *et al.*, 1990). Hence it is clear that new approaches to transfer function development must be explored. The following section outlines one such approach.

8.2. Transfer Functions for Future Precipitation Estimation

The precipitation information needs for hydrologic purposes are many and varied. It is likely that the most refined needs are for daily precipitation on a space scale of a few kilometers. Hence it is important to provide scenarios on this scale. Indeed, it is desirable even if longer-term, larger-scale values are needed since longer-term averages or totals can mask vital changes in precipitation regime. For one GCM which gave monthly precipitation totals similar to those observed, Robinson *et al.* (1993) found that the model indicated a large number of days with light precipitation rather than the few days of heavy precipitation which were actually observed. Thus the potential for misinterpretation when applying model results, for applications such as erosion control, is great. The difficulties and uncertainties in modelling daily precipitation arise primarily from the difficulty in adequately specifying in the models the complex nature of the processes leading to the production of an individual precipitation episode. Hence a major objective in developing a transfer function for precipitation must be to incorporate these precipitation production processes and their effects.

In the most general terms the precipitation production process, leading to variations in the amount, intensity, duration, and spatial extent of precipitation, depends on both the amount of water vapour in the atmosphere and on the mechanisms for releasing that water. Thus, over uniform terrain in the Intertropical Convergence Zone the amount of vapour available depends on whether the upwind trajectory of the Trade Winds was over a rapidly evapo rating water surface or a dry land one. The speed of the winds influences the amount of convergence, the intensity of uplift, and the rate of precipitation production. Similarly, the previous trajectory of a mid-latitude depression influences both the amount of moisture and the energy it contains, and thus the subsequent frontal precipitation. Since topography influences precipitation production directly, minor changes created by trajectory variations may have major consequences for precipitation distribution in mountainous areas.

These considerations emphasize the importance of airflow patterns for the production of individual precipitation episodes. Thus, if transfer functions relating daily precipitation to atmospheric circulation can be developed, they can be used with circulation estimates from GCM outputs in the current and future climate to provide the necessary scenario information. Although published GCM results still commonly emphasize temperature and precipitation, they are increasingly capable of showing and allowing analysis of future upper level pressure fields or atmospheric flow patterns (e.g. Crane and Barry, 1988). Thus, in mid-latitudes, it is becoming possible to consider the location, frequency, number and strength of the planetary waves with some degree of confidence. It remains to identify the necessary transfer functions using the observational data.

On a monthly time scale, observational connections between precipitation amounts and the upper atmosphere conditions have been indicated in several regions. In southern Africa, Taljaard (1986) used long-term average conditions to demonstrate that temporal and spatial precipitation changes were associated with both the 850 hPa and 700 hPa height fields. Using only the 700 hPa field for the western United States, Klein and Bloom (1992) developed predictive equations for monthly precipitation using a straightforward multiple regression approach. In a similar vein, there have been a considerable number of investigations of relationships with various circulation indices. Building on the global-scale work of Ropelewski and Halpert (1987, 1989), the influence of the Southern Oscillation Index has been investigated for numerous regions. In mid-latitudes the Pacific-North American Index (PNA) has been found to be a useful indicator of precipitation in some locations and seasons (Yarnal and Leathers, 1988; Leathers et al., 1991). However, none of these works explicitly considers global change issues, most existing studies (e.g. Ayers et al., 1993; Kalkstein et al., 1990; Gleick, 1986; Cohen, 1987) relying on simple transfer functions derived by statistical manipulation of monthly values. It is difficult to refine them further. The opportunity exists, however, to refine the circulation related results to develop transfer functions based on daily information.

For small space and time scales, Robinson and Henderson (1992) and Henderson and Robinson (1994) demonstrated that changes in atmospheric flow patterns influence the number, duration and intensity of individual precipitation events. Although these studies demonstrated spatial coherence for the results, the information was restricted to specific points. They used hourly precipitation data which had a network density so sparse that no inter-station relationships could be constructed, and no information on the variability of the spatial scale or spatial pattern of the individual events as a function of circulation could be developed. By relaxing the time scale to daily values, however, a much denser observational network is available and explicit consideration of the spatial scale of precipitation episodes is possible.

The present paper investigates the characteristic scale of daily precipitation episodes as a function of location and topography. The relationship of the results to circulation patterns is also explored. Since this type of investigation has not been undertaken previously, a full analysis followed by explicit attempts to develop transfer functions is not warranted. Rather, the feasibility of the approach for characterizing daily precipitation is tested through a case study, and the results are used to suggest opportunities for extension leading to transfer function development.

The spatial characteristics of precipitation on this daily scale can be established by considering the conditional probability of precipitation occurrence at each station pair in a network of stations. This approach is similar to that taken by several previous workers (e.g. Berndtsson, 1989; Sharon, 1979; Sumner, 1983). However, a major difference for the present study was that there was no concern with the amount of precipitation at either station. Rather, a binary relationship between rain and no rain was established. There is little guidance in the literature as to the nature of the spatial relationships in this binary mode, but it can be anticipated that it is, as with the amount, a function of location, topography and season in addition to circulation patterns. Thus it was desirable to consider and compare results for adjacent areas with similar climatic regimes but dissimilar topography as well as for widely separated regions. This paper considers the topographic and seasonal situation in Colorado and North Carolina, U.S.A., regions with contrasting climate but where there are areas of relatively flat terrain separated from areas with diverse relief by a marked cliff-like feature. Thereafter, winter conditions in North Carolina are considered as a function of circulation.

8.3. Study Area

In order to use a wide range of climatic and topographic conditions while utilizing similar data, two states within the United States of America having contrasting climate but similar topography, and where mountain precipitation is of special importance, were chosen for analysis. One, Colorado, lies at 37°-41°N, 102°-109°W, approximately in the middle of the North American continent. The other, North Carolina, extends 34°-37°N, 76°-84°W, and is located on the eastern seaboard of the continent. Climatically they are at similar latitudes and can be regarded as being influenced in similar ways by the planetary waves (Bryson and Hare, 1974). Thus the prevailing wind is from the west for both. However, the continental interior location of Colorado leads to a distinct continental temperature regime of cold winters and warm summers while North Carolina's temperatures are moderated because of proximity to the ocean. Location also influences the precipitation regime. Colorado has a dry summer only relieved by occasional isolated thunderstorms or the rare passage of a depression. In winter, however, depressions coming from the west, or generated in the area, provide considerable precipitation. North Carolina, in contrast, has precipitation throughout the year. In winter it is primarily associated with depression passage, while in summer rain comes from both depressions and from thunderstorms.

As far as terrain is concerned, the two states have similarities in that each has a mountainous west and a relatively flat east, so that each could be divided into a 'mountain' and a 'plain' zone. For Colorado the boundary between the two was the eastern edge of the Front Range of the Rocky Mountains, which runs approximately North-South at 105°W. To the east is a region of relatively flat land at about 2000 m above sea level. This is the western portion of the High Plains region, and is an irrigated agricultural area. Water requirements early in the growing season are often met by *in situ* snowmelt. However, most of that needed in summer comes from winter snowfall in the mountains, where it is stored either as snowpack or in reservoirs until required. The mountains themselves stretch west from the Front Range, where the highest peaks exceed 4000 m, across a plateau slowly losing altitude westward to an altitude near 2500 m at the state boundary. This plateau is highly dissected with some canyon-like valleys and many broad basins. Much of the area is desert, with water being abundant only along the water courses or where orographic uplift enhances precipitation. Indeed, much of the precipitation occurs within the general area of the Front Range, rather than farther west. Recreation, rather than agriculture, is the major economic activity. Skiing is particularly important, and adequate snowpack is vital.

In North Carolina the topographic division is similar to that of Colorado, but somewhat less clear-cut. The eastern edge of the Blue Ridge portion of the Appalachian Mountains provides the primary separation. This edge is a sharp escarpment, some 300 m high, running Northeast-Southwest from about 81°W in the north to 82°W in the south. The mountains are a series of interlinked ridges with local relief in excess of 500 m. To the east is a foothill zone sloping gently eastward, from an altitude near 300 m above sea level at the Blue Ridge to about 100 m some 200 km away at the junction with the coastal plain. Although the rolling hills, with local relief around 20 m, contrast somewhat with the virtually level, 200 km wide, coastal plain, they are distinctly flat in comparison to the western mountains. This whole lowland region has abundant precipitation throughout the year and major climate-related problems with water supply for agricultural and urban use are rare. It is unlikely that climatic changes will have a significant impact. In the mountains the situation is very different. The forested rolling hills are rapidly expanding with retirement communities and water demand is increasing. In addition, expanding lowland cities, especially Atlanta Georgia, some 150 km to the south, are looking to the area for additional water supply. Hence there is pressure which could well be intensified by climatic change.

8.4. Data

Daily precipitation observations have been made for upwards of one hundred years by several thousand volunteers in the United States. The data are contained in 'Climatological Data' published by the National Climatic Data Center (Hatch, 1983). Observation times vary, but tend to cluster around morning (usually within 30 minutes of 7 a.m. local time), afternoon (around 5 p.m.) and midnight observations. The first are by far the most numerous and are the only ones considered here. However, some comparative analyses were undertaken, which tended to support the current conclusions but, with a smaller sample size, with less statistical confidence.

The National Climatic Data Center has divided the U.S.A. into a series of small, climatically homogeneous regions known as Climate Divisions and has assigned each observing station to a single division. For North Carolina few foothill valleys penetrate into the mountains, climatic divisions follow topographic boundaries, and it is possible to assign stations unambiguously to the mountains or the plains. In Colorado the Climate Divisions follow drainage basins rather than topography, and several plains basins penetrate into the mountain region. Stations were categorized for the present work based on location with respect to the Front Range. Those identified as being west of it were classified as mountainous, those to the east as plains. For both states the plains stations can reasonably be regarded as representative of the whole topographically homogeneous area. However, the stations in the mountains are overwhelmingly in the valleys. Hence the results are strictly appropriate only for valley sites, and some of the orographic effects are obscured.

The season was the basic analysis unit, with summer (June, July, August) and winter

(December, January, February) being used here. The year for the winter is that of the appropriate December. Stations with more than 4 missing daily observations for the season were excluded from the analysis. Days with a 'trace' of precipitation were included as wet days. For the analysis of seasonal and topographic influences, data for the 1980-1985 period were used.During this period the number and location of stations available for analysis was rather stable, allowing the development of summary statistics and comparisons between years (Table 8.1). Further, the period provided a wide range of wet and dry periods. Rankings (Table 8.2) were determined from comparisons of the data with the seasonal climatic division averages for the 1890-1990 period provided by the National Climatic Data Center. They were expressed in deciles to give general indications of precipitation amounts for each of the regions.

TABLE 8.1.   Number of station pairs used in the development of the logarithmic relationship for each year, season and topographic region.

Year	Colorado				North Carolina			
	Summer		Winter		Summer		Winter	
	Mt.	Pl.	Mt.	Pl.	Mt.	Pl.	Mt.	Pl.
1980	2396	2056	2564	2041	930	2756	1056	2862
1981	2548	1988	2250	1404	1122	2756	1122	2442
1982	2780	1410	2622	1778	1056	2756	1104	2652
1983	2810	1614	2910	1566	1122	2860	1190	2848
1984	3256	1944	2598	1658	1190	2746	1116	2528
1985	2528	1670	2984	1480	1122	2862	980	2540

Circulation patterns were characterized by the Pacific-North America Index (PNA) (Wallace and Gutzler, 1981). The PNA has been found to be correlated with precipitation events in North Carolina, especially in winter (Henderson and Robinson, 1994). Positive values indicate meridional flow with a strong Aleutian Low, a ridge in the west and a trough in the east of the United States. Negative values indicate predominantly zonal flow, with a weak ridge and trough structure. Historically the range of winter values is from approximately 1.3 to -1.5, while in summer the weaker and more diffuse flow regime makes meaningful assessment more difficult, and the range is much more restricted. Further, the index is more appropriate for the southeast United States than for the mid-continent (Wallace and Gutzler, 1981), so the present analysis was restricted to North Carolina winter conditions. For the period used for the seasonal and topographic analysis, the early 1980's, there were few extreme conditions of atmospheric flow. Consequently for the investigation of the influence of circulation,

TABLE 8.2. PNA values and deciles of precipitation amounts derived from 100 years
of record for years used in analysis of topographic and seasonal influence on
precipitation (1 = driest 10%, 10 = wettest 10%).

| Year | Colorado | | | | North Carolina | | | | PNA | |
| | Summer | | Winter | | Summer | | Winter | | Summer | Winter |
	Mt.	Pl.	Mt.	Pl.	Mt.	Pl.	Mt.	Pl.	-	-
1980	1	3	1	1	1	1	1	1	-.52	.86
1981	9	5	5	4	1	5	10	10	.09	-.31
1982	6	8	4	6	5	5	7	8	.26	.83
1983	8	5	8	8	1	1	9	9	.01	.52
1984	9	3	6	5	6	5	2	4	.05	-.08
1985	4	3	3	6	8	8	1	1	.02	.77

five years with high and five years with low (highly negative) PNA values were selected
from the 1948-1990 period. Those chosen were years with both extreme values and a
relatively large number of stations available (Table 8.3). Since there was not a
completely consistent set of stations available and the total number used varies greatly
from year to year, the results, particularly those summarizing all years, must be used
with caution.

TABLE 8.3. Winter PNA values and number
of North Carolina station pairs used in the
analysis of extreme PNA years.

| Year | PNA | No. of Observations | |
		Mt.	Pl.
1976	1.29	2862	992
1969	1.07	2162	1122
1963	0.88	1892	928
1980	0.86	2862	1056
1982	0.83	2652	1104
1986	-0.53	2070	1122
1956	-0.59	1722	756
1951	-0.78	992	650
1955	-0.82	1560	600
1971	-0.88	2450	930

8.5. Methodology

The first step in the analysis was the identification of the conditional probability of precipitation for each station pair. For each season the number of days with precipitation at both stations, p_b, the number of days with precipitation at station 1 only, p_1, and the number at station 2 only, p_2, was determined. The probability of precipitation at station 2, given that it is raining at station 1, P_{12}, was given by:

$$P_{12} = p_b/(p_b + p_1) \qquad (8.1)$$

Similarly:

$$P_{21} = p_b/(p_b + p_2) \qquad (8.2)$$

For each region all station pairs were analyzed as a function of distance d between stations. Distance was determined using standard techniques from the latitude and longitude coordinates contained with the station history in the Climatological Data.

The nature of the relationship between the probability P_{ij} of precipitation at station j, given that precipitation is occurring at i, and the station separation distance d_{ij} was not well established. Some tentative results based on correlations of daily amounts had been suggested (Sumner, 1983, Berndtsson, 1989, Hevesi et al., 1992), but there was no recognized model for the binary rain/no rain approach required here. An analysis using data from a wet, a dry, and an intermediate season from the early 1980s for the mountain and plains regions in both Colorado and North Carolina indicated that a logarithmic relationship was most appropriate (Robinson, 1994). Not only did this reproduce the distance decay shape at short distances, but it also indicated a nearly constant probability at large distances. In addition, this latter probability began to approach values around 0.3 in Colorado and 0.5 in North Carolina. Without any consideration of conditional probabilities, individual stations have a probability of precipitation of approximately .25 and .30 in Colorado and North Carolina respectively. Thus the conditional probabilities at large distances suggested by the logarithmic model appear reasonable. Consequently a model of the form:

$$P_{ij} = a + b \ln(d_{ij}) \qquad (8.3)$$

was used for the analysis of the full dataset.

8.6. Results

On the broadest scale, there is a distinct difference in the results for Colorado and North Carolina irrespective of topography and season (Figure 8.1). In general, the higher the probability, the greater the frequency of precipitation over much of the region, while a gentle slope on this logarithmic plot indicates a predominance of large-scale widespread events, most likely connected with the frequent passage of extra-tropical cyclones. The steeper the slope, the more important are small-scale localized events characterized by convective activity. Thus the lower overall probabilities in Colorado were indicative of more localized precipitation, but with marked seasonal and topographic variations.

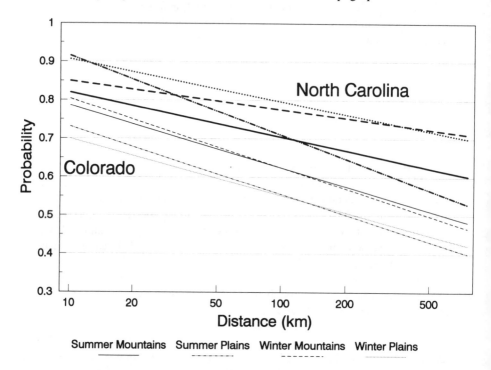

Figure 8.1. Relation between the conditional probability of precipitation and distance as a function of season and topography for Colorado and North Carolina.

In Colorado the plains had lower probabilities than the mountains in both seasons, while seasonal differences were minor. These results suggest that the processes in both areas were similar, but that topographic effects in the former created more frequent precipitation. The situation in North Carolina was rather more complex. The highest probability at short distances, and the steepest slope, occurred for the summer plains, indicating frequent local convective storms. The winter plains also had high probabilities, but a less steep slope, suggesting some convective activity superimposed on cyclonic precipitation. The results for the mountains in the same season suggested slightly less convective activity, while the summer mountain results showed considerably fewer events but with a similar distribution to that of winter. These suggestions are in agreement with the known synoptic climatology of both states, and indicate that the present method is suitable for providing a quantitative estimate of the spatial dimension of daily precipitation.

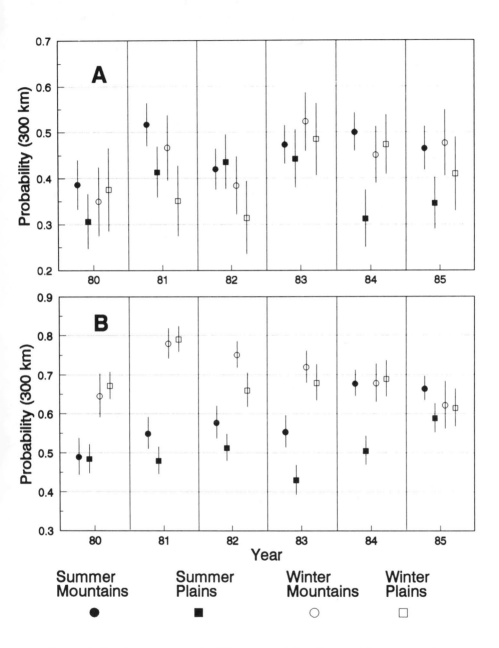

Figure 8.2. Probability of precipitation 300 km from a station where precipitation is occurring, as a function of season, region and year, for (A) Colorado and (B) North Carolina. Vertical lines indicate the range created by the standard errors of the logarithmic regression.

The year-to-year differences were investigated using the results of the model for a specific inter-station distance. An example is a spatial scale of 300 km, near the midpoint of the actual observed distances. There was a marked difference between the summer and winter probabilities in North Carolina. Thus the probability of precipitation at a point given that it is raining 300 km away was near 0.5 in summer, 0.7 in winter (Figure 8.2B). In all summers probabilities were higher in the mountains than on the plains. Several differences were considerably greater than the range of P_{ij} given by incorporating the standard error of the estimates of a and b, indicating a highly significant topographic contrast. The least difference occurred in 1980, a major dry year in both areas, and the only year with a significant zonal flow (Table 8.2). The differences between topographic regions in winter were much less marked, and varied in direction. Only in 1982 was it significant, with the plain once again having a lower probability. The overall probability values for Colorado (Figure 8.2A) were smaller than for North Carolina, with summer and winter values clustered around 0.4. Topographic differences were such that 5 of the 6 summer cases and 4 of the 6 winter ones indicated smaller storms in the plains than in the mountains. As with North Carolina, however, no connection either with precipitation totals or with circulation was apparent from this data sample.

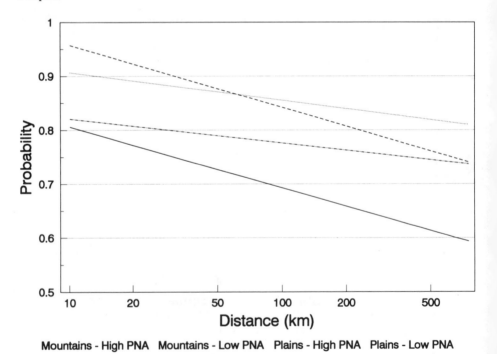

Mountains - High PNA   Mountains - Low PNA   Plains - High PNA   Plains - Low PNA

Figure 8.3. Relation between the conditional probability of precipitation and distance as a function of the Pacific-North America Index and topography for winter conditions in North Carolina.

The connection with circulation, using a wider range of PNA values (Table 8.3), was undertaken for the winter conditions in North Carolina. Composites of the five highest years and the five lowest years for each region were used initially (Figure 8.3). The plains showed a major difference between high and low conditions. High PNA conditions, with flow primarily from the southwest, indicated frequent occurrence of precipitation, with most being highly localized. Low (highly negative) conditions, however, gave lower probabilities and indicated more widespread rainfall events. The same pattern, but with a much less marked difference, was apparent in the mountains. Further, even the low PNA conditions for the plains gave a steeper slope than either of the mountain conditions.

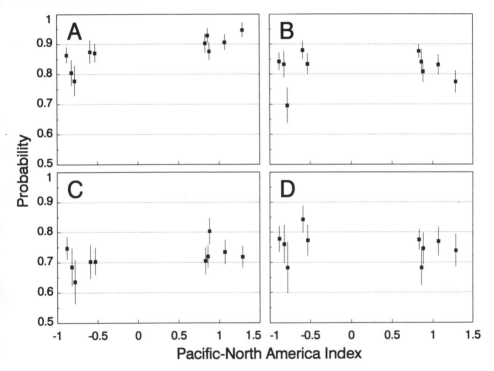

Figure 8.4. Conditional probability of precipitation as a function of the Pacific-North America Index
for North Carolina stations separated by (A) 20 km in the plains, (B) 20 km in the mountains,
(C) 500 km in the plains, and (D) 500 km in the mountains.

The relationships between PNA and precipitation probabilities for individual years were highly variable (Figure 8.4). The general patterns anticipated from Figure 8.3 were clearly displayed. Nevertheless there was great variability in the means from year to year suggesting that factors other than PNA play a significant role in the spatial scale of the precipitation. In addition, the large range of uncertainty each year, as indicated by the standard errors, suggests that the differences may not be highly significant. Certainly the results should be treated with caution, and further analysis incorporating more years and different areas is required before definitive statements can be made.

8.7. Conclusions

The development of conditional probability distributions for the occurrence of precipitation is a straightforward process. The present results indicated that there was a logarithmic relationship between the conditional probability and inter-station distance. The relationship for Colorado indicated lower overall probabilities than for North Carolina, indicating a greater preponderance of small-scale events. The results varied little with season for Colorado, but North Carolina displayed a lower probability in summer than in winter. Topographically, the Colorado mountain region had higher overall probabilities and a greater concentration of large-scale widespread events than the plains area. North Carolina was somewhat more complex but generally had the opposite tendency. However, for large inter-station distances of about 300 km, both states in both seasons had higher probabilities in the mountainous regions than in the plains, suggesting a strong orographic influence on the results.

The influence of circulation patterns on the spatial scale of precipitation for North Carolina winters was investigated using the Pacific-North America Index. There were general indications for the plains region that a high PNA value, meridional flow, was associated with more frequent and localized storms than were low values. While there was a similar relationship for the mountains, the differences were much less marked. Examination of conditions for individual years and consideration of the errors associated with the method indicated that the results were only preliminary and that much more work is needed before definitive statements can be made.

This conditional probability approach represents a departure from traditional methods of precipitation analysis, but the results as a whole indicate that it is possible to use this simple straightforward method to summarize the spatial scale of precipitation on a daily basis, to explore the influence of topography, season, and location, and to link daily precipitation with circulation. As such, it has the potential to form the basis for a transfer function relating the detailed observational evidence to model projections and thus provide estimates of precipitation regimes under a changed climate. Further testing, including use of other regions, circulation indices, and observational years, is needed. More explicit linkage to model output is also desirable and possible. Although at present the method deals only with the spatial scale of precipitation events, it is capable of much refinement and expansion. Amalgamation of the approach with statistical models such as those of Cox and Isham (1988) appears particularly promising, so that extensions to allow simultaneous estimates of precipitation amount, event duration, storm size, and thus ultimately the volume of water falling onto a watershed, should be possible. These would allow detailed analyses of local precipitation regimes in both the present and any future climatic conditions.

Acknowledgements

This work was supported by Grant No. DMS-9115750 from the National Science Foundation, and was undertaken as part of Cooperative Agreement No. CR-816237 between the Geography Department, University of North Carolina and the United States Environmental Protection Agency.

References

Ayers, M.A., Wolock, D.M., McCabe, G.J., and Hay, L.E. (1993) *Sensitivity of water resources in the Delaware River basin to climate variability and change*, Open File Report 92-52, United States Geological Survey, West Trenton, New Jersey, 68 pp.

Berndtsson, R. (1989) Topographical and coastal influence on spatial precipitation patterns in Tunisia, *International. J. Climatology* 9, 357-369.

Bryson, R.A., and Hare, F.K. (1974) The Climates of North America, in R. A. Bryson and F. K. Hare (eds.), *Climates of North America*, Amsterdam-London-New York, Elsevier, 1-48.

Chen, R.E., and Robinson, P.J. (1991) Generating scenarios of local surface temperature using time series methods, *J. Climate* 4, 723-732.

Cohen, S.J. (1986) Impact of CO_2-induced climatic change on water resources in the Great Lakes basin, *Climatic Change* 8, 135-153.

Cox, D.R., and Isham, V. (1988) A simple spatial-temporal model of rainfall, *Proceedings Royal Society London*, A 415, 317-328.

Crane, R.G., and Barry, R.G. (1988) Comparison of the MSL synoptic pressure patterns of the Arctic as observed and simulated by the GISS general circulation model, *Meteorology Atmospheric Physics* 39, 169-183.

Dickinson, R.E., Errico, R.M., Giorgi, F., and Bates, G.T. (1989) A regional climate model for the western United States, *Climatic Change* 15, 383-422.

Giorgi, F., Bates, G.T., and Nieman, S.J. (1992) Simulation of the arid climate of the southern Great Basin using a regional climate model, *Bulletin American Meteorological. Society* 73, 1807-1822.

Gleick, P.H. (1987) Regional hydrologic consequences of increases in atmospheric CO_2 and other trace gases, *Climatic Change* 10, 137-161.

Hatch, W.L. (1983) *Selective guide to climatic data sources*, Key to meteorological records documentation 4.11, National Climatic Data Center, Asheville, NC. np.

Henderson, K.G., and Robinson, P.J. (1994) Relationships between the Pacific/North American teleconnection patterns and precipitation events in the south-eastern United States, *International J. Climatology* 14, 307-323.

Hevesi, J.A., Istok, J.D., and Flint, A.L. (1992) Precipitation estimation in mountainous terrain using multivariate geostatistics. Part I: Structural analysis, *J. Applied Meteorology* 31, 661-676.

Houghton, J.T., Jenkins, G.J., and Ephraums, J.J. (1990) *Climatic change: The IPCC scientific assessment*, Cambridge University Press, Cambridge, 365 pp.

Kalkstein, L.S., Dunner, P.C., and Vose, R.S. (1990) Detection of climatic change in the western

North America Arctic using a synoptic climatological approach, *J. Climate* **3**, 1153-1167.

Kim, J.-W., Chang, J.-T., Baker, N.L., Wilks, D.S., and. Gates, W.L (1984) The statistical problem of climate inversion: Determination of the relationship between local and large-scale climate, *Monthly Weather Review* **112**, 2069-2077.

Klein, W.H., and Bloom, H.J. (1992) Specification and prediction of monthly and seasonal precipitation amounts in California and Arizona river drainage basins, *International. J. Climatology* **12**, 721-732.

Leathers, D.J., Yarnal, B., and Palecki, M.A. (1991) The Pacific/North American teleconnection pattern and United States climate. Part I: Regional temperature and precipitation associations, *J. Climate* **4**, 517-528.

Robinson, P.J. (1994) Precipitation regime changes over small watersheds, in V. Barnett and K.F. Turkman (eds.) *Statistics for the environment, volume 2*, Chichester, John Wiley and Sons, 43-59.

Robinson, P.J., and Finkelstein, P.L. (1991) The development of impact-oriented climate scenarios, *Bulletin American Meteorological. Society* **72**, 481-490.

Robinson, P.J., and Henderson, K.G. (1992) Precipitation events in the South-east United States of America, *International. J. Climatology* **12**, 701-720.

Robinson, P.J., Samel, A.N., and Madden, G. (1993) Comparisons of modelled and observed climate for impact assessments, *Theoretical Applied Climatology* **48**, 75-87.

Ropelewski, C.F., and Halpert, M.S. (1987) Global and regional scale precipitation patterns associated with the El Nino/Southern Oscillation, *Monthly Weather Review* **115**, 1606-1626.

Ropelewski, C.F., and Halpert, M.S. (1989) Precipitation patterns associated with the high index phase of the Southern Oscillation, *J. Climate* **2**, 268-284.

Sharon, D. (1979) Correlation analysis of the Jordan Valley rainfall field., *Monthly Weather Review* **107**, 1042-1053.

Sumner, G.N. (1983) The use of correlation linkages in the assessment of daily rainfall patterns, *J. Hydrology* **66**, 169-182.

Taljaard, J.J. (1986) Change of rainfall distribution and circulation patterns over southern Africa in summer, *J. Climatology* **6**, 579-592.

Wallace, J.M., and Gutzler, D.S. (1981) Teleconnections in the 500mb geopotential height field during the Northern Hemisphere winter, *Monthly Weather Review* **109**, 784-812.

Yarnal, B., and Leathers, D.J. (1988) Relationships between interdecadal and interannual climatic variations and their effect on Pennsylvania climate, *Annals Association American Geographers* **78**, 624-641.

P.J. Robinson
Department of Geography
University of North Carolina
Chapel Hill, NC 27599, USA

9 USE OF ARTIFICIAL NEURAL NETWORKS IN PRECIPITATION FORECASTING

HSIANG-TE KUNG, L. YU LIN and S. MALASRI

Abstract

The purpose of this paper is to study the potential uses of Artificial Neural Networks (ANNs) in estimating monthly precipitation for the City of Memphis, Tennessee. One of the popular models, the backpropagation algorithm (BP), was developed and installed in an internal program. Annual precipitation data from the City of Memphis provides the basis data to train the networks. The trained networks were then used to verify and predict precipitation for the City of Memphis. The results indicated that ANNs can recognize the precipitation pattern and provide an analogous precipitation trend to the existing precipitation data. One of the major advantages of this method is that its performance does not require a large quantity of data in the trend analysis.

9.1. Introduction

Precipitation is one of the major hydrological processes in the natural water cycle. It has been a major concern that precipitation has changed because of the changes in the global climate. As a result, it has increasingly complicated the difficult task of precipitation prediction. In the past, the prediction of precipitation pattern has relied primarily on the use of radar detection, time series analysis, and stochastic models. Although those methods provide a fairly reliable short-term result, large quantities of data were required.

An alternative approach is the use of Artificial Neural Networks (ANNs). ANNs is a modeling technique which has been widely applied in a number of domains, such as pattern recognition (Rumelhart *et al.*, 1986; Windrow 1988; Malasri *et al.*, 1991), noise cancellation (Hassoun and Spitzer 1988), and forecasting (Khotanzad 1991; Djukanovic 1991; Malasri and Lin, 1992; Williams *et. al.*, 1992). Among various ANN models, the backpropagation algorithm (BP) is the most popular network due to its ability to solve a wide range of problems. Additionally, it can be easily simulated by using a computer algorithm. The objective of this paper is to examine the potential use of ANNs for precipitation pattern recognition. Annual precipitation data recorded from the City of Memphis was used to train the ANNs. The results were verified and used to forecast the city's future precipitation.

J.A.A. Jones et al. (eds.), Regional Hydrological Response to Climate Change, 173–179.
© 1996 *Kluwer Academic Publishers. Printed in the Netherlands.*

9.2. Materials and Methods

The backpropagation algorithm was first discovered by Werbos in 1974. The architecture of the backpropagation model consists of three layers, i.e., input layers, output layers, and hidden layers (Figure 9.1). Within the networks, the algorithms emulate the learning process of a human brain and can be trained to recognize a pattern based on a given training set. In general, the learning processes of the backpropagation model is to minimize the differences between the computed (predicted) value and the desired (actual) output value for all training pairs during the training process.

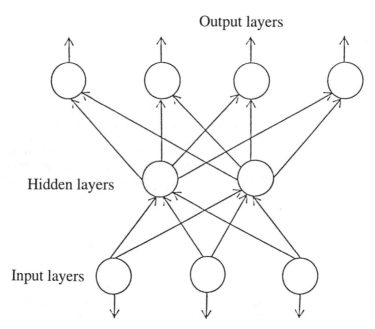

Figure 9.1. Architecture of a three-layer backpropagation.

A number of training pairs with desired input (X) and desired output (T) are initially presented to the networks. Once the number of hidden layers and accuracy threshold have been defined, the input data pairs are transferred to the hidden layers. Following the feed forward process, the results of hidden layer (H) and output layer (Y) are calculated using the following equations:

$$H_i = f\left(\sum_h X_h W_{hi} + \theta_i \right)$$
(9.1)

and

$$Y_j = f\left(\sum_i H_i K_{ij} + \lambda_j\right)$$ (9.2)

where θ_i is the ith hidden layer threshold value, λ_j is the jth output layer threshold value, and f( ) is the logistic sigmoid threshold function defined as:

$$f(x) = (1 + e^{-x})^{-1}$$ (9.3)

The weights (W) and biases (K) ranging from -1 to +1 are randomly assigned to adjust the values for output layer. A value ranging between 0 and 1 is assigned to the training rate coefficient (η) and the momentum coefficient (α). The training rate coefficient and the momentum coefficient are two constants which are used to adjust the weight and biases.

The discrepancy, an error (δ), between the computed and desired values of hidden layer and output layer is calculated using the following equations:

$$\delta_i^h = H_i\left(1 - H_i\right)\sum_j K_{ij}\delta^o$$ (9.4)

and

$$\delta_j^o = Y_j\left(1 - Y_j\right)\left(T_j - Y_j\right)$$ (9.5)

In most cases, the weights and biases of the hidden layer and the output layer have to be adjusted in order to meet the designated accuracy threshold. Equations (6), (7) and (8) are used to compute the hidden layer value. Equations (9), (10) and (11) are used to compute the output layer value. These processes are repeated until the discrepancy between the computed value and the desired value equal to the designated accuracy threshold. Therefore, it indicates that the training pattern has been recognized by the networks. The resulting model is able to be applied to verify and predict the future precipitation pattern.

$$\Delta W_{ij} = \eta \times_i \delta_j^h$$ (9.6)

$$W_{ij} = W_{ij} + \Delta W_{ij} + \alpha \Delta W_{ij}$$ (9.7)

$$\theta_{ij} = \theta_i + \eta \delta_j^h$$ (9.8)

and

$$\Delta K_{ij} = \eta H_i \delta_j^o$$ (9.9)

$$K_{ij} = K_{ij} + \Delta K_{ij} + \alpha \Delta K_{ij}$$ (9.10)

$$\lambda_i = \lambda_j + \eta \lambda_j^o \tag{9.11}$$

The ANNs were written using the BASIC language and installed as an internal program operating in MS-DOS environment. Three major modules were included: (1) Data Analysis Mode, (2) Training Mode, and (3) Prediction Mode. The data analysis mode was designed to manipulate and display the input data. For this study, the annual precipitation data in the city of Memphis were comprised as the input data. Based on the previous study (Malasri and Lin, 1992), a trial and error approach was conducted to select the optimum number of training pairs and hidden layers. To prevent excessive biases to the output values, the output values were normalized to a value ranging from 0 to 1.

The training mode provides the learning domain for the networks. During the training session, the training coefficient, the momentum coefficient, and the discrepancy threshold were initially set at 0.9, 0.9 and 0.05, respectively. As the training mode was completed, the networks automatically save the learning results. In the final stage, the prediction mode can be used to calculate the future precipitation based on the trained networks.

9.3. Results and Discussions

Figure 9.2 shows the annual precipitation data for the city of Memphis from 1930 to 1989. The maximum rainfall was 1802.89 mm (70.98 inches) in 1979 and the minimum rainfall was 775.72 mm (30.54 inches) in 1940. Mean annual precipitation was 1285.49 mm (50.61 inches).

Trial and error procedures were carried out to determine the optimum input layers and hidden layers before the networks start to train. The previous studies (Malasri and Lin, 1992; Malasri *et.al.*, 1991) revealed that the shorter training time usually provides the higher accuracy of networks. In addition, the training pairs equal to the hidden layers can support the higher efficiency of the networks. As the training coefficient or the momentum coefficient decreases, training time increases as a result of the low degree of accuracy. Theoretically the training time depends on several factors, such as the number of hidden layers, the training coefficient and the momentum coefficient. During the course of this study, various pairs of precipitation data were supplied to train the neural networks. The results showed that twenty input layers took the least training time. It implied that twenty input layers were the local optimum for this study. Therefore, 30 pairs of precipitation data with twenty input layers and twenty hidden layers were applied to train the neural networks.

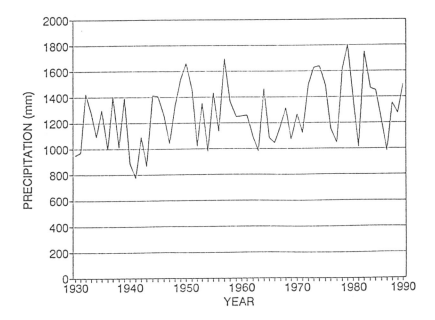

Figure 9.2. Annual precipitation in the city of Memphis.

The backpropagation model associated with the training coefficient and the momentum coefficient during the training process. The sigmoid function controls the output. It allows a smooth transition of output from any given set of input layers. The net took less than 5000 iteration cycles to learn the precipitation pattern.

The verification of the networks for the precipitation data in the City of Memphis demonstrated that the networks have achieved a recognition of the precipitation pattern. The difference is mainly because of the accuracy threshold. The prediction mode was then conducted after the training session was completed. Comparing the actual precipitation data to the computed data using the ANNs (Figure 9.3), the results indicate that the networks most likely followed the trends of actual precipitation data except for 1986. Although there was a difference between predicted precipitation and actual precipitation observed, the nets still provided an analogous trends, especially the precipitation in the last several years.

It was theoretically expected that more input layers can induce the higher level of accuracy in forecasting. However, it was not necessarily true in this study because more input layers will increase the training time. Consequently, a higher discrepancy value is generated. The choice of an optimum number of input layers is still debatable. A trial and error approach appears to be the only solution. Other parameter studies have also indicated that higher accuracy may be achieved by increasing the number of iterations during the training mode. If this is the case, the networks should be terminated only when a perfect solution is achieved.

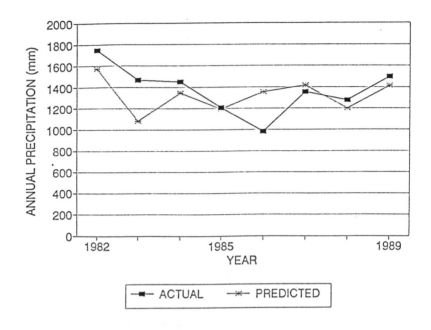

Figure 9.3. Comparison of actual and predicted precipitation.

9.4. Conclusions

In recent years, major global climatic changes have had a profound impact on precipitation patterns. In this study, it has been demonstrated that ANNs can be applied to pattern recognition and forecasting. However, the selection of the parameters such as input layers, hidden layers, the training coefficient, and the momentum coefficient is still highly debatable.

Some conclusions can be derived from this study:

(a) a reliable artificial network depends on the proper selection of input layers, hidden layers, and output layers;

(b) the number of hidden layers and the number of input layers should be equal to increase the performance of the networks;

(c) a shorter training time is desirable to increase the efficiency of the networks for pattern recognition and forecasting;

(d) the higher the training coefficient and momentum coefficient, the higher efficiency of the network is likely to be true;

(e) despite the highly variable precipitation pattern of the input data, the networks were able to memorize the pattern and make a reliable forecast; and

(f) further study is required to derive an optimum combination of parameters involved.

References

Djukanovic, M., and Babic, B. (1991) Unsupervised/supervised learning concept for 24-hour lead forecasting, *Proceedings of ANNIE (Artificial Neural Networks in Engineering)*, 819-827.

Hassoun, M., and Spitzer R. (1988) Neural network identification and extraction of repetitive super-imposed pulses in noisy 1-D signals, *Neural Networks Supplement* **1**, 1443.

Khotanzad, A., and Fowler M. (1991) Neural network based time series forecasting, *Proceedings of ANNIE (Artificial Neural Networks in Engineering)*, 813-818.

Malasri, S., and Lin L. Y.(1992) Forecasting with artificial neural networks, *Proceeding of Arkansas Academy of Science*, 1-7.

Malasri, S., Madhavan K., and Lin L. Y.(1991) Soil classifier: expert systems + neural nets, *Proceedings of ANNIE (Artificial Neural Networks in Engineering)*, 807-812.

Rumelhart, D.E., Hinton G.E., and Williams R. J. (1986) Learning internal representations by error propagation, MIT Press, Cambridge, Mass., USA, **1**, 318-362.

Werbos, P.J. (1974) *Beyond regression: new tools for prediction and analysis in the behavioral sciences*, Unpublished Ph.D. Dissertation, Harvard University, Cambridge, Mass., USA.

Williams, T.P., Khajuria A., and Balaguru P. (1992) Neural network for predicting Concrete Strength, *Proceedings of the 8th Conference on Computing in Civil Engineering*, 1082-1088.

Windrow, B., and Winter R. (1988) Neural nets for adaptive filtering and adaptive pattern recognition, *Proceeding of IEEE Computer*, 25-29.

Hsiang-Te Kung
Department of Geography
University of Memphis
Memphis, TN 38152, U.S.A.

L. Yu Lin and S. Malasri
Department of Civil Engineering
Christian Brothers University
Memphis, TN 38104, U.S.A.

10 GENERATION OF SEQUENCES OF AIR TEMPERATURE AND PRECIPITATION FOR ESTIMATION OF THE HYDROLOGICAL CYCLE IN CHANGING CLIMATIC CONDITIONS IN POLAND

MALGORZATA GUTRY-KORYCKA and PIOTR WERNER

Abstract

Downscaling experiments are undertaken using kriging and distance-decay function interpolation to produce high resolution maps of temperature and precipitation in Poland from the GFDL and GISS macroscale scenarios. Measurements of precipitation and temperature at 58 stations are used to test the applications and to calculate the effects of the scenarios on areal averages. Average annual precipitation in Poland would rise from 639 mm to 749 mm under the GISS scenario and to 655 in the GFDL scenario. The spatial median mean air temperature would rise from 7.7°C to 10.5°C and 11.9°C respectively.

10.1. Goal of Research and Empirical Data

Changes in the course of global geophysical processes may affect the size and time schedule of elements of the hydrological cycle in a catchment. Water balance, which constitutes a quantitative approach to this cycle at the scale of the particular catchment, is highly sensitive to global climate changing of the air temperature and precipitation that determine vertical and horizontal water exchange.

Using uncertain scenarios of changes that regard a global climatic warming as the main reason for unstationary circulation of water in a catchment, a simulation was carried out of the mean annual values of the air temperature and precipitation in Poland.

Simulation of the effect of global climatic changes may be carried out in two ways:

1) The expansion of the air temperature field into the Taylor sequence, $T = f(\lambda, \phi, H)$, in relation to three position coordinates, i. e. latitude and longitude, as well as elevation above sea level, H (Stopa-Boryczka et al., 1990), permits a description of the deformation of the field of air temperature as a result of hypsometry. The resulting empirical model of simulation of spatial distribution of the air temperature and the knowledge of basic geographical factors can be used for time and spatial forecast of global climatic changes elements in Poland.

2) The grid model which was used in these investigations takes into account the rhythm of changes of the fields of the air temperature and precipitation in relation to geographical coordinates. Thus it was assumed that spatial distribution of selected climatological elements (real and generated field) due to hypsometry and interpolation

J.A.A. Jones et al. (eds.), Regional Hydrological Response to Climate Change, 181–196.
© *1996 Kluwer Academic Publishers. Printed in the Netherlands.*

are in the quantitive proportion. In other words, the relative quantitative system of real and generated isotherms does not change.

The air temperature field was generated on the basis of average values for 30 years (1951-1980) from 58 meteorological gauges. In order to generate the field for precipitation, average yearly values of total precipitation for 90 years (1891-1980) from 57 gauges were used (Chrzanowski, 1988).

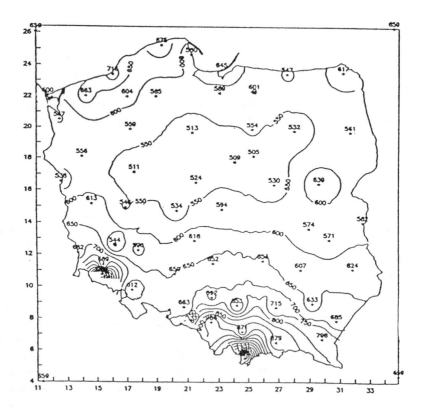

Figure 10.1. Normal annual precipitation in Poland (1891-80) in millimetres - Scenario 0.

10.2. Generation of Normal Values of the Air Temperature and Precipitation with Assumed Scenarios of Climatic Changes

Forecasts of climate changes are usually related to the level of double concentration of CO_2 and the greenhouse effect. Climatologists point to a positive trend of the air temperature, but it is more difficult to estimate a precipitation change trend due to its great stochastic variability both in time and space.

The sensitivity of a catchment to climatic changes in water balance is essentially dependent on precipitation. Models of global changes of climate developed in recent years, particularly in the USA and UK, permit the creation of scenarios of change based on these particular meteorological elements.

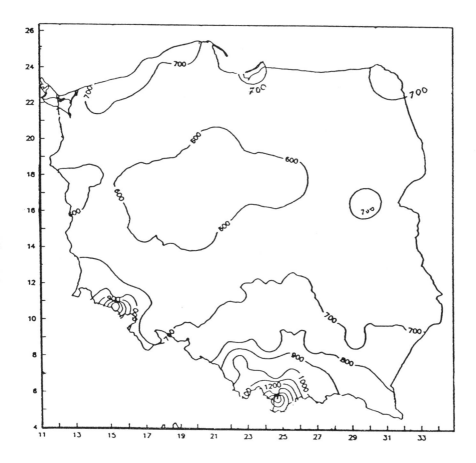

Figure 10.2. Generation of normal precipitation in millimetres - Scenario ($2 \times CO_2$), according to GFDL model.

In the Division of Water Resources of the Institute of Geophysics of the Polish Academy of Sciences, Prof. Z. Kaczmarek has been conducting research for several years with a view to developing and verifying selected climatic scenarios for the territory of Poland.

The first approach involves the simulation of hydrological processes using physically justified mathematical models of catchments and assumed scenarios of change in meteorological elements. The other approach is aimed at defining the degree of sensitivity of the particular processes of the hydrological cycle to climatic warming due to the greenhouse effect.

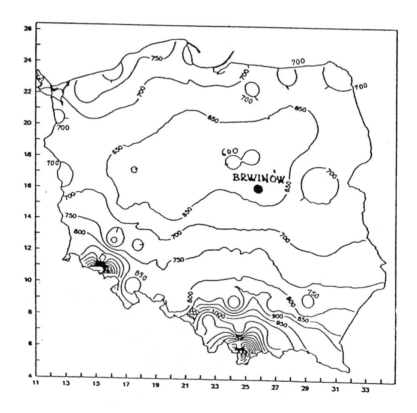

Figure 10.3. Generation of normal precipitation in millimetres - Scenario (2 x CO_2), according to GISS model.

Two scenarios were chosen corresponding to double concentration of CO_2 in the atmosphere calculated for Europe according to two global models of circulation, the General Circulation Model GFDL, constructed and verified at the Geophysical Fluid Dynamic Laboratory, and the GISS model, developed at the Goddard Institute for Space Studies. Results for these models for the region of Central Europe were obtained from the National Center for Atmospheric Research (NCAR) in Boulder, USA, from the pre-1989 model (Institute of Geophysics, 1992). Both global GCMs produce scenarios of climatic changes which envisage increase in the air temperature as well as in precipitation. Output data from these global models appear as files of values for temperature and precipitation corresponding to the nodes in the data net for scenario zero (1 x CO_2), that is to natural conditions, and data resulting from a double value (2 x CO_2). Numerical values of assumed climatic changes were defined for nodes of the geographical grid at intervals of 1 degree of latitude × 1 degree of longitude.

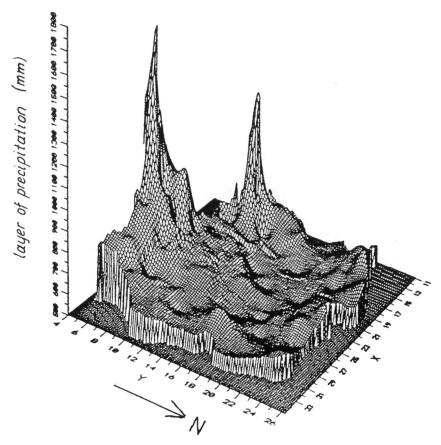

Figure 10.4. Three-dimensional distribution of normal annual precipitation in Poland (1891-1980) in millimetres - Scenario 0.

The region of Central Europe that was adopted is bounded by 45° and 55°N and by 10° and 25°E. The area outside Poland was divided into 150 unit entry files of 1 degree of latitude x 1 degree of longitude. To carry out the so-called downscaling of climatic information from the GCM models in the territory of Poland, several numerical procedures need to be applied.

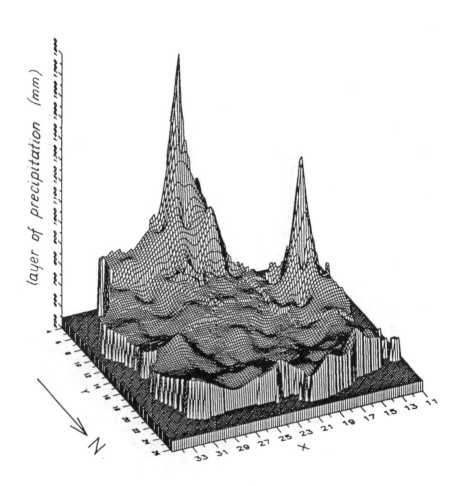

Figure 10.5. Three-dimensional distribution of normal annual precipitation in millimetres - Scenario (2 x CO_2), according to GFDL model.

The base map of Poland used was 1: 3 million in quasi-stereographical secant projection in the system type of GUGiK 1980 (Grygorenko, 1985). Transformation of the geographical system into the system of regular (Cartesian) coordinates was made using the ADS procedure in the programme PC ARC/INFO. The base of meteorological data was created using the programme Dbase IV. From this data base, the values of geographical locations of climatological stations were generated using the procedure PROJECT, which permits a transformation to quasi-stereographical projection. Next, using the procedure TRANSF/RM the location of station was transformed into the system of cartographical coordinates. The SURFER package was chosen for interpolation and presentation of the results obtained.

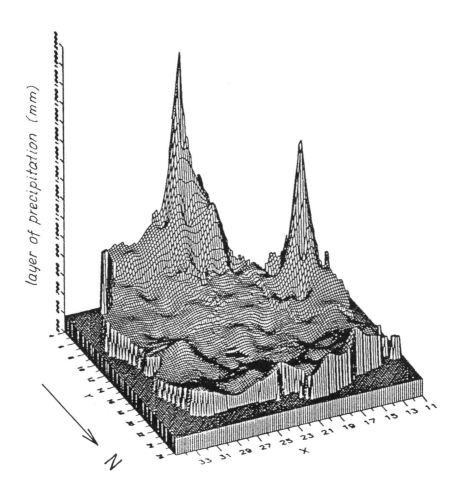

Figure 10.6. Three-dimensional distribution of normal annual precipitation in millimetres
- Scenario (2 x CO_2), according to GISS model.

As an interpolation net for the spatial model of climatological elements a square grid of cells 2 x 2 mm on the same scale as the base map was used (thus, the dimension of the field or cell of interpolation amounted to 6 x 6 km). In the interpolation, the area outside the territory of Poland was 'covered up'. The method of kriging for spatial interpolation was applied to construct the field of the air temperature. The method of kriging is based on the assumption that expected value is equal to the difference $[z(x)-z(x+h)^2]$, where $z(x)$ is the value of a meteorological variable at the station, while h indicates the distance between the stations. The method is confined to estimation of coordinates (z) at the nodes of the grid on the basis of values at the measuring points. The method assumes that the mean value of errors during estimation amounts to zero, while the variance is the least possible.

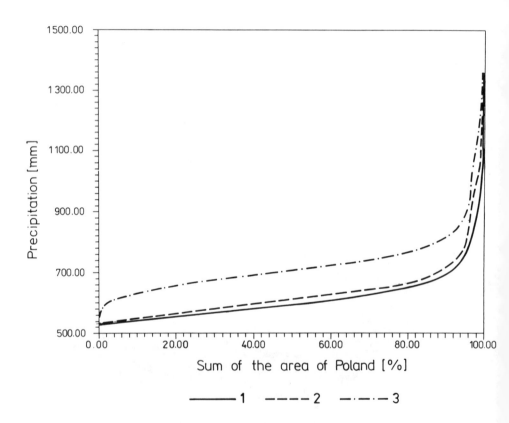

Figure 10.7. Areal distribution of precipitation in Poland as percentage of normal. 1. Period 1951-1980; 2. Generation according to GFDL - Scenario (2 x CO_2); 3. Generation according to GISS - Scenario (2 x CO_2). X axis - cumulative area of Poland covered by the given isopleth (%); Y axis - average annual precipitation (mm).

For interpolation of precipitation, the method of reverse distances between climatological gauges was used (Tanski, 1990). The interpolation procedures permitted
the creation of simulation maps in the two dimensional form of isopleths (isohyets and isotherms) that we can see in Figures 10.1-10.3 and 10.8-10.10 and in the three dimensional block diagrams representing the spatial distribution of the meteorological elements (Figures 10.4-10.6 and 10.11-10.13).

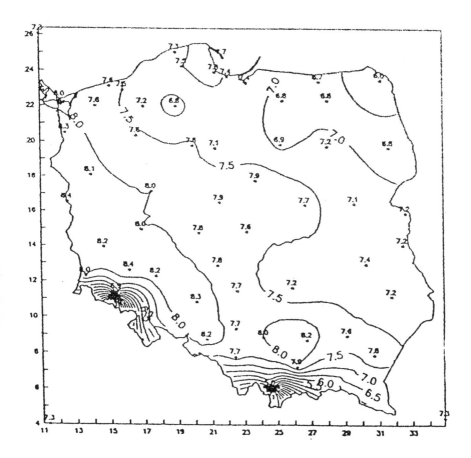

Figure 10.8. Average air temperature in Poland (1951-80) in degrees Celsius - Scenario 0.

The numerical and graphical areas obtained also made it possible to construct charts representing the relation of the numerical value of the meteorological element to the area limited by the isoline of the same value as the above-mentioned isoline. Curves of total annual precipitation distribution illustrate the share of Poland enclosed between the successive isohyets or isotherms of mean annual values (Figures 10.7 and 10.14). Finally, the maps produced for scenarios 0 and 1 were compared.

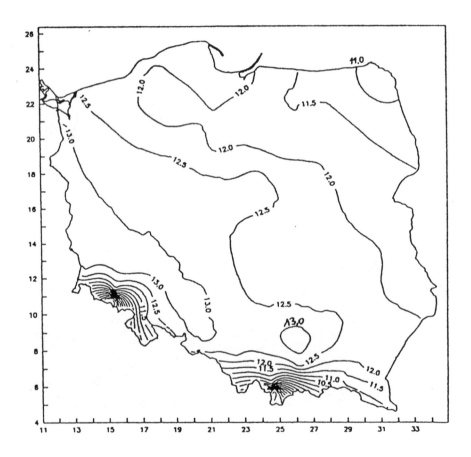

Figure 10.9. Generation of average air temperature in degrees Celsius - Scenario (2 x CO_2), according to GFDL model.

The disadvantages of the simulation is that the grid model takes into consideration mathematical but not geographical interpolation. Therefore in the areas with high gradients of meteorological elements, especially in the mountains and uplands, there may be large differences between the empirical value and those obtained from linear interpolation. Thus, deviations will appear in areas with a low density of measuring stations.

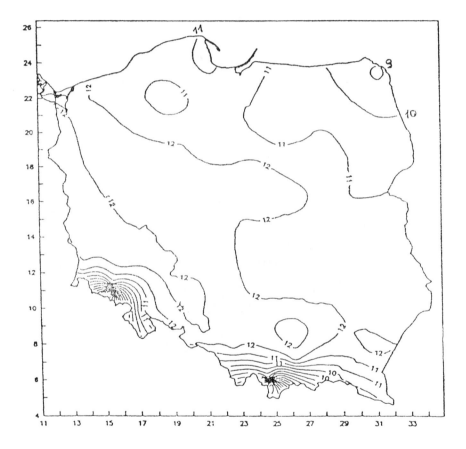

Figure 10.10. Generation of average air temperature in degrees Celsius - Scenario (2 x CO_2), according to GISS model.

The graphical and numerical areas obtained were taken as a basis for finding a method of transforming GCM-based predictions to subscaling models. The input simulation models obtained from GFDL and GISS models differ slightly. The selected scenarios of climatic changes envisage an increase in air temperature by about 4°C and an increase in precipitation over most of the year by 10 mm (except for September).

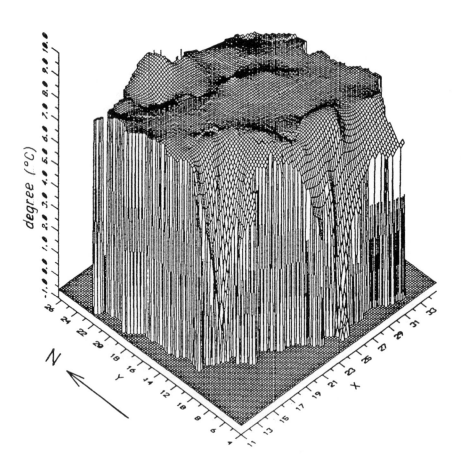

Figure 10.11. Three-dimensional distribution of average air temperature in Poland (1951-1980) in degrees Celsius - Scenario 0.

In the GISS model, the change in precipitation ranges from +1.1 mm to -13.2 mm, while the GDFL model expects the increase of precipitation to be slightly below 10 mm, with the exception of August, when the increase is greater than 30 mm, and June, September and October, when precipitation decreases by approximately 5 mm (Ozga-Zielinska, 1992).

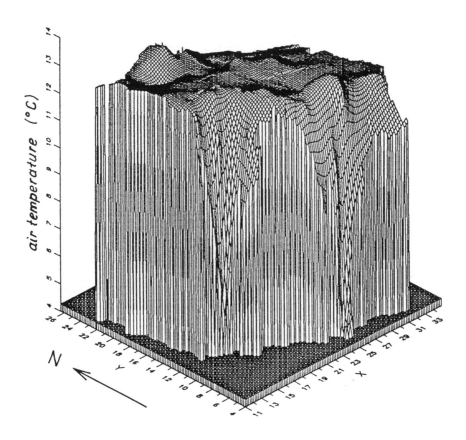

Figure 10.12. Three-dimensional distribution of average air temperature in Poland (1951-1980) in degrees Celsius - Scenario (2 x CO_2), according to GFDL model.

The interpolated fields of changes in air temperature and precipitation compared with normal real precipitation and simulated precipitation provided grounds for estimation of differences in value of these elements obtained from global GFDL and GISS models.

The real average annual precipitation total, or the spatial median value (i.e. half of the area of Poland has a value above this), in natural conditions amounts to 639 mm. According to the GISS model it will reach 749 mm, while according to to the GFDL model annual precipitation is expected to amount to 655 mm (Figure 10. 7).

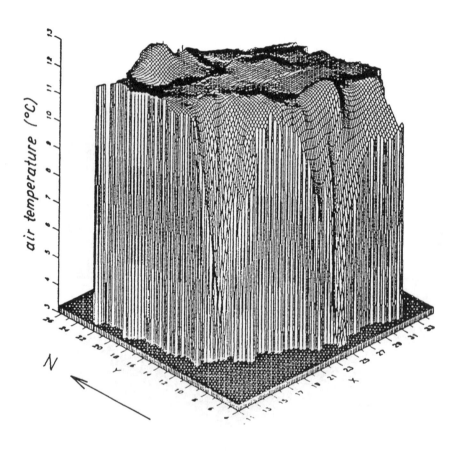

Figure 10.13. Three-dimensional distribution of average air temperature in Poland (1951-1980) in degrees Celsius - Scenario (2 x CO_2), according to GISS model.

Currently, 50% of Poland experiences a mean annual temperature of 7.7°C and above, while according to the scenario of doubled CO_2 contents in the GISS model it will amount to 10.5°C, and in the GFDL model it will reach 11.9°C (Figure 10.14).

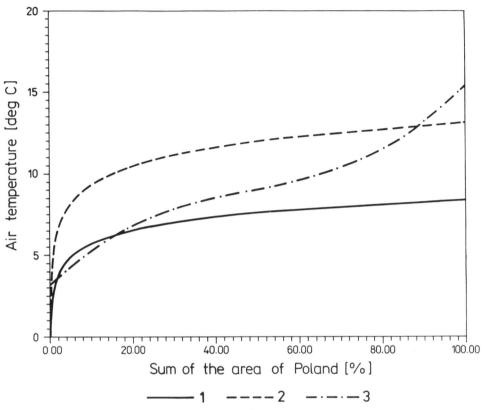

Figure 10.14. Areal distribution of average air temperature in Poland as percentage of normal. 1. Period 1951-1980; 2. Generation according to GFDL - Scenario (2 x CO_2); 3. Generation according to GISS - Scenario (2 x CO_2). X axis - cumulative area of Poland covered by the given isopleth (%); Y axis- average air temperature (°C).

10.3. Conclusions

Climate is described according to the statistics of meteorological phenomena observed over the period of at least 30 years. Research on the time and spatial structure of the natural fields of the air temperature and precipitation (scenario 0) has satisfied this requirement.

This research may be a starting point for combining GCMs used to forecast climatic changes with hydrological models. It paves the way for developing global models for grid nodes on a smaller scale. The stochastic methods applied permitted a determination of the local spatial structure of daily meteorological elements coupled with large-scale information.

The spatial distribution of isohyets obtained as a result of simulation, though it provides an approximate picture, could be a point of departure for modelling hydrological processes and their distribution under the assumed scenarios of climatic changes. This, however, requires the construction of spatio-temporal model linking local changes influenced by predicted global transformation of atmospheric circulation.

Methods of simulation using 90-year and 30-year sequences provide sufficient statistical material for a relatively precise estimation. Both approaches to simulation of fields of meteorological variables in long and short time intervals will be applied to combine global climate models with large-scale and small-scale models, given the GCM scenarios.

References

Chrzanowski, J. (1988) Regionalizacja i klasyfikacja dobowych opadów w Polsce (Regionalization and classification of daily precipitation in Poland), *Wiadomoœci IMGW* **11**, 32, No. 1-2.

Grygorenko, W. (1985) Uklad wspólrzêdnych i krój map topograficznych do celów gospodarczych w odwzorowaniu quasistereograficznym GUGiK 1980 (The system of coordinates and sheet division of topographical maps for economic purposes in a quasi-stereographic projection GUGiK 1980), *Przeglad Kartograficzny* **27**, No. 2.

Institute of Geophysics (1992) Scenariusze zmian klimatycznych wg modeli globalnych GFDL i GISS (Scenarios of climatic changes according to global models GDFL and GISS), Archives of the Institute of Geophysics, Polish Academy of Sciences.

Ozga-Zielinska, M. (1992) Efekt cieplarniany, jego perspektywy i skutki dla gospodarki wodnej w Polsce (Greenhouse effect, its prospects and consequences for water mamagement in Poland), *Gazeta Obserwatora IMGW*, No. 4-6.

Stopa-Boryczka, M. *et al.* (1990) *Atlas wspólzale¿noœci parametrów meteorologicznych i geograficznych w Polsce, (Atlas of relationships between meteorological and geographical parameters in Poland)*, Editions of University of Warsaw.

Tanski, T. (1991) *Surfer: przewodnik uzytkownika (Surfer: a user's guide)*, PLJ, Warszawa, *Temperatura powietrza - œrednie miesiêczne i roczne oraz ekstremalne (1951-1980), 1986) (The air temperature: monthly and annual as well as extremum means (1951-1980))*, Materia³y do Atlasu Klimatycznego Polski, IMGW, Warszawa (Archives of the Institute of Meteorology and Water Management).

Werner, P. (1992) *Wprowadzenie do geograficznych systemów informacyjnych (Introduction to Geographical Information Systems)*, Editions of University of Warsaw.

Werner, P. (1992) *Atlas GIS: Przewodnik uzytkownika (The Atlas of GIS: a user's guide)*, Editions of University of Warsaw.

Malgorzata Gutry-Korycka and Piotr Werner
Faculty of Geography and Regional Studies
University of Warsaw ul. Krakowskie
Przedmiescie 30, 00-927 Warsaw, Poland.

11 SOME ASPECTS OF CLIMATIC FLUCTUATION AT FOUR STATIONS ON THE TIBETAN PLATEAU DURING THE LAST 40 YEARS

MASATOSHI YOSHINO

Abstract

Making use of 40 years data, 1951-1990, at four stations (Lhasa, Changdou, Deqin and Ganzi) on the Tibetan Plateau, the fluctuations of air temperature and precipitation were analyzed. From the 1950s to the 1980s a warming and drying tendency is predominant, particularly in summer and in mid-winter. A seasonal change of different tendencies from March, April to May occurs in a diametrically opposite way from October, November to December. In the second part of the study, anomalies of air temperature and precipitation were analyzed in the cases of El Niño and La Niña. Distributions of the anomalies were shown by season for air temperature and precipitation separately. It was indicated that the anomalies at Lhasa are of different sign in most cases from those at the other three stations located in the eastern part of the Plateau in summer. The anomaly conditions are the same from October to December. This can be attributed to the influence of monsoonal atmospheric circulation systems in summer. It is also noted that the distributions of precipitation anomaly in the El Niño case are not opposite to those in the La Niña case. In the last part of study, combinations of negative/positive anomalies of air temperature/precipitation were clarified. Anomalies with combinations of negative (air temperature) and negative (precipitation) dominated in the cases of El Niño summer and La Niña winter, but combinations of positive (air temperature) and positive (precipitation) dominated in the cases of El Niño winter and La Niña summer, and a combination of anomalies with negative (air temperature) and positive (precipitation) predominated in the case of La Niña winter.

11.1. Introduction

It is said that, due to the increasing greenhouse gases caused by human activities, global temperature will increase $3 \pm 1.5°C$ in the 21st century. This is, however, a state averaged globally and it is still uncertain regionally. There is little information on such change or fluctuation of climate on the Tibetan Plateau.

J.A.A. Jones et al. (eds.), Regional Hydrological Response to Climate Change, 197–211.
© *1996 Kluwer Academic Publishers. Printed in the Netherlands.*

The Tibetan Plateau plays an important role for climatic change processes not only for Asia, but also for the Northern and Southern Hemispheres through the monsoonal circulations. Its details, however, have not yet been analyzed. A previous description showed a period of low mean annual air temperature during the first half of the 1960s and of January mean air temperature during the later half of the 1960s, based on the data obtained at Lhasa during the 20 years 1956-1975 (Academia Sinica Integrated Scientific Study Group for Tibetan Plateau, 1984). Since the second half of 1960s, the air temperature has been increasing. It was suggested that the secular change of the Tibetan high pressure at the 200 hPa level relates to this change in surface temperature and precipitation.

Du (1993) demonstrated the secular changes of air temperature and precipitation for the annual mean, January, April, July and October from 1951 to 1990. His figures demonstrated that the differences in the secular changes of air temperature and precipitation are rather great regionally and seasonally over the Tibetan Plateau. Wu and Lin (1981) and Lin and Wu (1984) reported that during the last 100 years over the Plateau the years of 1947-1962 were a rainy period and the years of 1963-1980 a dry period, both of which are included in the years of the present study. They also found that amplitudes of variability are larger than those in East China, relatively larger in the north and west Plateau, and the higher the altitude, the greater the amplitude. Yatagai (1991) pointed out the negative linear trend of the annual as well as summer precipitation in the Tibetan Plateau to the east of the Plateau. She found a decreasing trend of annual mean, spring mean and fall mean air temperature in the region under discussion. But, on the other hand, an increasing tendency was detected for summer. Chen *et al.* (1991) reported that, contrary to the warming tendency of the world, the eastern part of the Tibetan Plateau is getting cooler and the precipitation is decreasing.

A decreasing tendency of precipitation since the 1950s was also confirmed by the annual precipitation curve reconstructed from tree-rings in the Tibetan Plateau (Academia Sinica, Integrated Scientific Study Group for Tibetan Plateau, 1984; Zhang, 1991).

Based on these previous results, the present paper aims to clarify the air temperature and precipitation fluctuations in the Tibetan Plateau. The four stations are: Lhasa (91°08'E, 29°40'N, 3648.7 m), Changdou (97°10'E, 31°09'N, 3306.0 m), Deqin (99°10'E, 28°39'N, 3592.9 m), and Ganzi (100°00'E, 30°37'N, 3393.5 m). The data used cover 40 years from 1951 to 1990.

11.2. Trends between the 1950s to the 1980s

In this part of study, air temperature and precipitation tendencies in the Tibetan Plateau from the 1950s to 1980s are dealt with. Using a climatic table compiled by Du (1993), comparisons of air temperature and precipitation in the 1950s with those during the 1980s are shown in Figures 11. 1, 11.2 and 11.3. The Figures clearly show that trends from the 1950s average to the 1980s average, that is, differences between the 1950s and

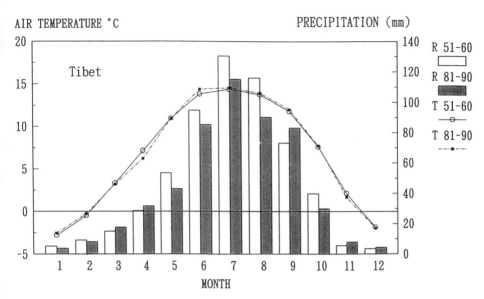

Figure 11.1. Monthly changes of air temperature and precipitation in Tibet for the 1950s and 1980s. Average of six stations: Lhasa, Changdou, Deqin, Ganzi, Maduo and Yushu. R51-60 is the ten year average precipitation, 1951-1960, and R81-90 is that for 1981-1990. T51-60 is the ten year average air temperature, 1951-1960, and T81-90 is that for 1981-1990.

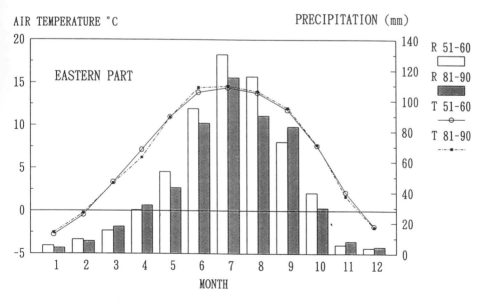

Figure 11.2. Same as Figure 11.1, but average of three stations, Changdou, Deqin and Ganzi, in the eastern part of the Plateau.

TABLE 11.1. Monthly trends in air temperature and precipitation
at four stations in the Tibetan Plateau from the 1950s to the 1980s.

Month	Trend from 1950s to 1980s Temperature	Precipitation	Magnitude of change	Related facts or estimated causes
1	+	-	**	Because of weaker winter monsoon in a global scale
2	+	-	**	during recent years.
3	0	+	*	Air temperature decrease caused by increasing cloud cover counter-balances global warming. Increased frequency of local invasion of moist air causes wetter conditions, though differences very small.
4	-	+	**	Air temperature decrease caused by increasing cloud cover exceeds global warming. Chance of precipitation increases, but amount is small.
5	0	-	*	Air temperature unchanged. Local convection is still weaker, because the Plateau is not yet warm enough (< 12°C monthly mean), so does not play any role in historical change. Regional moist air invasion becomes weaker, causing drier conditions.
6	+	-	***	Global warming effects, but summer monsoon in South Asia became weaker. SOI is low and Indian
7	+	-	***	summer rainfall shows drier tendency. Thus, the transport of water vapour from the Bay of Bengal
8	+	-	***	became weaker, resulting in drier conditions.
9	+	+	***	Summer season temperature conditions continue, so warmer historically. Local convection becoming stronger than before, resulting in wetter conditions.
10	0	-	*	Same as May
11	-	+	*	Same as April
12	0	+	*	Same as March

Note: Temperature trends: + denotes warmer, - cooler; Precipitation trends: + denotes wetter, - drier.
Magnitude of change is expressed as follows: *** the clearest, ** clearer, * clear, but Lhasa shows a
different tendency.

the 1980s, are not the same month to month. Their monthly status is summarized in the left-hand columns of Table 11.1. The facts to be noted are:

(1) Warming and drying in summer, June, July and August, are predominant.

(2) Warmer, but wetter conditions are noticeable in September, which is the most peculiar month in the year.

(3) Warming and dry status is found also in winter, January and February.

(4) Seasonal change of monthly tendencies from March, April to May occurs in an opposite way from October, November to December.

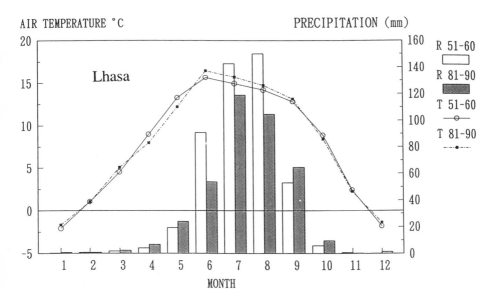

Figure 11.3. Same as Figure 11.1, but for Lhasa.

The related facts and/or estimated causes, referring to the previous studies (Luo *et al.*, 1983; Wang *et al.*, 1986; Zhu *et al.*, 1986), are listed on the right-hand side of Table 11.1. The conditions in summer and winter are considered to be related to the summer and winter monsoon activities during the 1950s and 1980s. Other months are transitional months. Thus, the changing conditions of cloud amount (longwave radiation cooling), transport of moist air from the Bengal Bay, and local convection, caused by the heated surface of the Plateau, could affect the various tendencies from the 1950s to the 1980s. It seems that the different heat balances caused by such warming/cooling and wetter/drier conditions from month to month resulted in the different tendencies. These should be further studied quantitatively.

TABLE 11.2. Air temperature and precipitation anomalies at
the four stations by month during El Niño and La Niña events.

	J	F	M	A	M	J	J	A	S	O	N	D
Lhasa												
Temperature (°C)												
El Niño	0.0	+0.1	-0.1	+0.4	+0.7	+0.2	+0.2	-0.1	-0.1	-0.8	-0.4	+0.2
La Niña	-0.1	0.0	+0.1	-0.3	-0.2	-0.2	-0.1	0.0	+0.1	+0.8	-0.2	-0.2
Precipitation (mm)												
El Niño	+1	0	0	+1	0	-1	+11	0	-5	+7	0	-1
La Niña	+1	0	0	+4	+2	+3	+1	-15	+10	-4	0	0
Changdou												
Temperature (°C)												
El Niño	+0.3	+0.3	0.0	-0.1	+0.3	-0.1	+0.3	-0.1	-0.4	-1.7	-0.1	+0.1
La Niña	-0.2	0.0	-0.2	-0.7	+0.2	-0.1	+0.1	-0.1	+0.5	+0.7	-0.3	-0.4
Precipitation (mm)												
El Niño	+1	0	-2	-3	-3	-4	-3	+14	-2	+5	-2	+1
La Niña	-1	-2	-1	+5	-1	+4	+5	-9	-8	-10	0	+1
Deqin												
Temperature (°C)												
El Niño	-0.3	-0.1	+0.2	-0.3	-0.2	-0.3	+0.5	-0.2	-0.3	-0.6	-0.1	+0.2
La Niña	+0.2	+0.4	-0.1	+0.4	+0.2	+0.3	+0.4	0.0	+0.4	+0.8	-0.2	-0.3
Precipitation (mm)												
El Niño	+3	+2	-5	+13	+5	-10	+9	+1	+1	-7	+1	+1
La Niña	-3	-9	+7	-4	-7	+7	+2	+2	-12	-12	+2	-2
Ganzi												
Temperature (°C)												
El Niño	+0.5	-0.1	0.0	-0.1	+1.0	-0.3	0.0	+0.2	-0.3	-1.1	0.0	+0.1
La Niña	-0.4	-0.1	-1.4	+0.2	-0.1	-0.1	+0.1	0.0	+0.4	+1.0	-0.2	-0.4
Precipitation (mm)												
El Niño	0	-1	-2	0	-8	-2	-10	-4	-9	-6	-2	+1
La Niña	+1	+1	+5	-4	+14	+8	+12	-5	+3	+1	-1	0
Southern Oscillation Index (SOI) (hPa)												
El Niño	2.7	2.6	3.4	1.7	1.3	0.5	-0.2	0.7	0.9	1.6	1.7	2.5
La Niña	5.6	5.9	4.7	2.8	2.1	1.8	1.5	3.0	3.9	3.8	3.9	4.5

TABLE 11.3. Summary of monthly air temperature and precipitation anomalies at the four stations in the Tibetan Plateau related to El Niño, La Niña and related phenomena.

			J+F	M+A+M	J+J+A+S	O+N+D
Lhasa						
Temperature:	monthly anomaly	El Niño	+0.1	+0.3	+0.1	-0.3
	(°C)	La Niña	-0.1	-0.1	-0.1	+0.1
Precipitation:	monthly anomaly	El Niño	+1	0	+1	+2
	(mm)	La Niña	+1	+2	0	-1
	seasonal total	El Niño	2	34	402	17
	(mm)	La Niña	2	34	396	7
Chandou						
Temperature:	monthly anomaly	El Niño	+0.3	+0.1	-0.1	-0.6
	(°C)	La Niña	-0.1	-0.2	+0.1	0.0
Precipitation:	monthly anomaly	El Niño	+1	-3	+1	+1
	(mm)	La Niña	-2	+1	-2	-3
	seasonal total	El Niño	7	60	372	39
	(mm)	La Niña	3	71	352	26
Deqin						
Temperature:	monthly anomaly	El Niño	-0.2	-0.1	-0.1	-0.2
	(°C)	La Niña	+0.3	+0.2	+0.3	+0.1
Precipitation:	monthly anomaly	El Niño	+3	+4	0	-2
	(mm)	La Niña	-6	-1	0	-4
	seasonal total	El Niño	35	166	379	71
	(mm)	La Niña	18	149	381	64
Ganzi						
Temperature:	monthly anomaly	El Niño	+0.2	+0.3	-0.1	-0.3
	(°C)	La Niña	-0.2	-0.1	+0.1	+0.1
Precipitation:	monthly anomaly	El Niño	-1	-3	-6	-2
	(mm)	La Niña	+1	+5	+4	0
	seasonal total	El Niño	11	109	426	48
	(mm)	La Niña	14	134	469	55
Indian rainfall*						
		El Niño	15	36	682	58
		La Niña	15	34	866	73
Southern Oscillation Index (SOI) (hPa)						
		El Niño	2.7	2.2	1.9	1.9
		La Niña	5.8	3.2	3.4	4.1

* Data from Parthasarathy *et al.* (1993)

11.3. Air Temperature and Precipitation Anomalies in the Cases of El Niño and La Niña

11.3.1. DATA ARRANGEMENT

It is of great importance to know whether the effects of El Niño/La Niña are negative or positive for the air temperature and precipitation anomalies at the stations located on the Tibetan Plateau. This is important not only because of the hydrological conditions of the Plateau from the standpoint of water resources, but also for the macro-processes of atmospheric circulations on the hemispheric scale.

In order to know its climatological relationships, the El Niño and La Niña periods (months) were taken from the table prepared by Tomita (1991). First, average monthly anomalies were calculated for each month in the cases of El Niño and La Niña respectively. The anomalies were obtained at the four stations, Lhasa, Changdou, Deqin and Ganzi, for monthly mean air temperature and monthly total precipitation. The numbers of months taken from Tomita's table are different month to month for the El Niño case and the La Niña case, but there were 12 to 17 months during the last 40 years, 1951-1990. The results of the calculated anomalies are given in Table 11.2. These are discussed below.

Next, seasonal changes of the anomalies are summarized. The seasons were divided as follows, taking into consideration the results in the previous part of the study: (i) January and February, (ii) March, April and May, (iii) June, July, August and September, and (iv) October, November and December. The results are shown in Table 11.3. For comparison, the precipitation amounts, including Indian rainfall, are also shown in Table 11.3. The values of Southern Oscillation Index (SOI) are shown at the bottom of Tables 11.2 and 11.3, to help in understanding the relationships between the El Niño/La Niña conditions and the air temperature precipitation anomalies.

11.3.2. DISTRIBUTION OF ANOMALIES

It is not enough simply to demonstrate the distribution of anomalies based upon the data at the four stations, and so this part of the study will discuss the distributions.

In Figure 11.4, the distribution of air temperature anomalies in the cases of El Niño and La Niña are shown for the respective seasons. In the case of El Niño, a positive anomaly dominates in mid-winter (January and February) and spring seasons (March, April and May), except at Diqin. At the eastern three stations, however, a negative anomaly appears and a positive anomaly remains only at Lhasa. A negative anomaly develops fully during October. In particular, an anomaly of -0.6 occurs at Changdou.

In the case of La Niña, the distribution is complex, but it does have a rather definite tendency. A negative anomaly develops at the northern three stations with a positive anomaly at Deqin located at the southeastern part of the study region, both in the mid-winter and spring seasons. The negative region retreats to the western part, i.e. Lhasa, and the positive region spreads over the eastern part represented at the three eastern

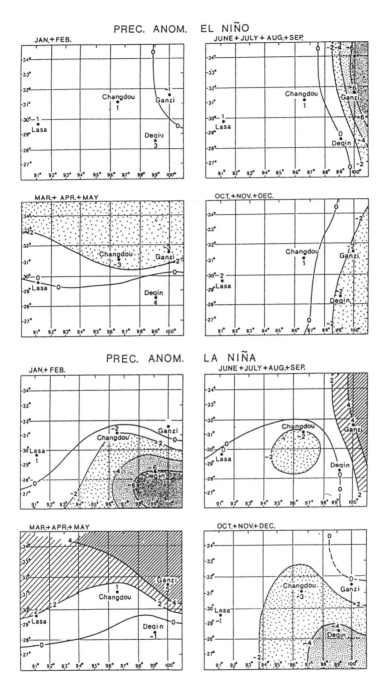

Figure 11.4. Distributions of air temperature anomaly (°C) in the case of El Niño (upper part) and in the case of La Niña (lower part).

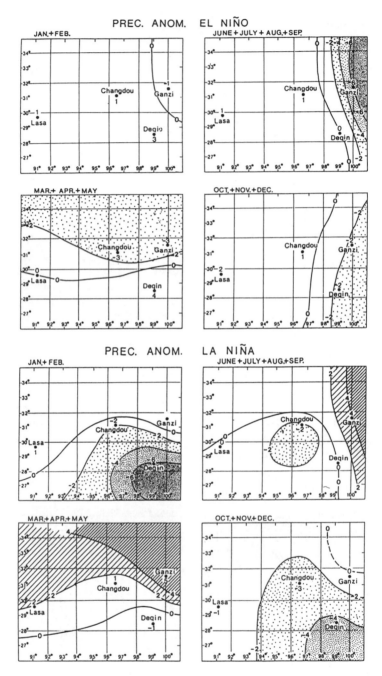

Figure 11.5. Distributions of precipitation anomalies (mm) in the case of El Niño (upper part) and in the case of La Niña (lower part).

stations in the La Niña case in June, July, August and September.

The positive anomaly covers the whole area in October, November and December. In conclusion, for the seasonal conditions, the following can be summarized: (i) the regional positive/negative anomaly conditions are roughly similar in mid-winter and the spring season; (ii) the positive/negative anomalies are different between Lhasa and the three eastern stations in the June, July, August and September season; (iii) the anomaly conditions are the same in the October, November and December season; (iv) the above-mentioned three conditions, (i), (ii) and (iii), are applicable both for the El Niño and La Niña cases; (v) at Lhasa, the western station, the positive/negative anomalies in winter and spring in relation to the El Niño/La Niña conditions continue until summer, i.e. the relationship is opposite only during October, November and December season; (vi) at Deqin, the negative/positive anomalies corresponding to the El Niño/La Niña cases appear in all seasons. This can be understood by the summer and autumn conditions continuing through to winter and spring.

The situation described in (v) can be attributed to the influence of atmospheric circulation systems, which are different between Lhasa and the eastern three stations. In other words, the position of Lhasa in the Tibetan Plateau causes different behavior at this station. The situation described in (vi) can be interpreted as due to the southerly flow all year around. These two are both caused by the topography of the Tibetan Plateau at the macro-scale.

Next, the distribution of precipitation anomalies shown in Figure 11.5 are dealt with. The negative anomalies appear in all seasons at Ganzi in the case of El Niño. This spreads over the northern part including Changdou in spring and the eastern part including Deqin in the October, November and December season in the case of the El Niño. This means that the tropospheric flow patterns, which cause precipitation, are weaker in the northeastern part during El Niño. In contrast, a positive anomaly dominates over most of the Plateau in winter, the southeastern part in spring and around Lhasa and Changdou in the second half of the year.

In the case of La Niña, a negative anomaly prevails at Changdou, the central part, and at Deqin, the southeastern part. This negative region is particularly enlarged in autumn and winter. The regions with positive anomaly in the case of La Niña appear at Lhasa and Ganzi in winter and this develops to cover the broad northern part in spring. In summer, the region becomes small occurring only at Ganzi, the northeastern part. This may be caused by the fact that the airflows bringing rainfall become inactive in the case of La Niña from the central to the southeastern part of the study region.

It must be stressed that the distribution of precipitation anomalies in the El Niño case are not completely opposite to the distributions in the La Niña case. They each show a unique distribution pattern.

11.3.3. COMBINATION OF AIR TEMPERATURE AND PRECIPITATION ANOMALIES

The combinations of monthly mean air temperature anomaly and monthly total precipitation anomaly were analyzed in this part of the study. First, the combinations were classified into four types, as defined in Table 11.4. That is, they are defined as negative or positive anomalies of air temperature and precipitation. Secondly, the frequencies of occurrence of Types A, B, C and D in the cases of El Niño and La Niña were counted from Table 11.2. The results are shown in Table 11.5.

TABLE 11.4. Definition of anomaly combination
types A, B, C and D.

		Air temperature anomaly*	
		Negative	Positive
Precipitation anomaly**	Negative	- - (Type A)	+ - (Type C)
	Positive	- + (Type B)	+ + (Type D)

* By monthly mean   ** By monthly total

It is very interesting to note that the frequencies of occurrence show a rather clear contrast between summer and winter, both in the El Niño and La Niña cases. The main points which should be noted are as follows:

(1) In the El Niño summer, Type A predominates, but in the El Niño winter, Type D.

(2) In contrast, in the La Niña summer, Type D predominates, but in the La Niña winter, Types A and B.

Because of this, it is suggested that we should estimate the air temperature and precipitation changes under global warming conditions separately for summer and winter.

The regionality of the monthly frequencies is as follows:

(1) In the case of El Niño, Type D dominates in Lhasa through the year. On the other hand, Type A dominates at the eastern station, Ganzi, and Type B at Deqin. There are no clear tendencies at Changdou, which is located at an intermediate position among the four stations under discussion.

(2) In the case of La Niña, Type B dominates at Lhasa and Ganzi throughout the year, but Type C at Deqin and Type A at Changdou. It is difficult to explain the causes of regionality at present, because of the sparse distribution of stations analyzed.

TABLE 11.5. Frequency of anomaly combination types, as defined in Table 4, for all four stations.

	Anomaly combination			
	Type A	Type B	Type C	Type D
El Niño				
Summer	8	4	3	2
Winter	4	3	3	6
La Niña				
Summer	2	4	3	7
Winter	7	7	2	0

* J+J+A+S   ** N+D+J+F

One of the problems to be studied further is a case study, for example, during the 1982/83 El Niño. The processes in such an extreme case (Arntz et al., 1991) should also be analyzed for the effect of the Tibetan Plateau. In the present study, only simultaneous relationships were considered. But the effect of snowcover over the Plateau on the summer monsoon in Asia is clear, as has been shown (Zhang et al., 1981), and the temperature increase in spring affects the onset date of summer monsoon in East Asia (Xu et al., 1985). Therefore, the lag correlation of El Niño/La Niña should be analyzed by taking into consideration the effect of the Tibetan Plateau in future studies.

11.4. Summary

In order to clarify climatic fluctuations in the Tibetan Plateau, climatic data at the four stations from 1951 to 1990 were analyzed as a preliminary approach. The following facts are of particular importance:
(1) Warming and drying tendencies are dominant in summer and in mid-winter.
(2) Seasonal changes in tendencies are opposite in spring and autumn.

(3) Negative/positive anomalies are similar at the three stations located in the eastern part of the Plateau, but different at Lhasa.

(4) The anomaly distributions are different from season to season. These can be attributed to the monsoonal circulations around the Plateau.

(5) Distributions of anomalies in air temperature and precipitation in the case of El Niño are not directly opposite to those in the case of La Niña.

(6) The combination of negative/positive anomalies of air temperature/precipitation revealed the following facts: a combination of anomalies with negative (air temperature) and negative (precipitation) dominated in the cases of El Niño summer and La Niña winter, but the combination of anomalies with positive (air temperature) and positive (precipitation) dominated in the cases of El Niño winter and La Niña summer.

(7) A combination of anomalies with negative (air temperature) and positive (precipitation) dominated in the case of La Niña winter.

(8) Lag correlation, interpretation through the atmospheric circulation systems and the physical processes causing the relationships mentioned above should be studied in future.

Acknowledgements

This study was partially supported by the Ministry of Education, Science and Culture (No. 05203108). Figures 11.1, 11. 2 and 11.3 were prepared by Dr. Du M.-y. and a chronological Table of El Niño and La Niña was provided by Mr. T. Tomita. I would like to express my sincere thanks to them for their assistance.

References

Academia Sinica, Integrated Scientific Study Group for Tibetan Plateau (1984) *Tibetan Climate*, Scientific Press, Beijing, 300 pp.

Arntz, W., and Fahrbach, E. (1991) *El Niño, Klimaexperiment der Natur*, Wissenschaftliche Buchgesellschaft, Darmstadt, 264 S.

Chen, Longxun *et al.* (1991) Preliminary analysis of climatic variation during the last 39 years in China, *Advances in Atmospheric Science* **8**, 3, 279-288.

Du, Mingyuan (1993) *Figures and Table 11.s of air temperature and precipitation variation in China*, Private publication, 645 pp.

Lin, Zhenyao, and Wu, Xiangdian (1984) Climatic change over the Qinghai-Xizang Plateau, in *Relationships between climate of China and Global Climate: Past, Present and Future*, Beijing International Symposium on Climate, Oct. 30-Nov. 3, 1984, I-14, 1-3.

Luo, Siwei, *et al.* (1983) The mean streamfield structures over the Qinghai Xizang Plateau and its neighbourhood in summer and winter, *Plateau Meteorology* **1**, 4, 60-73. (In Chinese with English abstract.)

Parthasarathy, B., Rupa Kumar, K., and Munot, A. A. (1993) Homogeneous Indian monsoon rainfall: Variability and prediction, *Proceedings Indian Academy of Science, Earth and Planetary Sciences* **102**, 1, 121-155.

Tomita, T. (1991) *The two successive-year El Niño and the single-year El Niño: an observational study*, Unpublished MS thesis, University of Tsukuba, Japan.

Tomita, T., and Yasunari, T. (1993) On the two types of ENSO, *J. Meteorological Society Japan* **71**, 2, 273-284.

Wang, Qiangian, *et al.* (1986) The effects of the QinghaiXizang Plateau on the mean summer circulation over East Asia, *Advances in Atmospheric Science* **3**, 1, 72-85.

Wu, Xiangdian, and Lin, Zhenyao (1981) Some characteristics of the climatic changes during the historical time of QinhaiXizang Plateau, *Acta Meteorologica Sinica* **39**, 1, 8997. (In Chinese with English abstract.)

Xu, Shuying *et al.* (1985) Seasonal change of heat conditions in the various regions and the summer monsoon activities, *Dili-Jikan* **15**, 105-114. (In Chinese.)

Yatagai, A. (1991) *Fluctuation of air temperature and precipitation during the 36 years, 1951-1986, in China*, Unpublished MS thesis, University of Tsukuba, Japan, 66 pp.

Zhang, Jiacheng (1991) *Climate of China. A series of climate for China*, Meteorological Publ., Beijing, 477 pp. (In Chinese with English abstract.)

Zhang, Lieting *et al.* (1981) Statistical analysis of early summer monsoon affected by the snow-cover in winter-spring over the Tibetan Plateau, in *Collected papers on Conference on Qing-zang Plateau Meteorology, 1977-1978*, 151-161. (In Chinese.)

Zhu, Qiangen, *et al.* (1986) A study of circulation differences between East Asian and Indian summer monsoons with their interaction, *Advances in Atmospheric Science* **3**, 4, 466-477.

Masatoshi Yoshino
Institute of Geography
Aichi University
Toyohashi, 441 Japan.

12 THE INFLUENCES OF THE NORTH ATLANTIC OSCILLATION, THE EL NIÑO/SOUTHERN OSCILLATION AND THE QUASI-BIENNIAL OSCILLATION ON WINTER PRECIPITATION IN IRELAND

S. DAULTREY

Abstract

Ireland's winter precipitation, while on average moderately high and very persistent, is highly variable. Three sources of this variability are examined: the North Atlantic Oscillation (NAO), the El Niño/Southern Oscillation (ENSO), and the Quasi-Biennial Oscillation (QBO). The effects of the NAO, ENSO and QBO, singly and jointly, on winter circulation patterns over the North America, the North Atlantic and Europe are described, with particular emphasis on their effects on the disposition of the westerlies and the tracks of extratropical cyclones, the principal sources of winter precipitation over Ireland.

Each winter (December-February) from 1957 to 1987 was categorised as to its phase of each of the NAO, ENSO and QBO, and from twelve synoptic stations Ireland's precipitation was categorised as above average, average or below average. Contingency tables of precipitation and oscillation phase were produced. These indicate that there was above average precipitation when the NAO was well positive (strong westerlies), below average precipitation when the ENSO was well positive ('High/Dry', but a more conventional ENSO classification yielded a much weaker association) and a tendency for above average precipitation during sunspot maxima particularly in the northern half of Ireland in QBO West phase. When ENSO was not well positive and during sunspot minima, precipitation in the north and west was above average when QBO phase was East and below average when QBO phase was West.

12.1. Introduction

The climate of Ireland is characterised by a lack of extremes: winters are mild and summers are cool; precipitation is moderate, but the number of rain days is relatively high. These characteristics derive from the predominance of oceanic influences. The interannual variability of temperatures is low, particularly in winter, but that of precipitation is quite high. The average water balance in the east shows moderate winter surpluses and small summer deficits, and in the west shows large winter surpluses and near zero summer deficits. Interannual variations in the water balance occur principally as a result of variations in precipitation, and these can be critical for water supply, particularly when a dry winter limits reservoir recharge in the more densely populated east of the country. These variabilities are also a result of oceanic influences, and it is

J.A.A. Jones et al. (eds.), Regional Hydrological Response to Climate Change, 213–236.
© 1996 *Kluwer Academic Publishers. Printed in the Netherlands.*

the purpose of this paper to investigate the association between winter precipitation in Ireland and three sources of variability in the atmospheric circulation over the North Atlantic Ocean, namely the North Atlantic Oscillation, the El Niño/Southern Oscillation and the Quasi-Biennial Oscillation of equatorial stratospheric winds.

12.2. The North Atlantic Oscillation (NAO)

The North Atlantic Oscillation (henceforth NAO) was first documented by Walker (1924) and Walker and Bliss (1932), and has more recently been investigated by van Loon and Rogers (1978), Rogers and van Loon (1979), Meehl and van Loon (1979), Rogers (1984) and Lamb and Peppler (1987). The extremes of the NAO were originally seen as the tendency for Greenland and Labrador to experience below average winter temperatures while those of northern Europe were above average ('Greenland Below'), and vice versa ('Greenland Above'), although some winters showed similar departures from the mean on both sides of the North Atlantic and were characterised as 'Both Below' and 'Both Above'. The 'Greenland Below' winters show strong westerly flow across the North Atlantic, while in 'Greenland Above' winters the westerly flow is weak. This allows the NAO to be more simply described in terms of the difference in anomalies of winter sea level pressure between the Azores and Iceland, a positive difference equating with 'Greenland Below' conditions and a negative difference with 'Greenland Above'.

The maps of van Loon and Rogers (1978) show Ireland to have above average sea level pressure in both 'Greenland Below' and 'Greenland Above' winters and a zero difference in pressure between the two types of winter. However, Rogers (1984), using Azores-Iceland sea level pressure anomalies to define the NAO, shows Ireland to have slightly higher sea level pressure during NAO positive winters and much higher 500 mb heights. Rogers and van Loon (1979) show that the North Atlantic immediately west of Ireland (10°W to 30°W) experiences fewer lows and fewer highs in 'Greenland Below' Januaries than in 'Greenland Above' Januaries. In both types of January the peak number of lows occurs at 60°N, to the north of Ireland, but in 'Greenland Above' Januaries there is a secondary peak at 35°N. In 'Greenland Below' Januaries the peak number of highs is observed from 35°N to 45°N, whereas in 'Greenland Above' Januaries there is a much higher peak at 35°N. Therefore the atmospheric circulation in the vicinity of Ireland shows slightly fewer but more intense depressions (and their fronts) passing to the north of the country in 'Greenland Below' winters, with weaker depressions passing to the north of Ireland and well to the south in 'Greenland Above' winters. This should imply more precipitation over and to the north of Ireland in 'Greenland Below' winters and more precipitation well to the south in 'Greenland Above' winters. Rogers and van Loon (1979) show the latter, and Lamb and Peppler (1987) explore this association for Morocco, but while the former authors show increased precipitation to the north of Ireland in 'Greenland Below' winters, they indicate decreased precipitation over Ireland.

12.3. The El Niño/Southern Oscillation (ENSO)

The Southern Oscillation component of the El Niño/Southern Oscillation (henceforth ENSO) was also documented by Walker (1924) and Walker and Bliss (1932), although it was well known before. It refers to a major transfer of atmospheric mass between the eastern and western Pacific Ocean which was first recognised by Bjerknes (1966) as being intimately linked with the southward extension of the austral summer El Niño warm current along the coast of Peru and the subsequent development of large areas of high positive sea surface temperature anomalies in the central and eastern equatorial Pacific Ocean (Rasmusson and Carpenter, 1982; Philander and Rasmusson, 1986; Graham and White, 1988). The usual measure of the state of ENSO is the Southern Oscillation Index (SOI), the diference in standardised sea level pressure between Tahiti and Darwin, Australia (Wright, 1984, 1989). Conditions vary between a high (positive) SOI, when pressure is high over the South Pacific, low over northern Australia and Indonesia, the southeasterly trades are strong and the eastern tropical South Pacific is generally dry ('High/Dry'), and a low (negative) SOI when pressure drops over the South Pacific, rises over northern Australia and Indonesia, the southeast trades weaken and in association with the spread of positive sea surface temperature anomalies, rainfall becomes more frequent in the eastern tropical South Pacific ('Low/Wet'). The ENSO cycle varies between 2 and 8 years with an average of about 3.5 years.

Not only does the climatic pattern of the whole tropical South Pacific change, but so too do aspects of all tropical climates (Yarnal and Kiladis, 1985) and of many extratropical climates (Yarnal, 1985; Hamilton, 1988). The Northern Hemisphere circumpolar vortex tends to contract following above average eastern equatorial Pacific sea surface temperatures (Angell, 1992), indicating general warming. Another major Northern Hemisphere response is for the upper westerly Rossby waves to increase in amplitude from the central Pacific to the western Atlantic following a 'Low/Wet' ENSO phase, with an intensified Aleutian low pressure zone, higher pressure with anomalously warm and dry conditions over western North America and lower pressure and anomalously cold and wet conditions over southeastern USA, a pattern known as Pacific/North American (PNA). Hamilton (1988) showed that this PNA pattern extended into the North Atlantic, with a negative sea level pressure anomaly centred at about 50°N 30°W, off the west coast of Ireland. This agreed with the earlier work of Angell and Korshover (1984), who related warm sea surface temperatures in the eastern equatorial Pacific (indicative of 'Low/Wet' ENSO phase) to below average central pressures of the Aleutian Low, Pacific High and Atlantic High and a southward displacement of the Icelandic Low. In their analysis of the displacements of the Northern Hemisphere circumpolar vortex at 300 mb the same authors (Angell and Korshover, 1985) show that the vortex is displaced toward Europe in seasons when eastern equatorial sea surface temperatures are below average (indicative of 'High/Dry' ENSO phase), but displaced toward the Western Hemisphere when sea surface temperatures are above average. This, they claim, is in accord with their earlier findings. Rasmusson and Mo (1993) present a detailed analysis of the period 1986-1989, covering the 1986 'Low/Wet' and the 1988 'High/Dry' ENSO phases. They show the winter following the 'Low/Wet' phase to have a southward displacement of depression tracks across the North

Atlantic and increased precipitation from the Gulf of Mexico northeast toward Ireland compared to the winter following the 'High/Dry' phase. The effects of ENSO over Europe are less regular and less pronounced, but Fraedrich (1990) showed that in central Europe the winters following a 'Low/Wet' ENSO phase have more cyclonic and less anticyclonic Grosswetter situations than those following a 'High/Dry' phase, and that the former are far more variable one to the next. Further analysis by Fraedrich and Müller (1992) demonstrated that following a 'Low/Wet' ENSO phase winter cyclone frequency is higher from southern Greenland to Ireland and thence across the North Sea and into the Baltic. In the winters following a 'High/Dry' ENSO phase cyclone frequency is higher from southern Greenland south of Iceland and into the Norwegian Sea. Hence winters in Ireland following a 'Low/Wet' ('High/Dry') ENSO phase should be slightly warmer (slightly colder) than average with higher (slightly lower) than average precipitation.

12.4. The Quasi-Biennial Oscillation (QBO)

The Quasi-Biennial Oscillation of equatorial stratospheric wind direction (henceforth QBO) is a more recently discovered atmospheric phenomenon, first reported by Ebdon (1961), Reed et al. (1961) and Veryard and Ebdon (1961). Since then there have been published many descriptions and updates (Ebdon, 1971; Holton and Tan, 1980, 1982; Labitzke, 1982; Naujokat, 1986), and there was a resurgence of interest in quasi-biennial weather cycles (Craddock, 1968). Although Berson and Kulkarni (1968) speculated about an association between the QBO and the double (Hale) 22-year sunspot cycle, it has only very recently been suggested that the QBO might modulate the relationship of the 11-year sunspot cycle to weather patterns (Labitzke, 1987; Labitzke and van Loon, 1988, 1989; Tinsley, 1988; van Loon and Labitzke, 1988, 1990; Barnston and Livezey, 1989, 1991; Venne and Dartt, 1990). The normal pattern of circulation in the middle and upper stratosphere is westerly in the winter hemisphere and easterly in the summer hemisphere, implying a transition from westerly to easterly at or near to the equator. Observations since the 1950s have shown that the transition moves from north of the equator to south and back in a very regular cycle with a period of 2.2 years. Hence the equatorial stratospheric wind varies from easterly to westerly and back approximately every two years, a Quasi-Biennial Oscillation with East and West phases. Labitzke (1987) showed that when the QBO is in West phase polar stratospheric warmings only occur when sunspot numbers are high, whereas in the East phase these warmings could occur at any sunspot number. Labitzke and van Loon (1988) substituted 10.7 cm solar flux for sunspot number and showed that in the East phase stratospheric warmings predominantly occur at low flux. Brier and Hanson (1989) used a non-parametric statistical test to confirm these findings. However, when Flueck and Brown (1993) reanalysed the original data and updated them to 1989, they found that up to 1972 only the West phase positive association between temperature and solar flux existed, but thereafter there was also a negative association in East phase.

Van Loon and Labitzke (1988) examined the associations of Northern Hemisphere surface temperature, sea level pressure and 700 mb height with winter 10.7 cm solar flux. For West phase winters strong positive correlations with sea level pressure were observed from the North Pole through North America to the Gulf of Mexico, with negative correlations in the subtropical North Atlantic and North Pacific Oceans; in the East phase winters the patterns tended to be reversed. They remark that 'In the east phase the main signal in the pressure is in an area centred on the Pacific Ocean, whereas in the west phase little goes on in the Pacific in comparison with North America and the Atlantic region.... The temperature signals show similar differences with respect to the two regions' (van Loon and Labitzke, 1988, 912). Their analysis of the 700 mb data indicated that in the West phase westerlies across North America into the North Atlantic must be appreciably weaker in solar maxima than in solar minima, with an opposite but less pronounced effect during East phase winters. Angell (1992) found a weak and non-significant correlation between the size of the Northern Hemisphere circumpolar vortex at 300 mb, with a generally contracted vortex in West phase except in the quadrant 90°E-180° where it was expanded. When the record contraction of the vortex in the high sunspot number East phase winter of 1989 was excluded, the vortex tended to be expanded at low sunspot numbers regardless of QBO phase, but at high sunspot numbers the vortex tended to be expanded in East phase and contracted in West phase. Tinsley (1988) showed that the mean latitude of winter storm tracks north of 50°N in the North Atlantic varies inversely with 10.7 cm solar flux by 6 degrees of latitude in the West phase, the opposite to what might be expected from the findings of Angell (1992); the relationship in the East phase is less clear. The implications for precipitation over Ireland are that the precipitation increase in West phase winters that should occur during solar minima with the intensified circulation could well be offset or even reversed by the poleward migration of the storm tracks.

12.5. Interrelationships between NAO, ENSO and QBO

The association between the NAO and the ENSO has been explored by Rogers (1984), who found a cospectral maximum between indices of the two at periodicities between 5 and 6 years. The association is clouded by the fact that the NAO index shows a decrease from the 1920s to the 1960s as the Icelandic low filled and the Azores high increased, but he noted a tendency of strong westerlies (a positive NAO index) to coincide with 'High/Dry' ENSO phases and for weak westerlies to occur during a 'Low/Wet' phase. Only over southeastern USA and the western North Atlantic is there a strong climate signal from both oscillations.

   Discussion of possible relationships between the NAO and QBO effects is limited. Tinsley (1988) observes that the NAO indices of van Loon and Rogers (1978) appear to show periodicities at both the QBO and solar cycle time scales: this is certainly the case for the 2.2 years of the QBO; but it is much less obvious for the 10-12 years of the sunspot cycle. Barnston and Livezey (1989) found that the spatial pattern of correlation between solar flux and 700 mb height during QBO West phase winters is well described by an unrotated principal component which is a combination of two modes of circulation as described by rotated principal components analysis (Barnston and Livezey, 1987). The first mode is the tropical/Northern Hemisphere (TNH) circulation pattern, with strong positive anomalies centred over eastern Canada and less strong negative anomalies centred over the subtropical eastern North Pacific, the subtropical western North Atlantic, Mongolia and eastern Europe. The second mode is the NAO circulation pattern counterposing Greenland/Iceland and the Azores. The Eurasian No. 1 (EU1) pattern (positive anomalies centred over Russia flanked by negative anomaly centres over central Siberia and southwest Europe) also shows significant association. In QBO East phase winters the EU1 pattern shows reversed and significant associations, while the NAO shows reversed and non-significant associations. Also important is the East Pacific (EP) circulation pattern of positive anomalies centred over northern Alaska and negative anomalies in the subtropical eastern North Pacific. These findings agree with the observation of van Loon and Labitzke (1988) that North America and the North Atlantic more clearly show QBO West phase responses.

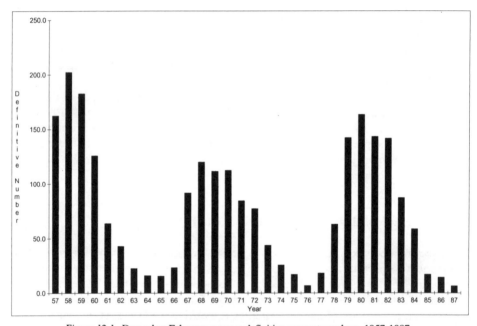

Figure 12.1. December-February average definitive sunspot numbers, 1957-1987.

The association between ENSO and QBO effects came under scrutiny from Barnston and Livezey (1991) when the expected QBO/climate relationship failed: with a high solar flux and the QBO in West phase, the normal tropospheric pattern of a positive temperature anomaly centred over the Canadian Arctic and negative anomalies centred over the subtropical eastern North Pacific, the subtropical western North Atlantic, the Baltic and northern China was reversed. The ENSO was in a 'High/Dry' phase, the first occasion since reliable stratospheric wind data have been available that this coincided with high solar flux in a QBO West phase. The suggestion was that the atmospheric temperature response was forced not by the QBO/solar association but by the ENSO phase. Barnston *et al.* (1991) extended this analysis, and suggested that the winter circulation responses to extremes of ENSO are modulated by the QBO. In QBO East phases the circulation anomalies correspond to the TNH and also to the West Pacific Oscillation (WPO) patterns of Barnston and Livezey (1987), with a warm anomaly over eastern North America and cold anomalies centred over southwestern USA and the central North Atlantic during periods with a negative SOI, and a reversed pattern during winters with a positive SOI (although still with cold anomalies over the central North Atlantic). In QBO West phases the circulation anomalies correspond to the PNA pattern, with strong cold anomalies centred over the Aleutians and southeastern USA during negative SOI winters, and strong warm anomalies centred over the subtropical western North Atlantic and western Russia and a strong cold anomaly centred over the Canadian Arctic. Figures 12.8b and 12.8d of Barnston *et al.* (1991) respectively suggest that during ENSO 'High/Dry' events cyclone tracks should be displaced well to the north of Ireland in QBO West phase and well to the south in QBO East phase, the former supported by Fraedrich and Müller (1992) but the latter not. Their Figures 12.8a and 12.8c both suggest weakened North Atlantic westerlies during ENSO 'Low/Wet' events, regardless of QBO phase and in accord with Rogers (1984).

12.6. Data and Methods

Winter phase of the QBO was determined up to 1985 by inspecting the equatorial stratospheric wind data of Naujokat (1986). These were found to be in accord with the phase identifications of Barnston and Livezey (1989) and of Flueck and Brown (1993), but not with those of Labitzke (1987), so the first authors' identifications were used after 1985. Data on 10.7 cm solar flux were not available to the author, but definitive sunspot numbers were (A. Koeckelenbergh, Sunspot Index Data Center, Brussels). December-February mean sunspot numbers were calculated and are shown for 1957-1987 in Figure 12.1. Initially, Labitzke (1987) was followed and sunspot numbers of at least 110 were classified as being High. This gave the 1965-75 cycle only three High values, so the threshold sunspot number was lowered to 80; numbers below 80 were classified as Low. The QBO phase and level of solar activity for each year are listed in Table 12.1.

Indices of the NAO were constructed from data derived from the Chadwyck-Healey and Climatic Research Unit, University of East Anglia, *World Climate Disc*. Following Rogers (1984) December-February mean sea level pressures for the Azores and Iceland were standardised about their 1895-1980 normals and the differences calculated. Figure 12.2 shows these data for 1957-1987; they are also shown in Table 12.1 categorised into positive (+) or negative (-) if the absolute value is at least unity, or intermediate (o) if it is not. Following van Loon and Rogers (1978), December-February mean temperatures for Jacobshavn (until 1970) and Egemende (after 1970), Greenland, and Oslo, Norway, were expressed as anomalies from the 1895-1980 normals. If the Greenland and Oslo anomalies were of opposite sign and the difference between the two was at least 4°C, the winter was categorised as 'Greenland Above' (GA) or 'Greenland Below' (GB); when the anomalies were of the same sign and each had an absolute value of at least 1°C, the winter was categorised as 'Both Above' (BA) or 'Both Below' (BB). These categorisations are shown in Table 12.1.

TABLE 12.1. Classification of each year 1957-1987 according to phase of
the QBO, sunspot numbers, Azores-Iceland index of the NAO, Greenland/
Oslo index of the NAO, Barnston and Livezey (1991) index of ENSO (SOI)
and the ENSO classification of Fraedrich (1990). For explanations, see text.

Year	QBO Phase	Sunspot Number	NAO Index	Year	SOI	ENSO
1957	East	High	+		o	o
1958	West	High	o		-	L/W
1959	East	High	-	GA	-	o
1960	West	High	-		o	o
1961	East	Low	+		o	o
1962	West	Low	o		+	o
1963	East	Low	-	GA	+	o
1964	West	Low	-		-	L/W
1965	West	Low	-	BA	o	H/D
1966	East	Low	-		-	L/W
1967	West	High	o		+	H/D
1968	West	High	-		+	o
1969	East	High	-	GA	-	o
1970	West	High	o	BB	-	L/W
1971	East	High	o	GB	+	H/D
1972	West	Low	o	GB	+	H/D
1973	East	Low	+	GB	-	L/W
1974	West	Low	+		+	H/D
1975	East	Low	+	GB	o	o
1976	West	Low	o		+	H/D
1977	East	Low	-	GA	o	L/W
1978	West	Low	-		-	o
1979	West	High	-	GA	o	H/D
1980	East	High	o	GA	-	o
1981	West	High	+		o	o
1982	East	High	-		o	o
1983	West	Low	+	GB	-	L/W
1984	East	Low	+	GB	o	o
1985	East	Low	-		+	o
1986	West	Low	-	GA	o	o
1987	East	Low	-	BB	-	L/W

The ENSO phase was described by two classifications. The first derived from an SOI
index constructed by Barnston and Livezey (1991). Those authors combined two
standardised indices: the frequently used Tahiti-Darwin sea level pressure difference,
but calculated as (Jan +Feb)/3 + (Dec + Mar)/6, and the Jan + Feb mean sea surface
temperature (multiplied by -1) for the equatorial Pacific, 10°S-10°N, 120°-160°W.

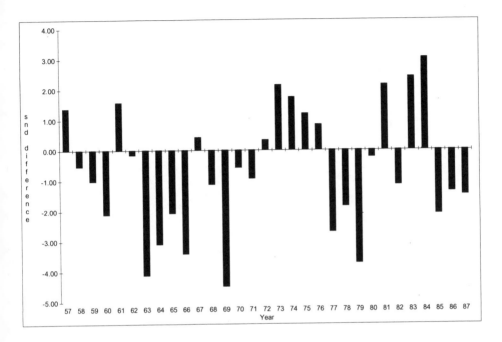

Figure 12.2. Azores-Iceland standardised sea level pressure difference index of the NAO, 1957-1987.

Following Barnston *et al.* (1991), absolute values of 0.5 or greater were categorised as positive (+) or negative (-), while other values were categorised as intermediate (o); the categorisations are shown in Table 12.1. The second classification was taken from Fraedrich (1990), who classified the December-February period according to its ENSO phase in the year of the December. He took his 'Low/Wet' years from Rasmussen and Carpenter (1983), adding 1982 and 1986; his 'High/Dry' years were taken from van Loon and Shea (1985), adding 1975. These categorisations are shown (thrown forward to the year of the January) in Table 12.1 as H/D, L/W or o (neither).

Precipitation data were the December-February totals for eleven synoptic stations in the Republic of Ireland, supplied by the Irish Meteorological Service, and for Belfast (Aldergrove), taken from the above-mentioned *World Climate Disc*. The locations of these twelve stations are shown in Figure 12.3. The period, 1957-1987, was chosen not for this exercise but for a parallel study using daily precipitation, temperature and sunshine data. For each station the seasonal total was expressed as a standardised anomaly from the 1958-87 mean (winter 1957 data were not available for Kilkenny and Malin Head); one such series, that for Belmullet, is shown as Figure 12.4. The seasonal anomalies were then categorised as: well above average (++), a value of more than +1.5;

Figure 12.3. Locations of the twelve stations used in the analysis.

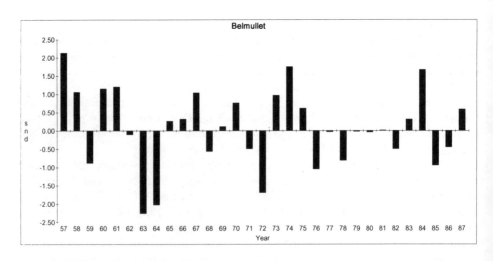

Figure 12.4. December-February precipitation, Belmullet, normalised about 1958-87 mean of 313 mm
and standard deviation of 72 mm.

above average (+), a value from +0.44 to +1.5 inclusive; average (o), a value between +0.44 and -0.44, which should yield 33% of the items in a Normal distribution; below average (-), a value from -0.44 to -1.5 inclusive; and well below average (--), a value of less than -1.5. These categorisations are shown in Table 12.2, together with an overall categorisation for Ireland.

TABLE 12.2. Classification of December-February precipitation for each station and for Ireland as a whole for each year 1957-1987. (++ > +1.5σ; +1.5σ > + > +0.44σ; +0.44σ > o > -0.44σ; -0.44σ > - > -1.5σ; -- < -1.5σ.)

Yr	MH	BA	Bm	Cn	Cm	DA	Br	SA	Kk	Rl	VO	RP	Irl
1957		o	++	+	++	o	+	+		o	o	+	+
1958	++	+	+	+	o	o	+	o	+	+	o	+	+
1959	-	-	-	-	-	-	-	-	-	-	-	-	-
1960	+	+	+	+	++	+	++	+	+	+	+	+	+
1961	+	o	+	+	+	+	++	++	+	o	+	o	+
1962	o	-	o	o	o	-	-	o	-	-	-	-	-
1963	--	-	--	-	--	o	--	--	-	-	-	-	-
1964	--	--	--	--	--	--	--	--	--	--	--	-	-
1965	o	o	o	o	o	o	+	o	o	o	-	-	o
1966	o	+	o	++	+	++	+	+	+	++	++	++	+
1967	+	+	+	+	-	+	+	o	o	-	-	-	o
1968	+	o	-	o	-	-	-	-	--	-	-	-	-
1969	o	o	o	+	+	+	+	+	+	+	o	+	+
1970	+	+	+	+	+	o	o	+	+	+	o	o	+
1971	-	-	-	--	-	-	-	-	-	-	-	-	-
1972	-	-	--	-	-	o	-	-	o	+	o	o	-
1973	o	o	+	o	+	-	-	o	o	-	+	o	o
1974	+	o	++	+	+	o	+	+	++	++	++	++	+
1975	o	o	+	o	+	-	+	o	o	-	-	-	o
1976	-	-	-	-	-	-	o	-	--	--	--		-
1977	o	+	o	+	o	+	o	o	o	+	+	++	+
1978	o	o	-	-	o	o	o	o	o	+	o	+	o
1979	o	++	o	o	o	++	o	o	+	+	+	+	+
1980	o	+	o	+	+	++	+	+	+	+	+	+	+
1981	+	o	o	o	o	-	o	o	-	-	-	-	o
1982	-	o	-	o	o	o	o	-	o	+	+	o	o
1983	+	-	o	o	+	o	o	+	o	o	o	o	o
1984	++	++	++	++	+	+	++	++	++	o	+	+	+
1985	-	-	-	-	-	o	-	-	-	o	o	-	-
1986	o	-	-	-	o	o	-	-	-	-	-	-	-
1987	+	-	+	-	o	-	o	o	-	o	o	+	o

(MH - Malin Head; BA - Belfast Aldergrove; Bm - Belmullet; Cn - Clones; Cm - Claremorris; DA - Dublin Airport; Br - Birr; SA - Shannon Airport; Kk - Kilkenny; Rl - Rosslare; VO - Valentia Observatory; RP - Roches Point; Irl - Ireland)

12.7. Results

12.7.1. ASSOCIATIONS BETWEEN NAO, ENSO, QBO AND IRELAND PRECIPITATION

None of the data series exhibited any statistically significant linear trend over time. Therefore in the first instance the categorisations of Ireland's precipitation (reduced to three categories) were cross-tabulated with the seasonal circulation classifications, using the FREQ procedure in SAS version 6 to derive descriptive statistics (SAS Institute, 1990). The study is most definitely an exploratory one, attempting only to identify possible associations. Therefore the descriptive statistics derived from the contingency tables, and their associated probabilities of occurrence under a null hypothesis of no association, are regarded as merely indicative of the possible existence of relationships (Flueck and Brown, 1993).

TABLE 12.3. Association between the Azores-Iceland index of the North Atlantic Oscillation and Ireland precipitation, December-February, 1957-1987.

Ireland precipitation	North Atlantic Oscillation			Total
	Negative (-)	Intermediate (o)	Positive (+)	
Above average (+)	5	3	4	12
Average (o)	4	1	4	9
Below average (-)	6	4	0	10
Total	15	8	8	31

Table 12.3 shows the association of the NAO index and winter precipitation. While any precipitation category was equally likely for negative and intermediate indices of the NAO, there was no occurrence of below average precipitation when the NAO was positive. Fisher's Exact Test yields a probability of occurrence under a null hypothesis of no association of 0.174. The Cochran-Mantel-Haenzsel (CMH) statistic against an alternative hypothesis of general association gives a similar result, with a probability of 0.213. However, the Uncertainty Coefficient Symmetric (UCS) is 0.124 with an asymptotic standard error (ASE) of 0.049, indicating a probability of occurrence under a null hypothesis of no association of 0.010.

A much more striking winter precipitation association is revealed by Table 12.4. This shows that there was only one winter with above average precipitation and a positive ENSO phase, and only two winters with below average precipitation and a negative ENSO phase. This association is reinforced by the fact that only 1 winter of 11 with an intermediate SOI showed below average precipitation. Fisher's Exact Test yields a probability of occurrence under a null hypothesis of no association of 0.025; the probability of the same from the CMH statistic is 0.018; and from the UCS statistic it is 0.051. However, this relationship is almost completely lost when a different classification of ENSO phase is used, that derived from Fraedrich (1990). The only evidence to be seen for it in Table 12.5 is that there is just one winter with below average precipitation associated with a 'Low/Wet' phase, and less convincingly that there are just two above average precipitation winters associated with a 'High/Dry' phase. The probability associated with the null hypothesis from Fisher's Exact Test is 0.745, from the CMH statistic is 0.731 and from the UCS is 0.599.

TABLE 12.4. Association between January-February Barnston and Livezey (1991) index of the Southern Oscillation and Ireland December-February precipitation, 1957-1987.

| Ireland | Southern Oscillation Index | | | |
precipitation	Negative (-)	Intermediate (o)	Positive (+)	Total
Above average (+)	5	6	1	12
Average (o)	4	4	1	9
Below average (-)	2	1	7	10
Total	11	11	9	31

The relationships between winter precipitation, sunspot number and QBO phase revealed by Table 12.6 are less definite. There appears to be an association between high sunspot number and high precipitation regardless of QBO phase in that 7 of the 14 high sunspot winters showed above average precipitation and only 3 recorded below average precipitation, although of the 12 above average precipitation winters, 7 had high sunspot numbers and 5 had low numbers. There is a suggestion that when sunspot numbers were low winter precipitation was modulated by QBO phase: only 2 of 9 East phase winters had below average precipitation whereas only 1 of 8 West phase winters had above average precipitation. However the probability of these frequencies arising by chance under the null hypothesis from Fisher's Exact Test is 0.338. There is also some evidence for a clearer relationship between sunspot number and precipitation during West phase winters: only 1 of 8 low sunspot winters had above average precipitation and only 1 of 8 high sunspot winters had below average precipitation. Fisher's Exact Test gives the probability of these frequencies occurring under the null hypothesis of no association as 0.145.

The details of the pairwise associations between the three atmospheric oscillations are shown in Tables 12.7-12.9. The associations of NAO phase with ENSO phase are shown in Table 12.7. No winter was categorised as Intermediate on both oscillations, and an intermediate score on one was associated with a positive or negative score on the other of a roughly equal likelihood. The probability of no association from Fisher's Exact test is 0.053, from the CMH statistic is 0.064 and from the UCS is 0.010. When the NAO was negative, ENSO was unlikely to be positive; when the NAO was positive, ENSO was most likely to be intermediate; when ENSO was negative, NAO was likely to be negative

TABLE 12.5. Association between the classification of ENSO events used by Fraedrich (1990) and Ireland December-February precipitation, 1957-1987.

Ireland precipitation	El Niño/Southern Oscillation			Total
	Low/Wet	Neither	High/Dry	
Above average (+)	4	6	2	12
Average (o)	3	4	2	9
Below average (-)	1	6	3	10
Total	8	16	7	31

TABLE 12.6. Association between sunspot number, phase of the Quasi-Biennial Oscillation and Ireland precipitation, December-February, 1957-1987.

Ireland precipitation	Sunspot Number				Total	
	Low		High			
	Quasi-Biennial Oscillation					
	East	West	East	West	East	West
Above average (+)	4	1	3	4	7	5
Average (o)	3	2	1	3	4	5
Below average (-)	2	5	2	1	4	6
Total	9	8	6	8	15	16

and unlikely to be positive; when ENSO was positive, NAO was likely to be intermediate. Table 12.7b shows the precipitation categorisations of the 31 winters, and
reveals a strong NAO/ENSO association with Ireland winter precipitation: when ENSO was positive and NAO was not positive, precipitation was never above average; when ENSO was not positive and NAO was not negative, precipitation was never below average. The CMH statistic gives a probability of 0.037 for the null hypothesis that there is no NAO/ENSO/precipitation relationship.

TABLE 12.7. Association between December-February Azores-Iceland index of the North Atlantic Oscillation and January-February Barnston and Livezey (1991) index of the Southern Oscillation, 1957-1987. (a) Simple frequencies. (b) Cell frequencies categorised as above average (wet), w, average, a, or below average (dry), d, winter precipitation.

(a) Southern Oscillation	North Atlantic Oscillation			Total
	Negative (-)	Intermediate (o)	Positive (+)	
Positive (+)	3	5	1	9
Intermediate (o)	6	0	5	11
Negative (-)	6	3	2	11
Total	15	8	8	31

(b) Southern Oscillation	North Atlantic Oscillation			Total
	Negative (-)	Intermediate (o)	Positive (+)	
Positive (+)	ddd	w a dddd		9
Intermediate (o)	www aa d		www aa	11
Negative (-)	ww aa dd	www	aa	11
Total	15	8	8	31

Table 12.8 indicates a slight propensity for negative rather than positive NAO phases to have occurred with high sunspots, regardless of QBO phase. Also, half the positive NAO phases occurred with low sunspots in QBO East phase. However the overall CMH statistic gives a probability of occurrence of these frequencies under the null hypothesis of no association as 0.605. Table 12.8b suggests a sunspot-precipitation relationship when the NAO is not positive and the QBO is in West phase: when sunspot numbers were low precipitation was never above average; when sunspot numbers were high only winter had below average precipitation. When precipitation is included in the crosstabulations, the probability of the null hypothesis of no association, as derived from the CMH statistic, falls to 0.283.

Finally, Table 12.9 reveals little or no association between sunspot number, QBO phase and ENSO phase, with the CMH statistic having a probability of occurrence under the null hypothesis of 0.587. There is, however, a slight suggestion from Table 12.9b of a sunspot-precipitation relationship when ENSO is not positive and the QBO is in West phase. When sunspot numbers were low no winter had above average precipitation, and when sunspot numbers were high no winter had below average precipitation. There is also the suggestion from Table 12.9b of a QBO modulation of the ENSO/precipitation association when ENSO was not positive and sunspots were low. Under these conditions no East phase winter recorded below average precipitation while no West phase winter recorded above average precipitation. The CMH statistic gives a probability of no association between ENSO phase, QBO phase, sunspot number and precipitation as 0.041.

TABLE 12.8. Association between sunspot number, phase of the Quasi-Biennial Oscillation and Azores-Iceland index of the North Atlantic Oscillation, December-February, 1957-1987.
(a) Simple frequencies. (b) Cell frequencies categorised as above average (wet), w, average, a, or below average (dry), d, winter precipitation.

(a)

North Atlantic Oscillation	Sunspot Number				Total	
	Low		High			
	Quasi-Biennial Oscillation Phase					
	East	West	East	West	East	West
Positive (+)	4	1	1	2	5	3
Intermediate (o)	0	3	2	3	2	6
Negative (−)	5	4	3	3	8	7
Total	9	8	6	8	15	16

(b)

North Atlantic Oscillation	Sunspot Number				Total	
	Low		High			
	Quasi-Biennial Oscillation Phase					
	East	West	East	West	East	West
Positive (+)	ww aa	w	w	aa	5	3
Intermediate (o)		ddd	w d	ww a	2	6
Negative (−)	ww a dd	aa dd	w a d	ww d	8	7
Total	9	8	6	8	15	16

12.7.2. ASSOCIATIONS BETWEEN NAO, ENSO, QBO AND STATION PRECIPITATION

Table 12.2 shows there to be considerable spatial variation in the receipt of winter precipitation over Ireland, with only 4 of the 31 winters having all twelve stations with the same three-level categorisation, and only 12 winters with at least nine of the twelve stations having the same three-level categorisation. Therefore the analysis described above was performed for each of the twelve stations to identify any regional response patterns that may exist.

TABLE 12.9. Association between December-February sunspot number, phase of the Quasi-Biennial Oscillation and January-February Barnston and Livezey (1991) index of the Southern Oscillation, 1957-1987. (a) Simple frequencies. (b) Cell frequencies categorised as above average (wet), w, average, a, or below average (dry), d, winter precipitation.

(a)	Sunspot Number				Total	
	Low		High			
	Quasi-Biennial Oscillation Phase					
Southern Oscillation	East	West	East	West	East	West
Positive (+)	2	4	1	2	3	6
Intermediate (o)	4	2	2	3	6	5
Negative (-)	3	2	3	3	6	5
Total	9	8	6	8	15	16

(b)	Sunspot Number				Total	
	Low		High			
	Quasi-Biennial Oscillation Phase					
Southern Oscillation	East	West	East	West	East	West
Positive (+)	dd	w ddd	d	a d	3	6
Intermediate (o)	www a	a d	w a	ww a	6	5
Negative (-)	w aa	a d	ww d	ww a	6	5
Total	9	8	6	8	15	16

When the NAO index was cross-tabulated with the station totals the pattern of frequencies was in most cases similar in that precipitation was rarely below average and usually above average when the NAO index was positive. The principal exception was Dublin Airport, which showed no pattern at all. The probabilities derived from the twelve Fisher's Exact Test are plotted in Figure 12.5a and isopleths drawn. With the exception of Rosslare the higher values are in the south and east and the lower values are in the north and west centred on Belmullet, Claremorris and Belfast Aldergrove.

When the ENSO index of Barnston and Livezey was cross-tabulated with the station totals the pattern of frequencies was much more consistent, with precipitation rarely above average and usually below average when ENSO was positive ('High/Dry'). Again the principal exception was Dublin Airport which showed just the rare above average precipitation when ENSO was positive. The probabilities from the Fisher's Exact test were much lower than for the NAO index, and their spatial distribution was different

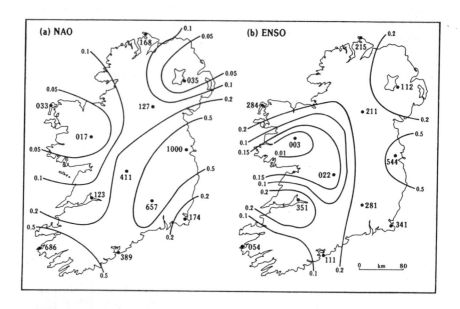

Figure 12.5. Probabilities (x1000) from Fisher's Exact Test of no association between station precipitation and (a) NAO phase, (b) ENSO phase (Barnston and Livezey index).

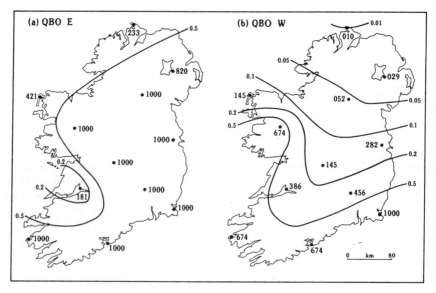

Figure 12.6. Probabilites (x1000) from Fisher's Exact Test of no association between station precipitation and sunspot number for (a) QBO East phase and (b) QBO West phase.

(Figure 12.5b). With the exceptions of Shannon Airport and Clones the lowest probabilites lay from southwest to northeast, with higher values in the nothwest and southeast.

To interpret the relationships between the QBO, sunspot numbers and winter precipitation, the values for QBO East and West phases were cross-tabulated separately. Figure 12.6 shows the distributions of the probabilities from the Fisher's Exact Tests. Figure 12.6a reveals that for the QBO East phase there is a northern and western fringe of lower probabilites. For the QBO West phase Figure 12.6b shows generally lower probabilities, with low probabilities in the north and much higher values in the south and west. The pattern of the frequencies within the contingency tables was described by Kendall's τ_b (tau b) coefficient, a rank correlation coefficient whose sign indicates the direction of association. The asymptotic standard errors of τ_b were used to calculate the probabilities of a null hypothesis of no association (a) for QBO East phase against an alternative hypothesis of negative association and (b) for QBO West phase against an alternative hypothesis of positive association, and these are plotted in Figure 12.7. The spatial patterns are similar to those of Figure 12.6, with the QBO East phase (Figure 12.7a) showing negative correlations between sunspot numbers and winter precipitation in the northwestern fringe, and in QBO West phase (Figure 12.7b) positive correlations across north and central Ireland, especially strong in the north.

The cross-tabulations of precipitation with NAO and ENSO phases (Table 12.7b), with NAO and QBO/sunspot phases (Table 12.8b) and with ENSO and QBO/sunspot phases (Table 12.9b) were repeated for precipitation at each of the twelve stations. For the NAO/ENSO/precipitation association, the pattern of cell frequencies was very similar to that seen in Table 12.7b except for Dublin Airport and Rosslare, and the spatial distribution of probabilities of no association was very similar to the southwest/northeast pattern seen in Figure 12.5b for the ENSO/precipitation association except that Shannon Airport's figure was not anomalously high. For the NAO/QBO/sunspot/precipitation relationships, the probabilities of no association tended to be lower in the north and west, somewhat similar to the NAO/precipitation pattern shown in Figure 12.5a. The suggested relationship of increased precipitation with increased sunspots when the NAO is not positive and the QBO is in West phase was strongly supported by the cell frequencies from Malin Head, Belfast Aldergrove and Clones, and less strongly from Belmullet, Dublin Airport, Birr, Shannon Airport, Kilkenny and Roche's Point. For the ENSO/QBO/sunspot/ precipitation relationships, the probabilities of no association were lowest in the middle west (Claremorris, Birr) and lower in the north (Malin Head, Belfast Aldergrove) and the southwest (Valentia Observatory, Roche's Point). The two suggested relationships when ENSO was not positive, namely a positive association between sunspot number and precipitation in QBO West phase, and increased precipitation in QBO East phase over West phase when sunspots are low, were strongly supported by the cell frequencies from Belmullet and Shannon Airport, and less strongly by those from Malin Head, Belfast Aldergrove, Clones, Claremorris, Kilkenny and Valentia Observatory.

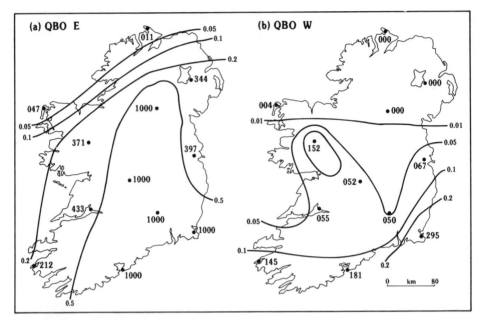

Figure 12.7. One-tail probabilities (x1000) from Kendall's tb statistic of no association between station precipitation and sunspot number (a) against an alternative hypothesis of negative correlation in QBO East phase and (b) against an alternative hypothesis of positive correlation in QBO West phase.

12.8. Discussion and Conclusions

The occurrence of average or above average winter precipitation when the NAO is in positive phase, i.e. when westerlies are strong, is to be expected. Sweeney (1985) calculated a mean daily precipitation over Ireland of 3.8 mm from Lamb Westerly airflow type, and 4.0 mm from Cyclonic type, both higher than from any other airflow category bar Southerly. That Rogers and van Loon (1979) showed precipitation increases to the north of Ireland but not over Ireland or Britain must be because they confined their attention to 'Greenland Below' and 'Greenland Above' situations. But Figure 12.5a does indicate that the NAO effect is greater in the northern and western half of Ireland.

The very strong association with below average winter precipitation and ENSO positive phase, and the consequent infrequency of below average precipitation in ENSO intermediate and negative phase situations is in complete agreement with the findings of Fraedrich (1990) of more cyclonic and more variable central European winters when ENSO is in 'Low/Wet' phase and of Fraedrich and Müller (1992) of cyclone tracks preferentially crossing Ireland and Britain into central Europe. What is remarkable is that the association is much weaker when those authors' ENSO phase classification is used, and bears testimony to the observation of Rogers (1984, 2004) 'that diagnostic studies of the SO are constrained by the definition of extreme events of the SO'. The Barnston and Livezey (1991) index uses contemporaneous (austral summer) Tahiti-

Darwin pressure difference and equatorial Pacific sea surface temperature rather than relying on Pacific Ocean events in the preceding austral winter, and it is tempting to speculate that its success in establishing a strong teleconnection is due to its lack of requirement of a half-year atmospheric 'memory'. The much weaker association of ENSO phase with winter precipitation in the southeast and in the northwest of Ireland (Figure 12.5b) bears investigation.

The tendency for winters with low sunspot numbers to be wetter in QBO East phase and drier in West phase is in part accord with the finding of Tinsley (1988) that cyclone tracks are well to the north of Ireland in low solar output QBO West phase winters. In better accord with Tinsley's results is the concentration of a positive association between precipitation and sunspot number in the northern half of Ireland in QBO West phase winters (Figures 12.6b and 12.7b): when sunspot numbers are high, cyclone tracks are just to the north of Ireland, producing precipitation especially over northern Ireland; whereas when sunspot numbers are low cyclone tracks are too far north of Ireland to produce precipitation there. However these, and the inverse relation between sunspot number and precipitation in QBO East phase winters in northwestern Ireland, appear to contradict the general association found by van Loon and Labitzke (1988) for 700 mb North Atlantic flow to be positively associated with 10.7 cm solar flux in QBO East phase and inversely associated in West phase.

As was described above, the three atmospheric oscillations do not operate independently of each other. At very least the coincidence of different phases emphasises certain circulation patterns, and possibly a phase of one oscillation may force or deny a particular phase of another. The sunspot number must be and the middle stratospheric equatorial wind should be independent of the NAO and ENSO. Table 12.7 indicates that the NAO and ENSO are unlikely to be independent of each other because none of the 31 winters is recorded as being intermediate on both oscillations. Presumably the ENSO phase is primary and the NAO phase secondary. When ENSO is in positive phase, the NAO is intermediate or in negative phase, and Ireland winter precipitation is below average. The only exception, 1974, showed the NAO in positive phase with the above average precipitation expected by that NAO phase. This contradicts the finding of Rogers (1984) that ENSO 'High/Dry' phases tend to occur when the NAO phase is positive. There is some support for the converse, also found by Rogers (1984), that the NAO tends to be in negative phase when ENSO is in 'Low/Wet' phase: such winters are variable in their precipitation amounts over Ireland.

The decreased winter precipitation during ENSO positive ('High/Dry') phases and the increased precipitation during NAO positive phases would appear to be the strongest controls, but when either ENSO or NAO was not in positive phase, there was some evidence for a direct association between precipitation and sunspot number. The additional tendency discussed above for QBO East phase winters to be wetter than West phase winters when sunspots are low and ENSO is not positive seems to be related to a QBO modulation of ENSO responses as suggested by Barnston *et al.* (1991), with middle North Atlantic cold pools present in East phases and subtropical North Atlantic warm pools present in West phases. It must be emphasised, however, that these solar and QBO effects are barely perceptible in the south and east of Ireland, where other controls must be operating.

In conclusion, it has been shown that winter precipitation in Ireland: (1) is strongly associated with ENSO phase, in particular being below average in positive phases; (2) is associated with NAO phase in that positive phases have above average precipitation; (3) is associated with sunspot activity with solar maxima showing above average precipitation particularly in QBO West phases; and (4) at solar minima and non-positive ENSO phases is modulated by the QBO to give above average precipitation in East phases and below average in West phases. These findings are frequently not in accord with North American analyses, suggesting that there is scope for further analysis from the eastern shore of the 'pond'.

References

Angell, J.K. (1992) Relation between 300 mb North Polar vortex and equatorial SST, QBO, and sunspot number and record contraction of the vortex in 1988-89, *J. Climate* **5**, 22-29.

Angell, J.K., and Korshover, J. (1984) Some long-term relations between equatorial sea-surface temperature, the four centres of action and 700 mb flow, *J. Climate and Applied Meteorology* **23**, 1326-1332.

Angell, J.K., and Korshover, J. (1985) Displacement of the north circumpolar vortex during El Niño 1963-83, *Monthly Weather Review* **113**, 1627-1630.

Barnston, A.G., and Livezey, R.E. (1987) Classification, seasonality, and persistence of low-frequency atmospheric circulation patterns, *Monthly Weather Review* **115**, 1083-1126.

Barnston, A.G., and Livezey, R.E. (1989) A closer look at the effect of the 11-year solar cycle and the quasi-biennial oscillation on Northern Hemisphere 700 mb height and extratropical North American surface temperature, *J. Climate* **2**, 1295-1313.

Barnston, A.G., and Livezey, R.E. (1991) Statistical prediction of January-February mean northern hemisphere lower tropospheric climate from the 11-year solar cycle and the Southern Oscillation for west and east QBO phases, *J. Climate* **4**, 249-262.

Barnston, A.G., Livezey, R.E., and Halpert, M.S. (1991) Modulations of the Southern Oscillation - northern hemisphere mid-winter climate relationships by the QBO, *J. Climate* **4**, 203-217.

Berson, F.A., and Kulkarni, R.N. (1968) Sunspot cycle and the quasi-biennial oscillation, *Nature* **217**, 1133-1134.

Bjerknes, J. (1966) A possible response of the atmospheric Hadley circulation to equatorial anomalies of ocean temperature, *Tellus* **18**, 820-829.

Brier, G.W., and Hanson, K. (1989) A nonparametric significance test of the solar-effect analysis of Labitzke and van Loon, *J. Climate* **2**, 1417-1418.

Craddock, J.M. (1968) Atmospheric oscillations with 'periods' of about two years, *U.K. Meteorological Office, Synoptic Climatology Branch Memo.* no. 22, 19 pp.

Ebdon, R.A. (1961) Some notes on the stratospheric winds at Canton Island and Christmas Island. Quart, *J. Royal Meteorological Society* **87**, 322-331.

Ebdon, R.A. (1971) Periodic fluctuations in equatorial stratospheric temperatures and winds, Meteorological Magazine **100**, 84-90.

Flueck, J.A., and Brown, T.J. (1993) Criteria and methods for performing and evaluating solar-weather studies, *J. Climate* **6**, 373-385.

Fraedrich, K. (1990) European Grosswetter during the warm and cold extremes of the El Niño/Southern Oscillation, *International J. Climatology* **10**, 21-32.

Fraedrich, K., and Müller, K. (1992) Climate anomalies in Europe associated with ENSO extremes, *International J. Climatology* **12**, 25-31.

Graham, N.E., and White, W.B. (1988) The El Niño cycle, A natural oscillator of the Pacific ocean/atmosphere system, *Science* **240**, 1293-1302.

Hamilton, K. (1988) A detailed examination of the extratropical response to tropical El Niño/Southern Oscillation events, *J. Climatology* **8**, 67-86.

Holton, J.R., and Tan, H.-C. (1980) The influence of the equatorial quasi-biennial oscillation on the global circulation at 50 mb, *J. Atmospheric Science* **37**, 2200-2208.

Holton, J.R., and Tan, H.-C. (1982) The quasi-biennial oscillation in the Northern Hemisphere lower stratosphere, *J. Meteorological Society Japan* **60**, 140-147.

Labitzke, K. (1982) On the interannual variability of the middle stratosphere during the northern winters, *J. Meteorological Society Japan* **60**, 124-138.

Labitzke, K. (1987) Sunspots, the QBO, and the stratospheric temperature in the North Polar region, *Geophysical Research Letters* **14**, 535-537.

Labitzke, K., and van Loon, H. (1988) Associations between the 11-year solar cycle, the QBO, and the atmosphere. Part I, The troposphere and stratosphere in the Northern Hemisphere in winter, *J. Atmospheric and Terrestrial Physics* **50**, 197-206.

Labitzke, K., and van Loon, H. (1989) Association between the 11-year solar cycle, the QBO, and the atmosphere. Part III, Aspects of the association, *J. Climate* **2**, 554-565.

Lamb, P.J., and Peppler, R. (1987) The North Atlantic Oscillation, Concept and an application, *Bull. American Meteorological Society* **68**, 1218-1225.

Meehl, G.A., and van Loon, H. (1979) The seesaw in winter temperatures between Greenland and northern Europe. Part III, Teleconnections with lower latitudes, *Monthly Weather Review* **107**, 1095-1106.

Naujokat, B. (1986) An update of the observed quasi-biennial oscillation of stratospheric winds over the tropics, *J. Atmospheric Sciences* **43**, 1873-1877.

Philander, S.G., and Rasmusson, E.M. (1986) The southern oscillation and El Niño, *Advances in Geophysics* **23**, 197-215.

Rasmusson, E.M., and Carpenter, T.H. (1982) Variations in tropical sea surface temperature and surface wind fields associated with the Southern Oscillation/El Niño. *Monthly Weather Review* **110**, 354-384.

Rasmussen, E.M., and Carpenter, T.H. (1983) The relationship between eastern equatorial Pacific sea surface temperatures and rainfall over India and Sri Lanka, *Monthly Weather Review* **111**, 517-528.

Rasmusson, E.M., and Mo, K. (1993) Linkages between 200-mb tropical and extratropical circulation anomalies during the (1986-1989 ENSO cycle, *J. Climate* **6**, 595-616.

Reed, R.J., Campbell, W.L., Rasmussen, L.A., and Rogers, D.J. (1961) Evidence of a downward propagating annual wind reversal, *J. Geophysical Research* **66**, 813-818.

Rogers, J.C. (1984) The association between the North Atlantic Oscillation and the Southern Oscillation in the Northern Hemisphere, *Monthly Weather Review* **112**, 1999-2015.

Rogers, J.C., and van Loon, H. (1979) The seesaw in winter temperatures between Greenland and northern Europe. Part II, Some oceanic and atmospheric effects in middle and high latitudes, *Monthly Weather Review* **107**, 509-519.

SAS Institute (1990) SAS procedures guide, version 6, third edition, SAS Institute, Inc., Cary, NC.

Sweeney, J.C. (1985) The changing synoptic climatology of Irish rainfall, *Transactions Institute British Geographers N.S.* **10**, 467-480.

Tinsley, B.A. (1988) The solar cycle and the QBO influences on the latitude of storm tracks in the North Atlantic, *Geophysical Research Letters* **15**, 409-412.

van Loon, H., and Labitzke, K. (1988) Associations between the 11-year solar cycle, the QBO, and the atmosphere. Part II, Surface and 700 mb in the Northern Hemisphere in winter, *J. Climate* **1**, 905-920.

van Loon, H., and Labitzke, K. (1990) Association between the 11-year solar cycle and the atmosphere. Part IV, The stratosphere not grouped by the phase of the QBO, *J. Climate* 3, 827-837.

van Loon, H. and Rogers, J.C. (1978) The seesaw in winter temperatures between Greenland and northern Europe. Part I, General Description, *Monthly Weather Review* **106**, 296-310.

van Loon, H., and Shea, D.J. (1985) The Southern Oscillation. Part IV, The precursors south of 15°S to the extremes of the oscillation, *Monthly Weather Review* **113**, 2063-2074.

Venne, D.E., and Dartt, D.G. (1990) An examination of possible solar cycle-QBO effects in the Northern Hemisphere troposphere, *J. Climate* **3**, 272-281.

Veryard, R.G., and Ebdon, R.A. (1961) Fluctuations in equatorial stratospheric winds, *Nature* **189**, 791-793.

Walker, G.T. (1924) Correlations in seasonal variations of weather, IX, *Memoranda Indian Meteorological Department* **24**, 275-332.

Walker, G.T., and Bliss, E.W. (1932) World weather V, *Memoranda Royal Meteorological Society*, **4**, 53-84.

Wright, P.B. (1984) Relationships between indices of the Southern Oscillation, *Monthly Weather Review* **112**, 1913-1919.

Wright, P.B. (1989) Homogenised long-period Southern Oscillation indices, *International J. Climatology* **9**, 33-54.

Yarnal, B. (1985) Extratropical teleconnections associated with El Niño/Southern Oscillation (ENSO) events, *Progress in Physical Geography* **9**, 315-352.

Yarnal, B., and Kiladis, G. (1985) Tropical teleconnections associated with El Niño/Southern Oscillation (ENSO) events, *Progress in Physical Geography* **9**, 524-558.

S. Daultrey
Department of Geography
University College Dublin, Belfield, Dublin 4, Ireland.

SECTION IV

IMPACTS ON SNOW, ICE AND MELTWATERS

SUMMARY

Modelling glacierized basins

■ A model is developed which combines energy balance, water balance and runoff components and is applied to a Tianshan basin. The simulated hydrographs average discharge are close to the observed.

■ Separate treatment of melt processes on the glacier and in the snowpack, separation of precipitation according to phase, and a tank model that treats different drainage processes discretely produce an effective physically-based model.

■ Runoff in the test basin is mainly controlled by the heat balance, although precipitation in the lower part of the basin occasionally causes poorer simulation results.

Glaciers of the Aral Sea basin

■ Previous analysis of summer temperatures in years of high global temperatures suggest a 0.5°C cooling in the Pamir and Tianshan mountains under 1°C global warming.

■ Equilibrium lines will descend by c.150 m leading to an 18% increase in the area of the cold firn zone and a commensurate reduction in the area of the ablation zone.

■ The area of glaciers with zero runoff will increase by 37% and total meltwater discharges are calculated to decrease in the order of 15%.

■ A marked decrease in icemelt will be partially compensated by increased snowmelt runoff from the lower slopes.

■ Artificial glacier melting will not be capable of making up the shortfall in runoff and the potential for artificial inducement will be slightly lower.

Climate sensitivity of the Central Asian cryosphere

■ Two-way feedbacks are envisaged between the surface cryosphere and the atmosphere: glaciers are indicators of climate change, but the Eurasian snowcover may also be a regulator of atmospheric circulation patterns.

■ Comparison of deviations in Central Asian snowcover, firn line altitudes, the zonality of circulation patterns and global temperatures shows zonal epochs during the periods of higher temperature in the 1930s and 1980s.

■ Trends in meridional temperature gradients, dominant circulation patterns and snow covered area show a series of lags. Peak temperature gradients precede peak zonality, which precedes maximum snow covered area, each with a lag of c.15-20 years.

■ Snow covered area is at a maximum at the end of a zonal epoch and proceeds to decrease throughout the subsequent meridional epoch, reaching a minimum at the end of that epoch. It then proceeds to gradually increase through the following zonal epoch.

■ The lag patterns are explained as follows:

1. Maximum snowcover causes cooling in Central Asia and increases the meridional temperature gradient between Central and Southern Asia.

2. This increases meridional airflow, especially in winter, but has a dampening effect on the Asian summer monsoon, for which higher temperatures in the continental interior are an important driving factor.

3. Weaker monsoons mean less vapour transport into Tibet from the Indian Ocean.

4. This is supplemented by weaker vapour transport from the Atlantic by the zonal winds, because they are obstructed by the meridional circulation in Central Asia.

5. These two effects combine to reduce snowcover in Central Asia.

6. As snowcover reaches a minimum, lower albedos increase net radiation receipts and less latent heat is consumed in melting, so that air temperatures rise.

7. Meridional temperature gradients reduce, zonal circulation regains ascendency.

8. Higher air temperatures maintain a period of lower snowcover for 7-14 years until increased snowfall from zonal airflow and the invigorated summer monsoon finally begins to expand the snowcovered area again.

■ Glacier mass balances depend more on circulation patterns, precipitation, cloudiness, solar radiation and local air temperatures than on global temperature changes.

■ 5 glacier systems studied show long-term negative balances corresponding to the predominantly meridional pattern of circulation this century.

■ Outlying glaciers are more sensitive to climate change than larger ice masses.

■ Firn lines are lowest around the end of a meridional epoch and throughout the first half of the zonal epoch. This is associated with a large increase in precipitation and cloudiness and a fall in local temperatures.

■ Firn lines are highest at the end of a zonal epoch when precipitation is at minimum.

Trends in Western China

■ There is widespread evidence of warming this century, especially during the winter and since the 1950s: January was 4.2°C warmer in North Xingjiang in the 1980s than in the 1950s.

■ Oxygen isotope evidence from the Dunde ice core taken in the Qilian Mountains indicates a temperature rise of about 1°C since the 1890s.

■ Evidence of a long-term drying trend since the early Holocene comes from proof of shrinking in over 1000 lakes, mostly on internal drainage systems in the Tibetan Plateau. Many lakes have become isolated since lake levels fell below the level of the outlet, as on Lake Yangzhouyong where the level is estimated to be still falling at 800 mm/100 years. The area of Lake Qinghai has reduced by 13% this century.

■ Mountain glaciers have continued in overall retreat during the 20th century, following the warming and especially the drying trends. Over the last 500 years, the total area of mountain glaciers in NW China has fallen by 7000 km^2, a reduction of 20% from maximum. Only 25% of glacier snouts are advancing.

■ Instrumental records from Lhasa show a decline in precipitation over the last 30 years approaching -200 mm per annum.

■ Historical documents and instrumental records indicate increased frequency of drought. Wet periods have decreased in duration and droughts have increased in length since 1883. The last wet period ended in 1962.

■ Global warming is predicted to intensify this drying trend.

13 RUNOFF FORMATION AND DISCHARGE MODELLING OF A GLACIERIZED BASIN IN THE TIANSHAN MOUNTAINS

KANG ERSI, SHI YAFENG, ATSUMU OHMURA and HERBERT LANG

Abstract

The experiments on energy and water balance of a Tianshan glacierized basin indicate that the runoff from the basin consists of glacier melt, snow melt and effective liquid precipitation. An energy-water balance model is developed to simulate the runoff from the glacier and snow covered area. The ice and snow melt is simulated by the parameterized energy balance model. The snow covered area is determined by the separation of precipitation form. The runoff produced from the basin is transformed by a tank model into the discharge at the basin outlet, which simulates the runoff processes on the glacial and non-glacial surfaces and in the intra- and subglacial drainage system and in the active layer of the permafrost. Finally, a runoff model of a Tianshan glacierized basin is developed to simulate the energy and water inputs to the basin and the discharge from the basin with the standard meteorological elements as the foremost inputs to the model.

13.1. Introduction

The mountain glaciers at the middle and low latitudes are not only sensitive to climate change, but also important in the formation and regulation of water resources. Therefore, the investigation of runoff formation and its link to the climatic variables in the vast glacierized basins of the inland mountainous area of China is very important both for the scientific inquiry into the change of regional water resources under the global warming and for runoff estimation and forecasting. During the years from 1985 to 1987, a comprehensive field experimental programme was carried out on energy and water balance in a glacierized basin of the Chinese Tianshan mountains (Kang *et al.*, 1992). Based on these field experiments, this paper is intended to study the runoff processes and their modelling in the glacierized basin and to link them to the energy and water balance in the basin, and then to link this to the standard meteorological elements.

The research basin, named the source basin, is located at the source area ($43°$ $06'$N, $86°$ $50'$E) of the Urumqi River. The underlying surface of the basin consists of 7 glaciers, bare rocks, moraines, alpine meadow and marsh ground. There are three hydrometric stations and a standard meteorological station (3539 m a.s.l.) within the

241

J.A.A. Jones et al. (eds.), Regional Hydrological Response to Climate Change, 241–257.

basic (Figure 13.1). The Total Control hydrometric station collects runoff for the entire source basin.

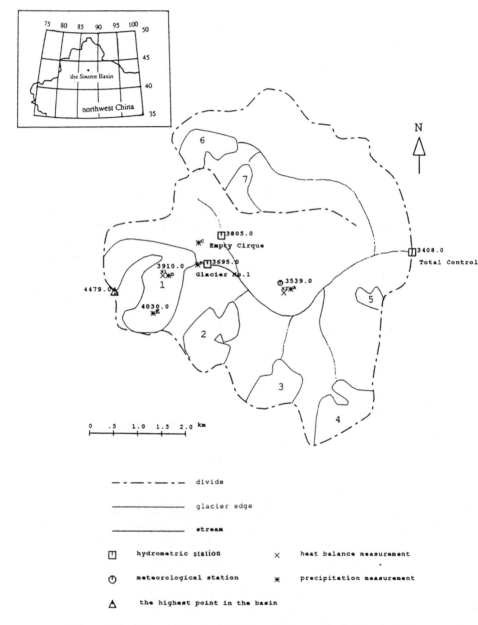

Figure 13.1. The source basin of the Urumqi River and its location in China. (Numbers on glaciers are based on the official numbering system of the Chinese Glacier Inventory.)

The two additional sub-basins referred to are the Empty Cirque basin and the Glacier No.1 basin (Table 13.1). The standard meteorological station, the Daxigou meteorological station, is located in the central part of the source basin.

On the glacier and snow surfaces, the runoff is simulated by the energy-water balance model. The melt water is simulated by the parameterized energy balance model. In the present study, two standard locations have been chosen to simulate the energy and water inputs to the basin. For the glacierized area, near the equilibrium line altitude, and for the non-glacierized area, near its central part. The precipitation form is separated by air temperature criteria and its daily range. The snow cover is approximated by the solid precipitation. The runoff produced both from the glacierized area and non-glacierized area of the basin is transformed into discharge at the basin outlet with the application of the tank model. The runoff formation processes and discharge at the three hydrometric stations are simulated and discussed.

TABLE 13.1. The hydrological basins of the research area.

Hydrometric station		Total Control	Empty Cirque	Glacier No. 1
Area (km²)	Total	28.9	1.68	3.34
	Glacierized	5.74	0.00	1.87
Altitude (m)	Maximum	4479	4393	4479
	Minimum	3408	3805	3695
	Mean	3817	4031	4065

13.2. Runoff Formation of a Glacierized Basin

13.2.1. RESULTS OF FIELD EXPERIMENTS

13.2.1.1. *Energy Balance*

Table 13.2 shows the simultaneously measured energy balance on Glacier No. 1 (Calanca and Heuberger, 1990) and the alpine tundra (Zhang *et al.*, 1992) during June to August of 1986. On the glacier surface, energy input is mainly net radiation followed by sensible heat flux; while energy consumption is mainly by the melt followed by evaporation. The energy balance on the alpine meadow is very different from that on the glacier surface. Net radiation is the sole energy source to the alpine meadow, amounting to more than twice that of the glacier surface, and all other components are energy sinks. The evaporation latent heat flux is by far the largest energy sink, while sensible heat flux and the heat for the snowmelt are small, and the heat conducted into the ground is the smallest sink.

TABLE 13.2. Energy balance on the surface of Glacier No. 1 and the alpine tundra at the source basin of the Urumqi River in summer (from July 6th to August 19th of 1986).

Underlying surface	Net radiation		Sensible heat		Latent heat		Melt heat		Subsurface heat	
	$W\,m^{-2}$	%	$W\,m^{-2}$	%	$W\,m^{-2}$	%	$W\,m^{-2}$	%	$W\,m^{-2}$	%
Glacier No. 1 (3910 m)*	+54.5	82.5	+11.6	17.5	-4.0	6.1	-62.1	93.9	0.0	0.0
Alpine tundra (3539 m)**	+122.4	100.0	-8.2	6.7	-104.2	85.2	-6.9	5.6	-3.1	2.5
Remarks	The positive sign symbolizes the heat flux to the surface, percentage is counted separately for heat input and output. The melt heat on the alpine tundra is owing to snow melt.									

* from Calanca and Heuberger (1990)   ** from Zhang *et al.* (1992)

13.2.1.2. *Water Balance*

During the years from 1983 to 1989, the annual water balance of the source basin (Kang, 1994) consists of precipitation of 546 mm, glacier net balance of -56 mm, evaporation of 170 mm and runoff of 432 mm. The glacier total melt is 150 mm, accounts for 35% of the total runoff of the source basin. That of the Glacier No. 1 basin consists of precipitation of 567 mm, glacier net balance of -158 mm, evaporation of 145 mm and runoff of 580 mm. The glacier total melt is 423 mm, accounts for 73% of the total runoff of the Glacier No. 1 basin.

13.2.2. PROCESSES OF RUNOFF FORMATION

Figure 13.2 shows the runoff formation processes in a glacierized basin. A glacierized basin is taken as an open system. The water input from the atmosphere to a glacierized basin is precipitation. Other water inputs, such as snow drift from outside of the basin and vapour condensation are neglected in this study. The liquid precipitation is directly transformed into runoff from both the glacierized area and non-glacierized area after evaporation. The solid precipitation on glacierized area consists of net accumulation, melt and evaporation, and that on the non-glacierized area consists of the redistribution by snow drift and avalanche to the glacier surface, melt and evaporation. The solid precipitation on the non-glacierized area is dealt with as snow cover. From the comparison of energy balance components (Table 13.2), heat input to the basin consists mainly of net radiation and sensible heat to the ice and snow surface, and of net radiation to the non-glacierized surface. Evaporation constitutes an energy sink. The net energy balance is the heat source for melt of ice and snow. Glaciers receive the net accumulated part of solid precipitation as storage, through mass change, and release water by net melt. The water produced from the glacierized part is transformed by glacierized runoff transformation, and that from non-glacierized part by non-glacierized runoff transformation, to form the discharge at the outlet of the basin.

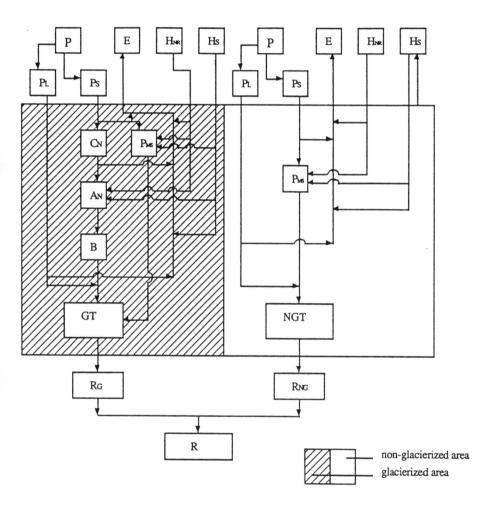

Figure 13.2. Energy - water balance and runoff formation in a glacierized basin.
P is precipitation, P_L liquid precipitation, P_S solid precipitation, E evaporation, H_{NR} net radiation, Hs sensible heat, C_N glacier net accumulation, A_N glacier net melt, B mass balance, P_{MS} melt part of snow-water equivalent, G_T glacierized runoff transformation, N_{GT} non-glacierized runoff transformation, R_G runoff from the glacierized area, R_{NG} runoff from the non-glacierized area, R runoff from the whole glacierized basin.

13.3. Runoff Modelling

13.3.1. ENERGY-WATER BALANCE OF A MELTING GLACIER

On the glacier surface during ablation season, the rain conducted heat and surface conducted heat are negligible (Lang, 1986; Ohmura *et al.*, 1990), the energy balance equation can be written as:

$$H_{NR} + H_S + H_L = H_M \tag{13.1}$$

where H_{NR} is net radiation, H_S sensible heat flux, H_L latent heat flux of evaporation, H_M melt heat of snow and ice. The ground water storage is neglected and the glacier mass balance is taken as the storage change of the basin, the water balance equation is:

$$P - E - R = B \tag{13.2}$$

where P is precipitation, E evaporation, R runoff, B glacier mass balance. Glacier mass balance equation is:

$$C_N - A_{NA} = B \tag{13.3}$$

where C_N is glacier net accumulation, A_{NA} glacier net ablation. Precipitation consists of solid part P_S and liquid part P_L:

$$P = P_S + P_L. \tag{13.4}$$

From Eqs. 13.1 to 13.4, the runoff from the glacierized area can be expressed as:

$$R = \frac{\tau}{L_M}(H_{NR} + H_S) - \frac{L_E}{L_M}E + P_L \tag{13.5}$$

where t is time period considered, L_M latent heat of melt, L_E latent heat of evaporation.
  If the subsurface heat and the heat from the liquid precipitation are neglected, Eq. 13.5 is also valid for the whole glacierized basin and on the snow covered non-glacierized area. Therefore, Eq. 13.5 is then considered as the equation of energy - water balance in a glacierized basin.

13.3.2. PARAMETERIZED ENERGY BALANCE MODEL

The heat flux from atmosphere to the glacier surface for melt is calculated with energy balance Eq. 13.1. In the practical application, the heat flux expressions of Eq. 13.1 can be parameterized using standard meteorological observations. The energy balance components measured at 3910 m near the mean altitude of equilibrium line are related to the meteorological elements at the Daxigou meteorological station by means of physical formulation and regression analysis (Kang, 1994). Eq. 13.1 is then parameterized as follows.

$$H_{NR} = R_G (1 - \alpha) + \varepsilon_e \varepsilon_a \sigma T_a^4 - \varepsilon_a \sigma T_e^4 \tag{13.6}$$

$$\frac{R_G}{R_{scd}} = 0.2319 + 0.5354 \frac{HD}{HD_c}, \tag{13.7}$$

$$\alpha = 0.82 - 0.03T_m - 1.74 \times 10^{-3}T_m^2 - 1.14 \times 10^{-4}T_m^3, \text{ (snow)} \tag{13.8}$$

$$\alpha = 0.27 - 0.01 \, T_m, \text{ (ice)} \tag{13.9}$$

$$\varepsilon_a = 0.69(1 + 0.42C) \tag{13.10}$$

$$\varepsilon_e = 0.97, \tag{13.11}$$

$$H_S = 0.97U_m T_m, \tag{13.12}$$

$$H_L = -2.73 \, U_m(6.11 - P_{me}), \tag{13.13}$$

where R_G is global radiation (W m^{-2}), α albedo, ε_a atmosphere emissivity, ε_e surface emissivity or absorption coefficient, T_a atmosphere temperature (K), T_e surface temperature (K), σ Stefan-Boltzmann constant, 0.5667 x 10^{-7} W m^{-2} K^{-4}, daily extra-atmospheric solar radiation, HD_c measured daily sunshine duration hours, calculated daily potential sunshine duration hours, C cover ratio of low cloud , C = 0.0 for clear sky, C = 1.0 for overcast skies, T_m (°C), U_m (m s^{-1}), P_{me} (hPa) are daily mean air temperature, wind velocity, and vapour pressure at the Daxigou meteorological station.

The daily energy balance at 3910 m a.s.l. near the mean equilibrium line of Glacier No. 1 in the basin is simulated during the ablation season of the years from 1986 to 1990 (Kang, 1994). During June to August of these years, the energy balance consists of 224 W m^{-2} global radiation, -121 W m^{-2} short-wave reflected radiation, 278 W m^{-2} long-wave incoming radiation, -308 W m^{-2} long-wave outgoing radiation, 13 W m^{-2} sensible heat flux , -5 W m^{-2} latent heat flux, and -80 W m^{-2} melt heat.

13.3.3. SEPARATION OF PRECIPITATION FORM

The water equivalent of snow cover is approximated by the separation of precipitation in the present study. Precipitation form is related to air temperature and its lapse rate (Lang, 1986). Most of the methods for separation of solid precipitation from the total precipitation are based on a critical temperature (WMO, 1986). Studies on the relationship between precipitation form and air temperature have been carried out since the early years of this century (Rohrer, 1989). The recent studies by Rohrer et al. (1990) show that the relative percentage of liquid, solid and mixed precipitation depends on the location, the hour of observation and sometimes on the season of observation. As precipitation can occur at any time with temperature T during the day, it makes the relationship of precipitation form to air temperature complex. In this study, the daily solid precipitation is separated by temperature criteria and the daily temperature variations. During the ablation season, when precipitation consists of both solid and liquid forms, the form is related to the temperature daily range (Kang, 1994).

When precipitation occurs during a day, there are two criterion temperature T_L and T_S, for liquid and solid precipitation separately. When the daily minimum air temperature $T_{min} \geq T_L$, precipitation is liquid, while when the daily maximum air temperature $T_{max} \leq T_S$ it is solid. When $T_{min} < T_L$ or $T_{max} > T_S$, it is possible for precipitation to have both liquid and solid forms (Figure 13.3a and b). The criterion temperature T_L and T_S can be determined by the field observations. Suppose the daily temperature variation is between T_L and T_S (Figure 13.3c), that is, $T_{max} < T_L$ and $T_{min} > T_S$, then for the precipitation over the 24 hours, suppose T_d is the daily mean air temperature, the following equation is set up:

$$\frac{P_S}{P} = \frac{1}{T_L - T_S}\left(T_L - \frac{T_d}{2} - \frac{T_{max}}{4} - \frac{T_{min}}{4}\right) \qquad (13.14)$$

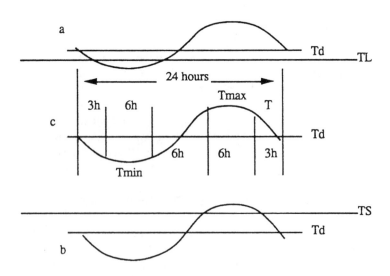

Figure 13.3. Determination of precipitation form by the critical temperature T_L for liquid precipitation, T_S for solid precipitation and daily temperature variation.

This is the basic equation to calculate the ratio of daily solid precipitation to the daily precipitation. Based on observations at the Daxigou meteorological station, the values for T_L and T_S are determined as

$$T_L = 5.5°C,$$

$$T_S = 2.8°C.$$

By this method, the daily solid precipitation can be obtained at any altitude of a mountain glacierized basin. The solid precipitation is dominant and increases with altitude in the source basin during the ablation season. During the periods from May to September of the years from 1986 to 1990, the averaged solid precipitation ratio is 0.67 at 3539 m a.s.l., 0.86 at 3900 m a.s.l. of the non-glacierized area, and 0.93 at 4010 m a.s.l. on the glacierized area of the basin.

13.3.4. RUNOFF FROM THE NON-GLACIERIZED AREA

On the ground surface without ice and snow cover, the general water balance equation should be used to calculate the runoff production instead of Eq. 13.5.

13.3.4.1. *Precipitation*

According to observations at the Daxigou meteorological station in the basin from 1959 to 1989, the mean annual precipitation is 429.3 mm with a standard deviation of 49.9 mm for the 31 years. The precipitation is concentrated in the summer months of June to August, accounting for 66% of the annual value, when the air temperature is high and the glacier ablation is rapid. The maximum monthly precipitation is in July when the air temperature is also the highest for the year (Kang et al., 1992).

There are systematic observation errors in the precipitation data (Sevruk, 1986). In order to investigate them in the Tianshan environment, a comparative field experiment was carried out with different precipitation gauges and at different altitudes in 1986 and 1987 in the Urumqi River basin (Yang et al., 1989). In the source area of the river, which is heavily glacierized, the systematic errors are studied of dynamic loss, wetting loss and evaporation loss for the Chinese standard rain gauge (Yang et al., 1992). It is found that dynamic loss creates the main part of the total observation error, accounting for 52.4%. The second largest error is due to wetting loss, 33.3%, and the smallest is due to evaporation loss, 14.3%. The correction factor for the total systematic error is 1.30. The monthly correction factor for the observed precipitation is also obtained.

It is found that precipitation increase with altitude in the Urumqi River basin occurs mainly in summer; other seasons do not have obvious increase with altitude (Yang et al., 1992). This can be explained by the continental monsoon climatic characteristics. The precipitation increase with altitude is caused by the enforced lifting of the air current containing precipitable water vapour. In the Tianshan area, winter is an extremely dry season controlled by the cold dry continental air mass, and the topographic lifting mechanism can not be set up well. According to the field experiments (Yang et al., 1992), the annual precipitation gradient is 18.3 mm/100 m in the source area of the Urumqi River. Based on the measurements at the Tianshan meteorological station and the two hydrometric stations, the Total Control and the Glacier No. 1, during May to August of 1983 to 1987, the following daily precipitation gradients $P_G(n, M)$ are obtained for May (M = 5) to September (M = 9),

$$P_G(n, 5) = 8.3\% \, P_{md}(n) \, / \, 100 \, m,$$
$$P_G(n, 6) = 5.5\% \, P_{md}(n) \, / \, 100 \, m,$$
$$P_G(n, 7) = 0.0\% \, P_{md}(n) \, / \, 100 \, m,$$
$$P_G(n, 8) = 7.2\% \, P_{md}(n) \, / \, 100 \, m,$$
$$P_G(n, 9) = 0.0\% \, P_{md}(n) \, / \, 100 \, m,$$

where, $P_{md}(n)$ is the error corrected daily precipitation at the Daxigou meteorological station, the day n belongs to the month M.

In July, when the altitude dependency of relationship is insignificant , the gradient is taken to be zero. The air temperature in July is the highest of the summer months (Kang et al., 1992). This may cause strong convection and the normal altitude dependency may be destroyed.

13.3.4.2. *Evaporation*
Based on the field measurements carried out on the alpine meadow (3539 m a.s.l.) in the source basin during the summer of 1986 by Zhang et al. (1992), the three methods are employed to estimate the evaporation: profile - flux relationship, weighing lysimeter and Bowen ratio. From the measured data, the daily evaporation rate Et (mm d^{-1}) from the alpine meadow is determined by the daily mean wind velocity U_m (m s^{-1}), vapour pressure P_{me} (mb) and saturation vapour pressure e_{sa} (mb) measured at the Daxigou meteorological station as follows,

$$Et = 0.33(U_m(e_{sa} - P_{me}))^{0.97}. \tag{13.15}$$

Eq. 13.15 is used for the estimation of daily evaporation from the alpine meadow during the period from May to September.

The annual evaporation from the non-glacierized area has been estimated using the water balance calculation. The evaporation value is distributed over the months of May to September based on the calculation of potential evaporation. Monthly evaporation ratio of non-glacierized area to the alpine meadow can be calculated. The daily evaporation ratio of non-glacierized area to alpine meadow is approximated by the relevant monthly ratio. Based on the calculated daily evaporation Et from the alpine meadow, the daily evaporation from the non-glacierized area is then calculated (Kang, 1994).

13.4. Discharge Modelling

The runoff produced in a glacierized basin is changed into the discharge at the outlet by the runoff transformation processes of the basin. On the glacier surface, the water production comes from snowmelt, glacial icemelt, and in a few case from rainfall. Snowmelt water runs through the snow layer or through the firn layer to the ice surface. The water on the glacier surface runs along the slope or surface ditches. Some of the meltwater runs into the glacier through tunnels or crevasses, then runs in the intra- and subglacial drainage system. All the water converges to the terminus, forming glacier runoff. On the non-glacierized surface, the water is produced from snow meltwater and rainfall. After evaporation, the water runs along the surface. This is controlled by the surface runoff processes. When the active layer is formed, the water penetrates and also runs through the active layer. The groundwater is neglected as mentioned before.

Because of the change of the conditions for both melt and runoff transformation in a glacierized basin, the ablation season is divided into a weak melt period and an intense melt period (Kang, 1991). The weak melt period designates the period with small discharge and dominated by seasonal snow melt, usually during the beginning and end of the ablation season. This period is generally in May, June and September in the glacierized basins of the Tianshan climatic and hydrological conditions. The intense melt period designates the period with large discharge, and glacial meltwater contributes to the runoff mainly during this period. This period generally consists of the months of July and August. Discharge modelling is carried out for the two periods separately in the present study.

13.4.1. APPLICATION OF THE TANK RUNOFF TRANSFORMATION MODEL

Based on these runoff transformation characteristics, the tank model developed by Sugawara (Sugawara *et al.*, 1984) is used to construct a runoff transformation model in a glacierized basin, which is taken the structure as shown in Figure 13.4 and consists of two vertically connected tanks. The upper tank with two side outlets and a bottom outlet is used to simulate the runoff transformation processes on the glacier surface and the ground surface, while the lower tank is to simulate those of the intra- and subglacial drainage system and the active layer of the permofrast. The runoff production is simulated based on Eq. 13.5 and taken as the net water input to the tank runoff transformation model, the discharge from the basin outlet is the sum of the discharges from the side outlets of the tank model. The parameters of the tank runoff transformation model are determined by calibration (Sugawara *et al.*, 1984).

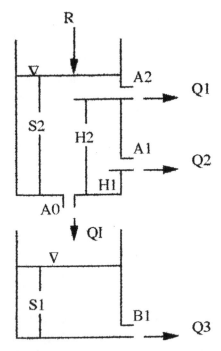

Figure 13.4. Tank runoff transformation model in a glacierized basin. R is the net water input (runoff production), S1 and S2 are water storage, H1 and H2 are the heights of the outlets of the upper tank, A1, A2 and B1 are drainage coefficient of the side outlets, A0 is the infiltration coefficient of the upper tank, Q1, Q2 and Q3 are discharge from the side outlets, QI is the infiltration rate, ∇ symbolizes the storage water level.

13.4.2. DISCHARGE MODELLING

13.4.2.1. *Runoff Transformation Processes*

Through the calibration for the years from 1986 to 1989, the parameters of the tank runoff transformation model are obtained for the three hydrological basins at the source area of the Urumqi River (Table 13.3).

The Empty Cirque basin represents the non-glacierized high mountain area of the Tianshan conditions. Except for some alpine meadow areas, there is no vegetation cover which is significant for runoff formation. The bare rock and moraine cover most of the ground surface. The snow cover is usually not deep but unstable, and because the precipitation season is also the ablation season, snow melts very quickly. These features make the heights of the two side outlets of the top tank always zero. This is identical with a linear reservoir model. During the weak melt period, the active layer is not yet well developed. Therefore, the infiltration is small, and the discharge from the active layer is negligible. During the intense melt period, the surface runoff is still the main part of the discharge. The ground active layer is well developed, and it plays some role in the runoff transformation. The larger value of infiltration coefficient and the drainage coefficient of the lower tank during the intense melt period can be explained also by the development of the active layer. During the weak melt period, the drainage coefficients of the outlets are mostly small. This could be attributed to the retention of runoff in the snow layer. During the intense melt period, the runoff transformation is quicker. The drainage coefficients are mostly larger. In the high mountain glacierized basins, the runoff from the non-glacierized area consists mainly of snow meltwater. The surface structure is complex, and the precipitation intensity is less as compared to the rainfall runoff basins. Therefore, even though the catchment area is small, it may take a few days to drain off the net water input of a day.

TABLE 13.3. Parameters of the tank runoff transformation model of the source basin of the Urumqi River.

Hydrometric basin	Melt period	H2	H1	A2	A1	A0	B1
Empty Cirque	Weak melt	0.00	0.00	0.043	0.064	0.015	0.004
	Intense melt	0.00	0.00	0.075	0.063	0.079	0.036
Glacier No. 1	Weak melt	5.00	0.00	0.247	0.191	0.021	0.004
	Intense melt	10.00	0.00	0.165	0.151	0.153	0.146
Source basin	Weak melt	0.00	0.00	0.184	0.117	0.016	0.008
	Intense melt	10.00	0.00	0.112	0.112	0.112	0.111

The runoff transformation of the Glacier No. 1 basin shows some nonlinear properties. When the net water input is more than a certain depth, the discharge can increase more rapidly. Along with the increase of meltwater, the surface albedo is often reduced, and the melt area increases. Consequently the drainage channels will be more developed. These may cause some of the nonlinear features of the increase in discharge. The height of the upper outlet should be related to the snow depth and the development of the surface drainage system. During the intense melt period, the infiltration coefficient and the drainage coefficient of the lower tank are large because of the development of the glacier drainage system. The drainage coefficients and the infiltration coefficient show rather identical values in an intense melt period. The meltwater runoff from the glacierized area is dominant during this period. The development of the glacier drainage system and the rather simple englacial drainage channels (Kang, 1994) may make the drainage conditions of the glacierized basin rather uniform.

In the weak melt period, the runoff transformation of the source basin shows the feature similar to that of the Empty Cirque basin. This indicates that the runoff formation is mainly at the non-glacierized part of the basin in this period. In the intense melt period, the runoff transformation shows the feature similar to that of the Glacier No. 1 basin. This indicates that the runoff formation at the glacierized part is important during this period. During the weak melt period, the runoff is mainly formed in the lower altitude zone of the basin. During the intense melt period, the runoff is mostly formed in the upper altitude zone of the basin, and the longer distance to the outlet may make the storage time longer. The infiltration coefficient A_0 and drainage coefficient B_1 represent the sub-surface runoff transformation both on the glacierized and non-glacierized area. The coefficients A_0 and B_1 are small during the weak melt period. During the intense melt period, the A_0 and B_1 values are rather large because of the development of the intra- and subglacial drainage system in the glacierized area and the development of the ground active layer in the non-glacierized area. The drainage coefficients and the infiltration coefficient show identical values in an intense melt period. Like the conditions of the Glacier No. 1 basin, the meltwater runoff from the glacierized area is dominant during this period. The development of the glacier drainage system, the rather simple englacial drainage channels and the development of the active layer may make the drainage conditions of the whole glacierized basin rather uniform.

13.4.2.2. Results

The discharge simulation is evaluated by the tank model evaluation criterion CR and the standard mean square error of discharge MSEQ (Sugawara *et al.*, 1984). Taking 1990 as a test year, the results of the simulation are evaluated in Table 13.4. The simulated mean discharge is close to the measured mean discharge (Yang, 1990). The evaluation criteria show very good values for the Empty Cirque basin and the source basin. The larger simulation error for the Glacier No. 1 basin of the test year can be explained by the more turbulent feature of the discharge as compared with the other years (Kang, 1994). This may be caused by the very changeable precipitation distribution and melt conditions both spatially and temporally at the high mountain glacierized area.

TABLE 13.4. The evaluation criteria for discharge modelling of the three hydrometric stations of the source basin for the test year 1990 (from May to Sept.).

Hydrometric station	Measured mean discharge (mm d^{-1})	Simulated mean discharge (mm d^{-1})	CR	MSEQ
Empty Cirque	3.1	2.8	0.307	0.409
Glacier No. 1	3.4	4.0	0.584	0.761
Total Control	3.3	3.5	0.328	0.367

Figure 13.5 shows the comparison of the simulated daily discharge (Q_c) hydrographs and the measured hydrographs (Yang, 1987, 1990) at the Total Control hydrometric station for the source basin. The daily course of the basic meteorological elements

which are measured at the Daxigou meteorological station and inputted to the energy - water balance model are inspected by the hydrographs. The basic characteristics of the measured hydrographs have been well simulated. In the source basin, the runoff consists of glacial melt runoff, snowmelt runoff and rainfall runoff. During the weak

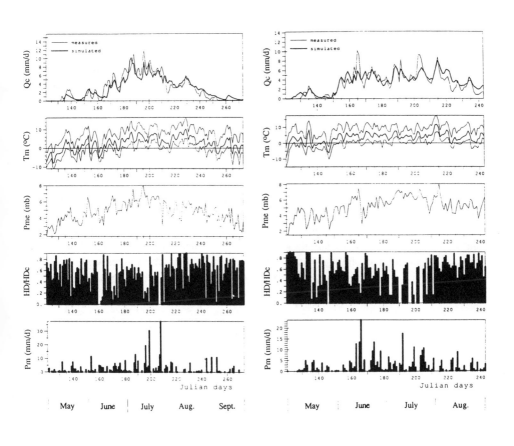

1987   1990

Figure 13.5. Daily discharge (Qc) simulation of the source basin hydrometric station (1987 is one of the years for the parameter calibration of the tank runoff transformation model, and 1990 is the year for the model test).

melt period of May, June and September, the snowmelt runoff is dominant, but in some case rainfall at the lower part of the lower zone may cause peak discharge. During the intense melt period of July and August, the daily discharge fluctuation coincides with those of air temperature, vapour pressure and sun duration with some time lags. Therefore, the runoff is controlled mainly by the heat conditions and consists mainly of glacial meltwater and snow meltwater. The precipitation at the lower part may at times affect the peak discharge. In this case, the peak discharge may not be well simulated by the model transformation. In the test simulation, the June 16th peak discharge shows this situation.

13.5. Conclusion

The hydrological discharge simulation of a glacierized basin can be divided into two steps. The first is the simulation of the runoff formation. It is carried out by the simulation of energy-water balance and all the parameters are physically determined. The second step is the simulation of the runoff convergence processes whereby the runoff is transformed into the discharge at the basin outlet. It is carried out using the tank runoff transformation model, the parameters of which need to be calibrated. The model presented here links the standard meteorological elements to the processes in a glacierized basin, then to the runoff from the basin. The good discharge simulation results identify the response of runoff to the climatic variables in a glacierized basin. Therefore, this model can be applied to analyze the sensitivity of changes in glaciers and runoff in mountainous regions to changes in climatic variables. Further research and field experiments on the simulation of energy - water balance and runoff should be carried out in the hydrological basins on the scale of mountain ranges with covers of glaciers, snow and permafrost, in order to investigate the changes of mountain climatology, hydrology and water resources under global climatic change.

References

Calanca, P. and Heuberger, R. (1990) Energy balance, in A. Ohmura, H. Lang, F. Blumer and D. Grebner (eds.) Glacial Climate Research in the Tianshan, *Zürcher Geographische Schriften*, Heft 38, ETH Zürich, 60-70.

Kang, E. (1991) Relationship between runoff and meteorological factors and its simulation in a Tianshan glacierized basin, IAHS Publ. No. 205, 189-202.

Kang, E. (1994) Energy-water-mass balance and hydrological discharge, *Zürcher Geographische Schriften*, Heft 57, ETH Zürich.

Kang, E., Yang, D., Zhang, Y., Yang, X., and Shi, Y. (1992) An experimental study of the water and heat balance in the source area of the Urumqi River in the Tien Shan mountains, *Annals of Glaciology* **16**, 55-66.

Klemes, V. (1990) The modelling of mountain hydrology: the ultimate challenge, IAHS Publ. No. 190, 29-43.

Lang, H. (1986) Forcasting meltwater runoff from snow-covered areas and from glacier basins, in Kraijenhoff and Moll (eds.), *River Flow Modelling and Forecasting*, D. Reidel Company, Dordrecht, 99-127.

Ohmura, A., Lang, H., Blumer, F., and Grebner, D. (eds.) (1990) Glacial Climate Research in the Tianshan, *Zürcher Geographische Schriften*, Heft 38, ETH Zürich.

Rohrer, M. (1989) Determination of the transition air temperature from snow to rain and intensity of precipitation, in B. Sevruk (ed.) *Precipitation Measurement*, WMO/IAHS/ETH, 475-481.

Rohrer, M. B., and Lang, H. (1990) Point modelling of snow cover water equivalent based on observed variables of the standard meteorological networks, International Association of Hydrological Sciences Publ. No. 193, 197-204.

Sevruk, B. (1986) Correction of Precipitation Measurements, in B. Sevruk (ed.), *Zürcher Geographische Schriften*, Heft 23, ETH Zürich, 14-23.

Sugawara, M., Watanabe, I., Ozaki, E., and Katsuyama, Y. (1984) Tank Model with Snow Component, Publ. No. 65, National Research Center for Disaster Prevention, Japan, HEC Press, Tsukuba, 163-247.

WMO (1986) *Intercomparison of models of snowmelt runoff*, Operational Hydrology Report No. 23, World Meteorological Organization, Geneva, No. 646.

Yang, D., Shi, Y., Kang, E., and Zhang, Y. (1989) Research on analysis and correction of systematic errors in precipitation measurement in Urumqi River basin, Tianshan, in B. Sevruk (ed.), Precipitation Measurement, WMO/IAHS/ETH, workshop on precipitation measurement, St. Moriz, ETH Zürich, 173-180.

Yang, D., Shi, Y., Kang, E., Zhang, Y., Shang, S., and Wang, X. (1992) Analysis and correction of the systematic errors from precipitation measurement, in: Y. Shi, E. Kang, G. Zhang and Y. Qu (eds.), *Formation and Estimation of Mountain Water Resources in the Urumqi River Basin*, Science Press, Beijing, 14-40.

Yang, X. (1987/1990) Illustration of the compilation of hydrological and climatic data at the source of Urumqi River, Annual Report on the work at Tianshan Glaciological Station 6 and 9, Lanzhou Institute of Glaciology and Geocryology, Academia Sinica, Lanzhou, China.

Zhang, Y. and Kang, E., and Yang, D. (1992) An experimental study on the evaporation in the source area of the Urumqi River, in: Y. Shi, E. Kang, G. Zhang and Y. Qu (eds.), *Formation and Estimation of Mountain Water Resources in the Urumqi River Basin*, Science Press, Beijing, 79-89.

Kang Ersi and Shi Yafeng
Lanzhou Institute of Glaciology and Geocryology
Academia Sinica
Lanzhou 730000, China.

Atsumu Ohmura and Herbert Lang
Swiss Federal Institute of Technology (ETH)
CH-8057 Zurich, Switzerland.

14 IMPACT OF FUTURE CLIMATE CHANGE ON GLACIER RUNOFF AND THE POSSIBILITIES FOR ARTIFICIALLY INCREASING MELT WATER RUNOFF IN THE ARAL SEA BASIN

A.N. KRENKE and G.N. KRAVCHENKO

Abstract

According to Kovaneva (1982), who investigated the surface air temperature distribution in years with high global temperature values, the summer temperature in the Pamirs and Tyan-Shan will fall by $0.5°C$ in the event of a $1°C$ global warming of mean annual air temperature. Precipitation will increase by 10-15% according to Groisman (1981). This will lead to the descent of the equilibrium line on the glaciers by 120-170 m. The boundary between the warm and the cold ice-formation zones and the upper limit of the runoff from the glaciers will descend by 90-150 m. As a result, the area of the cold firn zone will increase by 18% at the expense of the ablation zone. The decrease in the natural glacier runoff is estimated. The following data were taken into account: the altitudinal distribution of the glacier surface, the dependence of melting on the summer air temperature and the refreezing of melted water in the firn bodies in different ice-formation zones. The decrease in the total glacier runoff due to assumed climate change appears to be equal to 15%. Any artificial increase of glacier runoff could not balance this decrease in the natural runoff. The potential for such artificial increase would be less by 0.2 km^3. The share of accumulation area in the total runoff would be bigger than now.

14.1. Introduction

The total area of glaciers in the Aral sea basin over the territory of the FSU republics (Kirghizia, Tadjikistan, Uzbekistan) according to the USSR glacier catalogue (1968-1980), and the unpublished correction to it, is 10890 km^2. According to the unpublished inventory compiled by N.A. Urumbaev, the area of glaciers in the Afganistanian part of the Amu-Darya basin is an additional 3900 km^2. The glaciers occupy only 4% of the total Aral sea basin though they are responsible for 15% of the runoff in the basin. In the dry periods when the water in the basin is especially important the share of the glaciers in the runoff increases to 40%. Because of the lack of necessary data from the Afganistan section the following analysis is dealing only with FSU data. We hope our main conclusions will be valid for the Afganistanian glaciers as well.

The general scientific option is that global climate warming began towards the end of nineteenth century (e.g. Jones, 1988) and suggests its further prolongation by the anthropogenic 'greenhouse effect', although there are regional differences in the degree

J.A.A. Jones et al. (eds.), Regional Hydrological Response to Climate Change, 259–267.

and even in the sign of the temperature changes. During the last century some decrease in the summer air temperature has taken place in the Asian continental region including Aral sea basin (Predstojashchie izmenenija, 1991). The object of this paper is to evaluate the impact of supposed climate change on glacier runoff in the Aral sea basin.

14.2. Assumptions

According to Krenke and Khodakov (1966) the main factors influencing glacier runoff are the yearly solid precipitation and the summer air temperature. Judging from previous years with positive global temperature anomalies, in the mountainous part of the Aral sea basin a 1°C global warming corresponds to a decrease in summer air temperature by 0.5°C (Kovaneva, 1982) and an increase in the yearly precipitation by 15% in the Pamirs and the western Tyan-Shan and by 10% in the eastern and central Tyan-Shan (Vinnikov and Groisman, 1979; Groisman, 1981). At the height of the glaciers in the Aral sea basin practically all precipitation is solid. The changes are positive for the mass balance of the glaciers and negative for the runoff from the units of glacier surface. In this paper we will not take in the account the changes in the glacier surface and morphology. At the moment, the glaciers' balance is negative and its improvement is thought to be only capable of creating zero balance. Processes influencing glacier morphology are not well known and have not been put into the calculation.

The changes in the air temperature lead to changes in the melting/precipitation ratio, to changes in the water loss due to refreezing in the firn body and thus to changes in the glacier runoff coefficient. According to Shumsky (1955) and Krenke (1975) glacier runoff is equal to zero in the case of melting below a ratio of 0.1 from the solid precipitation. The melted water if it exist refreezes in the upper firn layer deposited in the last year. Thus we consider the '0.1' level as the limit of glacial runoff formation. If the melting/solid precipitation ratio is between 0.1 and 0.5 the 'cold firn zone' exists and the loss of water on refreezing is assumed to be equal to 0.5. The runoff coefficient will also be equal to 0.5. If the melting/solid precipitation ratio is between 0.5 and 1.0 the 'warm firn zone' or the 'superimposed ice zone' exist depending on the ice temperature during the summer. In both cases the water loss due to refreezing is assumed to be equal to 0.25 and the runoff coefficient equal to 0.75. The runoff coefficient is practically equal to 1.0 in the ablation zone, where the melting/solid precipitation ratio is higher than 1.0. Under the assumed climate change scenario, the opposite sign of change in air temperature and precipitation changes this ratio, the position of altitudinal limits between glaciological zones and their areas.

14.3. Methods

We have taken the above mentioned reconstruction for the global warming by 1°C as the climate change scenario (Groisman, 1981, Kovaneva, 1982). The same changes were assumed to apply at different altitudes. This is supported by the comparison of the data from stations situated at different levels. For example, the sign of interannual changes in summer temperature was the same in 34 cases out of 39 at the Kulyab (610 m a.s.l.) and Lednik Fedchenko (4169 m a.s.l.) stations. The Aral Sea basin was divided into 14 regions according to the river basins. For each basin the total glacierized area and its distribution over each 200 m altitudinal zone were determined. The so-called hypsographic curves of the glacierized areas were plotted for each basin. From these curves, the area of any altitudinal belt could be estimated.

Elsewhere, we have drawn the position of the equilibrium line on the glaciers of the region (Krenke, 1975, 1982). That is the level of balance between accumulation and ablation, which is close to the snowline. The summer air temperature and melting at the equilibrium line and the solid precipitation at the same level were thus determined by Krenke (1975, 1982). To estimate the summer ablation from the air temperature, we used the following formula (Krenke and Khodakov, 1966):

$$A = (T_s + 9.5) \tag{14.1}$$

in which A is the ablation in mm and T_s is the mean June-August air temperature in degrees C. We calculated the depths and the volume of ablated snow and ice between each 200 m of altitude from formula (1) and plotted the ablation/height curves for each of the 14 regions. The vertical summer air temperature gradient above the glacier surface, according to previous estimation based on field observation (Krenke, 1982), is assumed to be 0.65°C. Because of the lack of data, the solid precipitation at each level above and some hundreds of meters below the snowline was assumed to be the same as at the snowline height. Comparing the ablation/height curves with the amount of precipitation for each of the 14 basins, we estimated the contemporary position of the highest limit of glacier runoff and the height of the lower boundary of the cold firn zone. We then plotted the ablation/height curve for the future climate scenario. The summer air-temperature for all the regions and all the altitudinal belts was assumed to be 0.5°C lower than now. These new ablation/height curves were compared with the solid precipitation which were assumed to be 15% higher for the Pamirs, Guissar-Alay mountains (Alay, Turkestan, Zeravshan and Alay ranges) and western Tyan-Shan (Talas, Ugam, Pskem and Chatkal ranges). For all the ranges of Northern and Central Tyan-Shan, the solid precipitation were increased by 10%. Comparing the new ablation/height curve with the precipitation we have estimated the future altitudinal position of the limit of glacier runoff, the lower boundary of the cold firn zone and the equilibrium line over the glaciers.

The surfaces of all glaciological zones were estimated for both contemporary and future cases. We have estimated the runoff from each zone for both cases in the following way. We calculated the melted layer for each altitudinal belt inside the glaciological zone according to formula (1). Ablation under moraines was assumed to be equal to 0.5 from open surface, to which the formula was applied. The runoff was assumed to be zero above the glacier runoff limit, to be equal to 0.5 from the ablation in the cold firn zone and to 0.75 from the ablation beneath the lower boundary of the cold firn zone, though above the equilibrim line. We estimated the glacier runoff value by multiplying the runoff layer on the surface area for each altitudinal belt inside each glaciological zone.

For both scenarios, we have estimated the possibilities of artificial runoff generation by scattering coal dust over the glacier surface in order to change the albedo. The following rules were used in this calculation:

1. The coal dust had no influence on the runoff above the runoff limit and in the areas of the continuous moraines.

2. From the open ice in the ablation zone, according to our own (Dolgushin *et al.*, 1965) and other experiments, the increase in runoff due to the coal dust is assumed to be equal to 25% from the natural runoff.

3. In the cold firn zone, the albedo will drop from 0.7 to 0.4 because of the dust, and the melting will increase twofold. All the additional water will be added to the runoff.

4. In the cold firn zone, the albedo will drop because of the dust from 0.8 to 0.4 and the melting will increase threefold, though 50% of the additional water would refreeze in the firn body.

14.4. Results and discussion

The results of the estimation of the contemporary and future boundaries between the glaciological zones are given in the Table 14.1. Small changes in the altitudinal positions of the glaciological zones would lead to important changes in glacier runoff. The limit of glacier runoff on those glaciers which are high enough to have a part above it will expand downslope by 100-120 m. As a result, the glacier area without runoff, will increase by 37% from 663 km^2 to 909 km^2 (Table 14.2). The lower boundary of the cold firn zone will descend downslope by 130-150 m and its area will increase by more then 18% from 4000 km^2 to 4740 km^2. The equilibrium line will descend by 120-160 m. The warm firn and superimposed ice zones will descend by approximately 140 m. Its surface area will not changed very much, decreasing by 13% from 2150 km^2 to 2440 km^2. The ablation area will decrease by 25% (2800 km^2 instead of 3710 km^2), because of the lowering of the equilibrium line. The latter change will take place totally at the expense of the bare ice. The surface of the ice beneath the moraine will not change.

According to Table 14.2, the total glacier runoff will decrease by 14% from 12.8 km^3 to 11.0 km^3. Inspite of this, the runoff from the accumulation area will increase by 17% from 3.5 km^3 to 4.1 km^3. The runoff from the ablation area will decrease by 23% from 10.0 km^3 to 7.7 km^3, and the share of the accumulation area in the total ablation will increase from 27% to 37%. The most drastic changes in glacier runoff, with decreases exceeding 20%, could occur in the basins with a marked continental climate. This applies to the basins of the upper part of Naryn River in the highlands of inner Tyan Shan, Kyzylsu basin (Alay valley) and the Bartang basin (highlands in the Pamirs). Minor changes, with runoff decreasing by less than 12%, are expected in the basins with maritime climate on the western periphery of the mountains: the Zeravshan, Kashka-Darya, Obikhingou basins. The Moorgab basin is an exception, where the calculated changes are small inspite of its continental climate.

TABLE 14.1. The position of glaciohydrological boundaries on glaciers in the Aral Sea drainage system (meters above ocean level). 1 = contemporary position; 2 = position under assumed climate change scenario.

Basin name and number	Highest point on glacier	Upper limit of glacier runoff		Lower boundary of cold firn zone		Equilibrium line position		Lowest point on glacier
	1 and 2	1	2	1	2	1	2	1 and 2
1. Arys, Pskem, Chatkal and right tributaries of Naryn below Kokomeren	4440	4440	4440	4000	3870	3620	3470	2720
2. Kokomeren, Atbashi and Naryn tributaries	5420	5350	5250	4400	4290	4100	3980	3400
3. Left tributaries of Naryn down the Atbashi	5120	5000	4890	4380	4240	3950	3800	2760
4. Kara-Darya and left tirbutaries of Syr-Darya up to Aksu	5560	5270	5160	4750	4620	4100	3960	2750
5. Surkhan-Darya and Kafirnigan	5300	5100	4980	4200	4050	3800	3640	2720
6. Kyzylsu and Surkhob up to Obikhingou junction without Muksu	7000	5100	4980	4700	4560	4370	4220	2320
7. Fedchenko glacier	7400	5700	5590	4950	4820	4650	4510	2880
8. Muksu without Fedchenko glacier	7400	5700	5590	4950	4820	4650	4510	2820
9. Obikhingou and right tributaries of Pyange (Vanch to Vakhsh junctions)	6350	5870	5750	4620	4470	4180	4020	2700
10. Vanch, Yazgulem and right tributaries of Pyange (Vanch to Gunt junctions)	6590	5650	5530	4800	4660	4460	4300	2580
11. Bartang	6970	5650	5540	4800	4670	4600	4450	3400
12. Moorgab	6100	5780	5680	5250	5130	5000	4870	4260
13. Gunt and Pyange upstream of Gunt mouth	6780	5780	5670	5250	5120	5050	4900	3420
14. Zeravshan and Kashka-Darya	5620	5350	5240	4460	4320	3900	3790	2780

Naturally, in the case of cooling and increased precipitation the snow runoff outside the glaciers will increase. Thus, global warming will lead to a shift in the runoff structure in the Aral sea basin in favor of snowmelt runoff and at the expense of glacier runoff. As a result increased flow regulation will be necessary to supply water to the agricultural fields at the required time, and this has to be taken into account in water use planning.

TABLE 14.2. Changes in glacier runoff under the assumed scenario. 1 = contemporary conditions, 2 = future climatic scenario (Surface areas in km², Ablation and Runoff in km³)

Basin no.	Without runoff Surface 1	Without runoff Surface 2	Cold firn zone Surface 1	Cold firn zone Surface 2	Cold firn zone Ablation 1	Cold firn zone Ablation 2	Cold firn zone Runoff 1	Cold firn zone Runoff 2	Warm firm and Superimposed ice zone Surface 1	Warm Surface 2	Warm Ablation 1	Warm Ablation 2	Warm Runoff 1	Warm Runoff 2	Ablation Surface 1	Ablation Surface 2	Ablation/Runoff 1	Ablation/Runoff 2	Surface 1=2	In total Ablation 1	In total Ablation 2	In total Runoff 1	In total Runoff 2	Change Runoff 2-1	Change %
1	0	0	50	86	.04	.07	.01	.03	108	143	.20	.29	.14	.22	209	139	.76	.52	367	1.00	.88	.91	.77	-.14	-15
2	8	26	559	647	.13	.15	.06	.07	273	249	.22	.17	.17	.13	275	192	.48	.32	1110	.83	.64	.71	.52	-.19	-27
3	5	12	106	142	.06	.07	.03	.04	166	193	.24	.31	.18	.24	292	222	1.03	.78	569	1.33	1.16	1.24	1.06	-.18	-14
4	3	7	37	55	.01	.01	.01	.01	194	235	.21	.28	.15	.21	332	270	1.04	.84	566	1.26	1.13	1.20	1.06	-.14	-12
5	1	2	60	86	.07	.11	.04	.06	71	67	.21	.21	.15	.15	62	39	.38	.25	194	.66	.57	.57	.46	-.11	-19
6	136	154	435	487	.08	.08	.04	.04	249	228	.14	.15	.10	.13	259	211	.29	.21	1080	.51	.44	.43	.38	-.10	-23
7	62	84	297	339	.04	.06	.03	.03	145	137	.10	.10	.07	.08	255	200	.56	.45	760	.70	.61	.66	.56	-.10	-15
8	112	145	594	737	.11	.14	.06	.08	348	272	.22	.20	.17	.15	325	225	.64	.55	1380	.97	.84	.87	.73	-.14	-16
9	24	34	335	393	.14	.20	.07	.10	183	173	.32	.36	.25	.27	244	286	.90	.71	786	1.36	1.27	1.22	1.08	-.14	-12
10	65	80	223	264	.11	.17	.06	.08	126	124	.21	.24	.15	.18	317	264	1.19	.96	732	1.51	1.37	1.40	1.22	-.18	-13
11	58	92	634	712	.36	.42	.17	.20	173	146	.25	.22	.18	.17	247	163	.57	.36	1110	1.18	1.00	.92	.73	-.19	-21
12	38	61	211	252	.06	.07	.03	.04	138	133	.10	.11	.07	.08	189	129	.31	.24	574	.47	.42	.41	.36	-.05	-12
13	137	189	266	307	.06	.08	.03	.04	127	132	.08	.11	.07	.08	515	419	1.35	1.12	1050	1.49	1.31	1.45	1.24	-.24	-14
14	14	23	193	233	.08	.11	.04	.06	209	208	.34	.39	.26	.29	189	141	.50	.43	605	.92	.93	.80	.78	-.02	-2
Total	663	909	4000	4740	1.35	1.74	.67	.88	2510	2440	2.84	3.14	2.11	2.38	3710	2800	10.00	7.69	10 890	14.20	12.60	12.80	11.00	-1.83	-14

The shift in the runoff pattern in favour of the accumulation area increases the potential of artificial melting there. In contrust, in the ablation area these resources will decrease. The results of calculations made according to the rules described in section 14.3 are given in Table 14.3. The changes are not to great. Only in the westernmost part of the mountains (western Tyan-Shan, Zeravshan and Kashka-Darya basins) is the balance in favour of increasing the potential for artificial melting. In all other areas the potential for artificial melting will decrease, in total by 0.2 km³, which is in the margin of error of the prediction procedure. The cost of artificial melting will increase because of repeated summer snowfalls in the accumulation area. However, the ecological danger to the preservation of the glacier will be less because of the more healthy state of the glaciers.

TABLE 14.3. Changes in the potential artificial runoff
due to the application of coal dust

Number of river basin (see Table 14.1)	Potential artificial runoff (km³)		Difference between the future and present conditions	
	In present conditions	Under the future climate scenario	km³	%
1	0.31	0.33	+0.02	+6
2	0.30	0.26	-0.04	-13
3	0.39	0.39	0.0	0
4	0.32	0.34	+0.02	+6
5	0.26	0.24	-0.02	-8
6	0.08	0.07	-0.01	-12
7	0.16	0.14	-0.02	-14
8	0.32	0.28	-0.04	-12
9	0.42	0.40	-0.02	-5
10	0.33	0.32	-0.01	-3
11	0.38	0.33	-0.05	-13
12	0.16	0.16	0.0	0
13	0.37	0.33	-0.04	-11
14	0.32	0.34	+0.02	+6
Total	4.12	3.93	-0.19	-5

Budyko's scenario was based on the analysis of the relationship of the local climate changes to the global warming for the period 1881-1980. During the last warming in the 1980s the climate change features were opposite to the previous century. In Central Asia the summer was warmer and precipitation not greater than in the average. If the future changes are similar to this last example, the changes in glacier runoff will differ from those described. Even so, the described method of prediction could still be used.

14.5. Conclusions

The results suggest that global warming will lead to the shift of mountain runoff in the Aral Sea basin in favour of the snowmelt runoff at the expense of icemelt runoff. The amount of necessary seasonal artificial runoff regulation needed will increase. The decrease of glacier runoff under 1°C global warming is calculated to be about 15%. Glacier runoff patterns will change in favor of the accumulation area. The accumulation/ablation areal ratio will increase from 2.0 to 2.9 (i.e. by 45%), and the share of the accumulation area in the runoff will increase from 0.27 to 0.37 (i.e. by 37%). If global warming remains stable enough, glacier runoff could increase anew due to the increasing area of glacier surface.

Acknowledgements

The authors are thankfull to the Academy of Sciences of China Peoples Republic and to the Russian Fundamental Science Fund (Grant no. 93-05-9699) who supported the authors of this publication and its presentation at the Lhasa Conference.

References

Groisman, P.Ya. (1981) Empiricheskie ocenki svyazi processov globalnogo potepleniya s regimom uvlagnyonnosti territorii SSSR (Empirical evaluation of the relation between the global warming and cooling and the precipitation regime over the USSR territory.), *Izvestiya AN SSSR, Ser. Geographicheskaya* 5, 86-95. (In Russian.)

Dolgushin, L.D., Kemmerikh, A.O., and Krenke, A.N. (1965) Polevye issledovaniya po teme 'Razrabotka nauchnykh osnov iskusstvenogo usileniya tayaniya lednikov Sredneyi Azii' (The field studies according to the project 'Development of the scientific base for the artificial increasing of Central Asia glacier melting'), in *Materialy glyaciologicheskikh issledivaniyi, Khronika, Obsuzhdeniy* 11, 5-15. In Russian

Johns, P.D. (1988) The influence of ENSO on global temperature, *Climate monitor* 17, 3, 80-89.

Kovaneva, N. (1982) Statisticheskoe issledovanie sovremennykh poleyi izmeneniya temperatury prizemnogo vozdukha i atmosphernogo davleniya (Statistical study of the regularities in the contemporary changes of the surface air temperature and the pressure fields), in *Voprosy Hydrologii Sushi*, Hydrometeoizdat, Leningrad, 200-210. (In Russian.)

Krenke, A.N. (1975) Climatic conditions of present-day glaciation in Soviet Central Asia, in International Association of Hydrological Sciences Pub. No. 104, 30-41.

Krenke, A.N. (1982) *Massoobmen v lednikovykh sistemakh SSSR. (Mass exchange in the glacier systems of USSR)*, Hydrometeoizdat, Leningrad. (In Russian.)

Krenke, A.N., and Khodakov V. G. (1966) O svyazi poverkhnostnogo tayaniya lednikov s temperaturoyi vozdukha (About the relation between surface ablation of the glaciers and the air temperature), in *Materialy glyaciologicheskikh issledovaniyi, Khronika, obsuzhdeniya* 12, 153-164. (In Russian.)

Budyko, M.I. (ed.) (1991) *Predstoyashchie izmeneniya klimata (Future climate changes)*, Hydrometeoizdat, Leningrad, 250 pp. (In Russian.)

Shumskiyi, P. A. (1955) *Osnovy strukturnogo ledovedeniya (The basis of stuctural ice science)*, Izd. Akademii Nauk SSSR, Moscow. (In Russian.)

USSR (1980) *Katalog lednikov SSSR. (Inventory of the glaciers in the USSR)* (1968 - 1980), Tom (Vol.) 14, Vyp (Issue) 1, Chasti (Parts) 1-11, Vyp. 3, Chasti 1-20, plus unpublished supplementary inventory by N.A. Urumbaev. (In Russian.)

Vinnikov K. Ya., and Groisman P. Ya. (1979) Empiricheskaya model sovremennykh izmeneniy klimata (Empirical model of the contamporary climate changes), *Meteorologiya i Hydrologiya* **3**, 16-24. (In Russian.)

A.N. Krenke and G.N. Kravchenko
Institute of Geography
Russian Academy of Science
Staromonetny 29, Moscow 109017, Russia.

15 GLACIERS AND SNOWCOVER IN CENTRAL ASIA AS INDICATORS OF CLIMATE CHANGE IN THE EARTH-OCEAN-ATMOSPHERE SYSTEM

VLADIMIR B. AIZEN and ELENA M. AIZEN

Abstract

We propose a scheme of natural interactions in the earth-ocean-atmosphere system, where the seasonal snowcover of Eurasia serves as a possible regulator of circulation, and glaciers serve as indicators of global climate change. Standard hydrometeorological, expedition and satellite data collected since the 1890s are used to obtain quantitative estimates of present-day deviations of circulation indices, area covered by seasonal snow, and global air temperature. We estimate deviations of the mass-balance components for five mountain-glacier systems in remote areas of central Asia.

15.1. Introduction

This investigation concerns interactions between natural processes in the 'earth-ocean-atmosphere' (EOA) global system. The main goal is understanding of some mechanisms in this system and the determination of their maximum probable deviations from the normal state. Our work is based on the thesis that the EOA system is self-regulating (Gernet, 1981; Kotlyakov, 1987), i.e., where natural processes maintain this system in a steady state.

Taking the EOA system to be relatively stable against external effects on the present time scale (i.e., during the last century), we wish to estimate the probable range of variation of natural processes within the system. We consider catastrophic events such as the impact of large meteorites or great volcanic eruptions are to be external to the system we are studying.

Numerous investigations are dedicated to questions of snowcover and general circulation interaction, e.g. Karapetyance and Morozov (1983) and Kotlyakov (1987). Voeikov (1889) wrote about the cooling effect of snowcover on the atmosphere. Kumar (1988) and Barnett (1989) established that variations of snowcover extent on the Eurasian continent influence monsoon activities. Kukla (1981) shows that there are a great number of feedback relations in which snowcover is of great importance. Differences in average air temperature in April between days with snowcover and

J.A.A. Jones et al. (eds.), Regional Hydrological Response to Climate Change, 269–285.

without have a mean of 4.3°C (Pokrovskya and Spirina 1965). According to Krenke and Loktionova (1989), if there is no snowmelt, the heating of atmosphere would be in two or three times greater. Afanasieva and Esakova (1964, 1969, 1974) demonstrated a feedback relation between the Eurasian seasonal snowcover boundary and a zonal index of circulation, and found a correlation between this boundary and the zone of frontal activity. The interaction between snowcover and the development of the Siberian anticyclone was established by Kotlyakov (1987). Robinson (1992) discussed the critical role of snowcover in the global heat budget.

In this paper we are considering one part of the EOA system, where seasonal snowcover functions as a self-regulating mechanism. Because Eurasia is a much larger land mass than other continents, there are long periods when either zonal or meridional circulation patterns prevail there (Figure 15.1). There is a much stronger correlation between snow-covered areas in the Northern Hemisphere and in the Eurasia (r = 0.84) than between snow - covered areas in the Northern Hemisphere and in the North America (r = 0.23). Seasonal snow-covered areas for North America, Eurasia and the entire Northern Hemisphere during 1966-76 derived from satellite data (Matson, 1977) was used for these calculations.

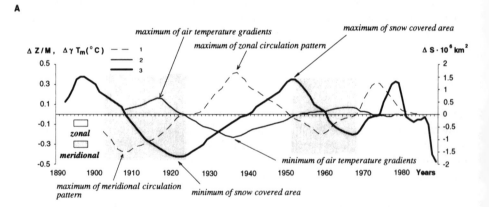

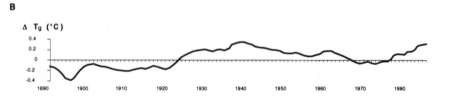

Figure 15.1. A) Changes in (1) the zonal index ($\Delta Z/M$), (2) meridional gradients of air temperature ($\Delta\gamma T_m$), according to Vinnikov (1986) and (3) snow-covered area, during zonal and meridional epochs. B) Changes in global air temperatures between latitudes $87.5°$ and $17.5°N$, after Vinnikov (1986).

15.2. Methods

15.2.1. SNOW-COVERED AREAS

To obtain quantitative estimates of the present oscillations of climate in the EOA system, we used archived data on the yearly position of seasonal snowcover boundary in the territory of the USSR from 1890 to 1975 (Loktionova, 1976, 1985). Yearly data on the maximum area covered by snow in Eurasia and in Northern Hemisphere were obtained from satellites for the periods 1966-1976 (Matson, 1977) and 1967-1988 (Barry, 1990). Annual deviations of snow covered areas from the mean in the Northern Hemisphere from 1972 to 1989 are from Folland et al. (1992). Unfortunately, because of the absence of data for the position of the seasonal snowcover boundary in winter months outside the USSR, we used data for April when this boundary is in the USSR. The area of maximum snowcover extent in Eurasia during the winter-spring period (Matson, 1977) is well correlated with its April extent in the USSR (r = 0.72). Therefore the April data for the area covered by snow inside the USSR were used as an index of maximum extent of seasonal snowcover for the Eurasian continent and for the entire Northern Hemisphere. We then calculated values of five-year moving means of the area covered by seasonal in the USSR during April from 1890 to 1975 (S_s), for the Northern Hemisphere during 1967-1989 and their deviations from mean ($\Delta S = S - S_s$) (Figure 15.1).

In the same way we obtained the deviations of all other parameters. We used the ratio (Z/M) between the duration of zonal circulation patterns (Z) and meridional circulation patterns (M) (Krenke, 1987). To plot the deviations of the zonality ratio ($\Delta Z/M$), we used the data for the period since 1900 to 1987. Deviations are the differences between the long-term mean of this index for the mentioned period (Z/M) and its moving five-year mean.

For the analysis of deviations from the mean meridional air temperature gradient ($\Delta \gamma$ $T_m/10°$lat.) and global air temperature ($\Delta T_g °C$) in the Northern Hemisphere, we used the data presented by Vinnikov (1986).

On the basis of our analysis (Figure 15.1) meridional circulation patterns predominates during two periods: 1903 - 1924, and 1952-1968, and two periods of zonal circulation patterns predominates from 1925 to 1951; and 1969-1987.

15.2.2. SNOW WATER EQUIVALENT

For the analysis of snow water equivalent we used annual data of 150 stations situated in the USSR for the period from 1936 to 1973:

1. The mean maximum snow water equivalent, $\overline{X}_o$, for the observed period 1936-1973;

2. The mean snow water equivalent values, $\overline{X}_z$, and $\overline{X}_m$, for each circulation epoch (zonal: 1936 - 1951, 1969-1973, meridional: 1952 - 1968);

3. The mean long term deviations from the mean for each epoch:

$$\Delta X_z = \frac{1}{N} \sum_{i=1}^{N} (X_{iz} - \overline{X_z}) \tag{15.1}$$

$$\Delta X_m = \frac{1}{N} \sum_{i=1}^{N} (X_{im} - \overline{X_m}) \tag{15.2}$$

where X_{iz} and X_{im} are the maximum snow water equivalent for year 'i', corresponding to zonal and meridional epochs, and N is the number of years for which there are data in each epoch.

4. The standard deviation for the entire period of observation σ_o, as well as for each epoch, σ_z and σ_m.

5.. The Fisher criterion, F, the level of significance of the difference between averages and variance of snow water equivalent for two circulation epochs.

The calculations were done for zonal and meridional epochs. In this case snow water equivalent, but not seasonal snow-covered area was considered.

15.2.3. GLACIERS

Considering the snowcover of Eurasia as a regulator of macro scale climatic changes, we would like to quantify the changes in the main indices of mass-energy transfer for some mountain-glacier systems located on the periphery of central Asia (the North and Central Tien Shan, Central Pamir, Southeast Tibet and North Himalayas). The main indices are mean values of summer air temperature at the mean long term altitude of firn line position, T_s, annual quantity of precipitation at the mean long term altitude of firn line position, P_r, altitude of the firn line position H_{fl}. We calculated values of five-year moving means of these main indices and their deviations from the mean ($\Delta T_s = T_s - T_{s_5}$, $\Delta P_r = P_r - P_{r_5}$, $\Delta H_{fl} = H_{fl} - H_{fl_5}$) (Figure 15.2) for five regions of glaciation in the central Asia. We used expeditionary and station data on the Golubina, Tuyuksu, and Inylchek glaciers since 1900, the Fedchenko glacier since 1934, the Hailuogou and Xixibangma glaciers since 1952, and standard meteorological information from 36 stations. These areas were chosen because outlying mountains are the first to be affected by changes in the general circulation flows of the atmosphere. Therefore, glaciation in these regions is less balanced and responds strongly to climatic changes. However, each of these areas has unique glacioclimatic conditions depending on orography, latitude and location in central Asia (Suslov,1980; Krenke,1984; Aizen, 1985, 1990; Aizen and Aizen, 1991, 1992, in press).

Mass balance was calculated using measured data and the calculating method of Konovalov (1979), Aizen (1988), and Glazirin (1991).

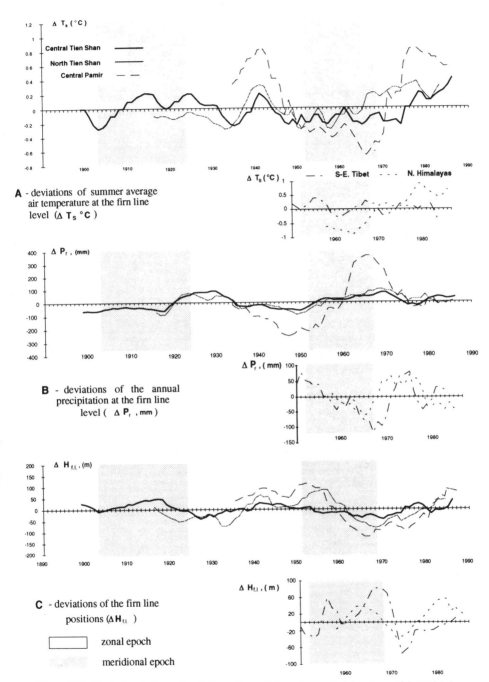

Figure 15.2. Deviations in the regime indices of mass balance in the glacial systems of Central Asia during zonal and meridional (shaded) epochs: A) Deviations in local summer temperature at firn line level; B) Deviations in annual precipitation at firn line level; C) Deviations in firn line position.

15.3. Results

15.3.1. SNOW-COVERED AREAS

We cannot suggest that there is a periodic recurrence, because our investigations are limited to the existing period of measurements. However, 5-year moving means do seem to show some trends in the development of oscillations. By self-regulating mechanism, we mean the capacity of snowcover to influence itself by feedback through atmosphere cooling, i.e. the interaction occurs through changes in the meridional gradient of global air temperature and consequently through changes in the prevalence of zonal or meridional patterns.

Calculated curves (Figure 15.1) established a clear interaction between the general circulation of the atmosphere, the meridional gradient of global air temperature, and snow covered area. For example, the maximum meridional index (around 1908 and 1959) occurs near to years when the meridional gradient of air temperature has maximal values. The extreme minimum and maximum conditions in snow covered area are observed in years (around 1922 and 1951) with average conditions in the development of meridional and zonal patterns, ($\Delta Z/M$ is near 0 and Z/M is near 1). Correspondingly, the average state in snow covered area is observed in years when the extreme development of any circulation form occurred (around 1909, 1940, 1961 and 1974). However, interaction happens continuously, not just in extreme years, i.e. any fixed condition of atmosphere circulation corresponds to a certain state in snow covered area, and *vice versa*. The direct influence of the general circulation of the atmosphere becomes more apparent in the years of changing sign in anomalies of the zonal index. For example, around 1924 the anomalies of the zonal index changed from negative to positive values, i.e. zonal forms of circulation begin to predominate and about this time the process of reduction in snow covered area stopped and anomalies in snow covered areas began to increase gradually. The curve of anomalies in the snow covered area changes its direction (Figure 15.1). The feedback loop becomes more sharply apparent in years of changing of sign in anomalies of snow covered area (around 1909, 1940, 1961 and 1974), because about this time the curve of zonal indexes changes its direction. When the area covered by snow becomes greater than average (around 1941 and 1974), the growth in the duration of zonal forms ceased and the duration of meridional forms begins to increase, because of the cooling effect of snowcover.

We can explain the lag of change in snow covered area in terms of snow water equivalent, which is not shown in Figure 15.1. In flat terrain, snow-covered area is determined by the location of the snow boundary. Its location is controlled by air temperature and the quantity of solid precipitation or snow water equivalent, like the snowline in mountains. For example, the development of meridional forms of circulation is accompanied by a global decrease in air temperature (Figure 15.1), which favours snowcover conservation and its greater areal extent. However, during these periods of lower air temperatures, snow water equivalent and the quantity of precipitation drop, resulting in an opposite effect in the extent of snowcover, especially in the southern latitudes (Figure 15.3A).

 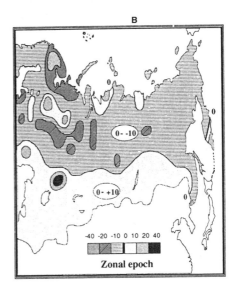

Figure 15.3. The distribution of snow water equivalent deviations in the former USSR during
(A) meridional (1952-1968) and (B) zonal (1936-1951, 1969-1973) epochs.

15.3.2. SNOW WATER EQUIVALENT

On the basis of annual snow water equivalent data, we have drawn maps of their
deviations from the mean during the zonal and meridional epochs of atmosphere
circulation (Figure 15.3). Using analysis of variance, we have established the correlation
between the values of snow parameters and the circulation conditions. The degree of
closeness of the connection depends on the latitude of the site (Karapetyance and
Morozov, 1983).

The statistics for several chosen stations are shown in the Table 15.1. According to
Fisher's variance ratio, the influence of atmospheric circulation conditions on snowcover
weakens from North to South. The greater stability of synoptic processes in zonal
epochs (Dzerdzeevsky, 1974) results in an evident growth in deviations of snow water
equivalent from the mean. The mean value of absolute deviations in snow water
equivalent for the zonal epoch is three times larger than the corresponding value in the
meridional epoch (Karapetyance and Morozov, 1983). The deviation maps also illustrate
the variation of snow water equivalent with the circulation regime, and the feedback
influence of snowcover on the general circulation of the atmosphere. An increase in
precipitation during the zonal epoch results in positive deviations in snow water
equivalent in southern Eurasia (Figure 15.2B). This effect should cause an increase in
the meridional gradient of global air temperature, a corresponding increase in meridional

TABLE 15.1. Statistical descriptions of snow water equivalent during meridional (m) and zonal (z) circulation epochs and for the whole period. Symbols explained below.

Station	$\varphi°N$	$\overline{X}_o$	σ_o	$\overline{X}_m$	$\overline{X}_z$	σ_m	σ_z	Δ_m	Δ_z	F	F'	DF
Bashanta	48	33	23	29	48	19	23	-4	15	3.75	3.38	25
Svatovo	49	40	25	35	56	21	30	-5	16	3.50	3.35	24
Aktiubinsk	50	70	28	66	88	24	39	-4	18	3.30	3.32	30
Shilka	52	21	11	19	35	11	12	-2	14	3.79	3.31	31
Nora	53	62	16	68	55	15	19	6	-7	3.30	3.33	27
Poltavka	54	55	15	55	71	17	11	0	16	4.11	3.28	34
Chulim	55	86	37	77	120	39	24	-9	34	5.03	3.32	30
Birsk	55	141	49	155	115	40	52	14	-26	4.71	3.35	27
Ishevsk	57	173	37	165	196	33	36	-8	23	4.45	3.34	28
Sobolevo	58	221	64	236	172	63	40	15	-49	5.10	3.42	23
Gonda	58	46	18	42	57	17	18	-4	11	4.42	3.34	28
Bereleh	62	84	22	89	74	24	6	5	-10	4.73	3.29	33
Arhangelsk	64	124	42	132	104	53	29	8	-20	4.05	3.27	35
Ain	66	81	16	87	68	14	13	6	-13	7.23	3.37	26

$\varphi°N$ - latitude; $\overline{X}_o$, $\overline{X}_m$ and $\overline{X}_z$ - average snow water equivalent for whole period and for meridional and zonal epochs respectively (mm); Δ_m and Δ_z - average deviations from long-term mean for meridional and zonal epochs (mm); σ_o, σ_m and σ_z - standard deviations for whole period, meridional and zonal epochs (mm); F - Fisher criterion; F' - limits; DF - degrees of freedom.

circulation and a weakening of zonal circulation and monsoon intensity. During the meridional epoch (Figure 15.3A) deviations have the opposite sign, which is explained by processes opposite to those which occur during the zonal epoch.

15.3.3. GLACIERS

The means of the mass-energy parameters for the mountain glacier systems studied and the characteristics of their variability are presented in Table 15.2. On average, the glacier systems have a negative mass balance, in concordance with data about the dominant meridional circulation, from 1900 to 1987. During this period the average value of the zonal index was 0.82. This form of circulation is not favorable for moisture transport inside Eurasia, resulting in negative mass balance values. The highest values of negative balance are typically found among glacier systems on the Northern periphery of central Asia, especially the glaciers of the Central Pamir, where glaciation conditions are determined by the influence of Atlantic air masses.

TABLE 15.2. Main statistical indices of five Central Asian glacial systems. Symbols explained below.

Glacial system	Prevalent aspect	φ °N	S_σ km²	$H_h - H_l$	T_s °C				P_r (mm)					H_{fl} (m)				B (g cm⁻²)			
					x̄	σ	min.	max	x̄	σ	min.	max.	d	x̄	σ	min.	max.	x̄	σ	min.	max
N Tien Shan	N, NW	42	1651	5020-3000	2.0	0.5	0.8	3.6	697	127	392	1058	5	3848	120	3560	4220	-15.0	32	-110	52
C Tien Shan	W	41	4320	7500-2600	-1.8	0.4	-2.7	-0.6	840	61	708	988	6	4476	47	4416	4646	-31.8	13	-47	16
C Pamir	N	38	3480	7120-2580	-0.2	0.9	-2.0	2.1	1192	284	728	2060	3	4780	186	4338	5187	-41.0	78	-207	150
SE Tibet	E, SE	29	90	7400-2900	4.9	0.7	2.1	6.2	2207	110	1758	2149	6	5200	171	4839	5647	-9.9	49	-123	99
N Himalayas	N	28	4840	8000-5500	2.4	0.4	1.7	2.9	558	94	436	901	8	5900	84	5700	'6300	-3.4	27	-84	111

φ°N - average latitude of glacial system; S_g - area of glacial system; $H_h - H_l$ - upper and lower levels of glaciations; T_s °C - summer temperature at firn line;
σ - standard deviations; P_r - annual precipitation at the firn line; $H_{f.l.}$ - position of the firn line; B - mass balance of the glacial system; σ - standard deviations;
d - month of maximum precipitation.

Glaciers in the Central Pamir and regions with monsoon type precipitation have the highest variability of mass balance and firn line position. In contrast, high stability in these characteristics is typical in the Central Tien Shan and on the northern slope of the Himalayas, which are located closer to the interior areas of Tibet.

15.4. Discussion

On the basis of these results, a scheme of interaction in the EOA system was developed. Certainly this scheme requires further investigation. Nevertheless, the evidence points to the important role of seasonal snowcover in climate change, and the necessity of accounting for it in the simulation of global atmospheric processes.

During the maximum development of meridional atmosphere circulation (around 1909 and 1959) the meridional gradient of air temperature reached its maximum values (around 1918 and 1964) (Figure 15.1). This prevented the penetration of moisture from the Indian and Pacific oceans into the regions that normally receive monsoon precipitation, such as the Himalayas and Tibet. During these periods, monsoon activity weakened, because a monsoonal gradient in air temperature directed opposite to the meridional one has low values. At the same time, during maximum development of meridional forms of atmosphere circulation, the penetration of western (zonal) moisture-carrying airstreams was prevented, reducing the delivery of moisture from the Atlantic Ocean. Thus, a decrease in precipitation and consequently in snow water equivalent (Figure 15.2), especially in southern latitudes (Figure 15.3A), results in a decrease in the area annually covered by seasonal snow (Figure 15.1). A few years after the maximum development of the meridional form of circulation (around 1921 and 1963) the seasonal snowcover occupied the smallest area. Thus, during maximum development of meridional forms of circulation (1906-1909, Figure 15.1) snow accumulation in Eurasia reached its lowest values (Figure 15.3A). The quantity of heat consumed in snow melting decreased and greater atmosphere heating began (air temperature from 1908 begins to increase, Figure 15.1), especially at high latitudes.

From this moment, a major shift in the circulation pattern took place. The decrease in the duration of the meridional pattern began and the gradual growth of zonal indices can be observed (Figure 15.1). The meridional air temperature gradient weakened and the increased zonal indices was accompanied by increased snow accumulation. However, the boundary of seasonal snowcover still moved north, because the global air temperature rose. Seven to fourteen years after the circulation shift began, increasing snow accumulation stopped the movement of the seasonal snowcover boundary towards higher latitudes. Starting in 1924, the beginning of a positive deviation in the zonal index, seasonal snow covers more and more area, gradually moving the snowcover boundary south again. At the end of the 1930s, snow accumulation was so great (Figure 15.3B) that from this moment the next decrease in global air temperature began (Figure 15.1), owing to the cooling influence of the growing snowcover. Therefore, at the end of the 1930s and the during the middle of the 1970s the transition took place towards a gradual decrease in the zonal indices and corresponding growth in the meridional atmospheric circulation pattern. At the same time, the decrease in air temperature lead to positive anomalies in snow covered areas. Seven to fourteen years after the maximum of the zonality indices the growth in the meridional air temperature gradient and the declining snow accumulation halt the increase in snow-covered area.

We postulate that the observed 7-14 year lag of correspondence in variations of atmosphere circulation and snow covered area (Figure 15.1) is the result of using data in snow covered areas rather than the amount of snow accumulation. The changes in the location of snow boundary reflect the influence of two opposite factors: air temperature and precipitation. We do not mean that this lag is the result of a feedback loop. Changes in snow covered area influence atmospheric circulation continuously. We simply observe this lag in the data and are unsure of its physical significance or its long term consistency because of the short period of record. This problem has not considered further in this paper, but is a subject for future investigations.

Analysis of annual data shows that the anomalies in global temperatures and indices of circulation are interconnected with the anomalies in snow-covered areas in the Northern Hemisphere and Eurasia. The degree and type of correlation is connected with corresponding periods of calculated data. During the predominance of one circulation pattern the highest positive correlation is observed for data with a 13-years lag. Linear regression equations used data with a 13 year lag for the zonal and meridional epochs, i.e., ΔT_g, $\Delta Z/M$ data from 1908 - 1938 and 1939 - 1957 were used with corresponding ΔS data from 1921-1951 and 1952 - 1970. The annual deviation from 5-years moving means were used for this calculation.

$$\Delta T_g = -0.06 + 0.26\Delta S \quad (r = 0.69) \qquad (15.3)$$

$$\Delta Z/M = 0.07 + 0.29\Delta S \quad (r = 0.69) \qquad (15.4)$$

The absolute extreme values in anomaly changes during considering period of years are for ΔT_g -0.49 to 0.56°C; for $\Delta Z/M$ -0.77 to 1.08; and for ΔS -6.9·10^6 to 12.8·10^6 km^2 The extreme 5-year moving values of anomalies are summarized in Table 15.3.

TABLE 15.3. Maximum deviations in climatic indices in the Northern Hemisphere.
Symbols explained below.

Epoch	Years	ΔT_o °C	$\Delta\gamma(T_m)$ °C	$\Delta Z/M$	ΔS 10^6 km^2	N'
Zonal 1	1924-51	0.40	-0.23	0.50	1.4	15-12
Zonal 2	1969-85	0.35	-0.04	0.34	1.3	10-7
Meridional 1	1902-23	-0.40	0.13	-0.47	-1.7	15
Meridional 2	1952-68	-0.20	0.08	-0.14	-0.7	7

ΔT_g ,$\Delta\gamma(T_m)$ - deviations of global air temperature and its meridional gradient; $\Delta Z/M$ - deviations of zonal indices; ΔS - deviations of snow-covered area; N' - number of years between the extreme values of $\Delta Z/M$ and ΔS.

We think that outlying areas of mountain glaciers in the central Asia mountains are more susceptible to climatic change than the interior ones. Analysis of glaciation in the EOA system shows that present-day fluctuations of global air temperature do not exert a direct influence on glaciers. According to Budyko (1974) and Vinnikov (1986), fluctuations of global air temperatures show up to a greater extent in winter and at high latitudes, i.e. North of present-day glaciation in central Asia. This can be explained by the fact that a rise of global air temperature is generally accompanied by an intensification of monsoon and zonal circulation increasing cloudiness and reducing insolation. According to our estimates, the present-day cloudiness accompanying monsoon activity reduces insolation to 1/6 of what is theoretically possible in these latitudes (Aizen and Loktionova, 1992). Theoretically, the amount of radiation to which this region could be exposed is great (at $25°$-$30°$ North, radiation $I = 0.0836$ MJ m^{-2} min^{-1} at the top of the atmosphere). Without atmospheric effects (reflection and absorption from clouds, etc.) during the warm half of the year irradiance equals 7315 MJ m^{-2} (40 MJ m^{-2} d^{-1} (Alisov *et al.*, 1952)). According to expeditionary measurements at 3400 m on Hailougou glacier during the summer-autumn monsoon of 1990, the average insolation intensity amounted 6.9 MJ m^{-2} d^{-1} (Aizen and Loktionova, 1992).

The intensification of monsoon activity results in a drop in local air temperature, especially at low latitudes. In contrast, a global drop in air temperature should be accompanied by an increase in solar radiation and an increase in local temperature, especially in low latitudes. For quantification of the interaction between local air temperature and the indices of circulation intensity, we calculated the coefficients of correlation between average summer temperatures and the summary values of annual precipitation (Table 15.4). The negative values of the coefficients confirmed this assumption. Although there is no clear inverse relation here (the values of correlation coefficients are small, Table 15.4), in all cases there is a tendency for a decrease in local air temperature to accompany a simultaneous increase in precipitation, and *vice versa*. This is especially apparent in the southern areas of glaciation in central Asia. In our opinion, these relationships are one of the pre-conditions for the existence of glaciers at low latitudes.

TABLE 15.4. Correlation coefficients for regime indices in Central Asian glacial systems.

Glacial system	$T_c°/P_r$	$T_c°/H_n$	P_r/H_n
N Tien Shan	-0.17	0.85	-0.35
C Tien Shan	-0.10	0.79	-0.59
C Pamir	-0.32	0.90	-0.60
SE Tibet	-0.45	0.73	-0.53
N Himalayas	-0.42	0.77	-0.58

$T_c°$ - summer air temperature at the firn line;

P_r - annual precipitation; H_n - firn line position.

The position of the firn line serves as one of the main indices of glacier existence (Krenke, 1982; Aizen, 1990; Glazirin, 1991). According to our investigations, its altitude is determined by predictors such as average summer air temperature and annual precipitation (Table 15. 4). We calculated the values of 5-year moving means of summer air temperatures, annual precipitation totals and firn line position and their deviations from the long term means.

Lowering of the firn line reaches its maximum at the end of meridional circulation epochs and the beginning of zonal ones (i.e., 5-7 years the before the maximum of zonal period precipitation increase). It occurs apparently before the maximum of zonality, because a decrease in meridional gradients shows up first in outlying alpine areas of central Asia. The reduced effect of the Siberian and Tibetan anticyclones at this time facilitates the penetration of the outlying mountain areas in central Asia by moisture-carrying airstreams from the Atlantic, Indian and Pacific oceans. This process occurs first in the Pamir, then, almost at the same time, in the North and Central Tien Shan, and finally in southeastern Tibet and on the northern slopes of the Himalayas.

The firn line is highest at the end of zonal epochs and the beginning of meridional ones, when minima of precipitation are observed. Thus, the most favorable time for enhanced glaciation in central Asia is at the end of meridional epochs and the beginning of zonal ones, when maximum lowering of the firn line is observed (Figure 15.2). If we compare these changes with variations in the position of the seasonal snow boundary in Eurasia (Figure 15.1), we see that the maximum lowering of the firn line on glaciers is observed 7-10 years after the extreme northern position of the seasonal snow boundary in Eurasia, and *vice versa*. The highest present-day values of anomalies in firn line depression are presented in Table 15.5 and Figure 15.4.

TABLE 15.5. Maximum deviations of regime indices in Central Asian glacial systems.

Glacial system	Beginning of zonal epoch			Beginning of meridional epoch		
	ΔT_s °C	ΔP_r mm	ΔH_n m	ΔT_s °C	ΔP_r mm	ΔH_n m
N Tien Shan	-0.30	100	-75	0.3	-90	70
C Tien Shan	-0.25	90	-35	0.2	-70	50
C Pamir	-0.65	360	-110	0.8	-250	100
SE Tibet	-0.25	75	-68	0.3	-120	40
N Himalayas	-0.80	60	-40	0.8	-60	40

ΔT_s - deviations of summer air temperature at firn line; ΔP_r - deviations of annual precipitation; ΔH_n - deviations in firn line position.

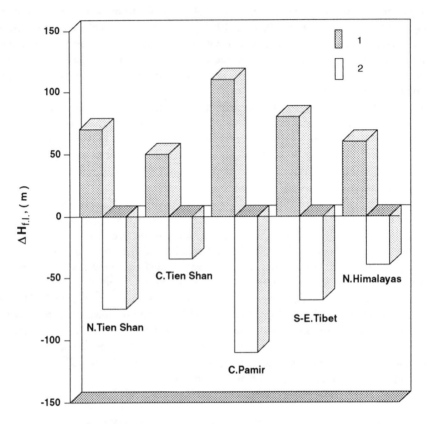

Figure 15.4. Maximum deviations in firn line altitude within the glacial systems
of Central Asia at the beginning of (1) meridional and (2) zonal epochs.

15.5. Conclusions

1. A scheme of interaction of natural processes in the earth-ocean-atmosphere system
was considered. In this system the variations in seasonal snowcover on the Eurasian
continent may be related to changes in the dominant of circulation pattern, which, in
turn, influence climatic conditions in central Asia.

2. Some general relationships and estimates of parameters in the general circulation of
the atmosphere, and snowcover were obtained.

3. The minimum area occupied by seasonal snowcover occurs after the maximum of
meridionality and *vice versa*. The maximum snowcover extent on the Eurasian continent
occurs after the maximum of zonality.

4. Using the available data, we calculated the relationship between anomalies in
global temperature, area occupied by seasonal snowcover, and the zonality indices.

5. Data on the mass transfer in five glacier systems of central Asia were used as reference indicators for changes in moisture conditions. The results of a covariance analysis have shown that global variations of air temperature do not exert a direct influence on the state of present-day glaciation. They become apparent through variations in the general circulation of the atmosphere, precipitation, cloudiness, solar radiation, and local air temperature.

6. The glacier systems on the average have a negative mass balance, which testifies to a general trend of decreased glaciation in central Asia, corresponding to the meridional character of the period from 1900 to 1987. This pattern is not favorable for moisture transport inside the Eurasian continent. The greatest of negative mass balance is typical of glacier systems on the northern periphery of central Asia, especially for the glaciers in the Central Pamir.

7. We have shown that the outlying mountain areas of central Asia respond first to changes in general circulation of the atmosphere. The lowest firn line elevations are observed at the end of meridional epochs and during the first half of zonal ones. During these periods a large increase in precipitation and drop in local air temperature occurs due to increased cloudiness and reduction of solar radiation at the surface. In contrast, the beginning of meridional epoch is accompanied by a rise of the firn line elevation. The 1980s and 1990s correspond to this period.

Acknowledgments

The authors wish to thank Dr.Tetsuo Ohata (Water Research Institute, Nagoya University) and his colleagues for many useful discussions during the course of this work. The review of the original manuscript by Prof. Alexander N. Krenke (Institute of Geography Russian Academy of Science) and Prof. Gleb E. Glazirin (Tashkent University) is gratefully acknowledged.

References

Afanasieva, V.B., and Esakova, N.P. (1964) Statistical relations among some weather characteristic anomalies, *Works of MGO*, Hydrometeo-Publishing, Leningrad, USSR, No. 165, 105 - 114. (In Russian.)

Afanasieva, V.B., and Esakova, N.P. (1969) About relation between planetary truncated frontal zone with snowcover boundary location, *Works of MGO*, Hydrometeo-Publishing, Leningrad, USSR, No. 236, 70-75. (In Russian.)

Afanasieva, V.B., Esakova, N.P., and Titov, V.M. (1974) Statistical relations between temperature and circulation conditions and snowcover boundary location, *Works of MGO*, Hydrometeo-Publishing, Leningrad, USSR, No. 298, 130-135. (In Russian.)

Aizen, V.B. (1984) The mass accumulation in the feeding area of Medvejii glacier during the period between glacier motion, *Data of Glacial Studies*, Moscow, USSR, No. 54, 131-136. (In Russian.)

Aizen, V.B. (1985) The mass balance of Golubin glacier during 1959/60 and 1981/82, *Data of Glaciological Studies*, Moscow, USSR, No. 53, 44-53. (In Russian.)

Aizen, V.B, (1987) *The glaciation and its evolution at the North Tien Shan periphery range*, Unpublished Ph.D thesis. Moscow University, Russia, 210 pp. (In Russian.)

Aizen, V.B. (1990) Nourishment conditions and mass exchange of some glaciers in the Tien Shan Mts, In *Proceedings of the Fourth National Conference of Glaciology and Geocryology*. Lanzhou, China, 86-94. (In Chinese.)

Aizen, V.B., Aizen, E.M., Nesterov, V.B., Sexton D.D. (1991) Heat and mass exchange processes in glacial system of Central Tien Shan, in *Glacier-Ocean-Atmosphere Interaction*, IAHS Publication No. 208, Wallingford, England, 329-337.

Aizen, V.B., and Aizen, E.M. (1993) The dynamic of hydrological cycle of subcontinental mountain-glacial system in the Central Tien Shan, *J. Glaciology and Geocryology* No. 3, Lanzhou, China, 442-459.

Aizen, V.B., Aizen, E.M., Melack, J.A., and Martmaa, T. (in press) Isotopic Measurements of Precipitation on central Asian Glaciers (Southeastern Tibet, northern Himalayas, central Tien Shan), *J. Geophysical Research*, USA.

Aizen, V.B., and E.M. Aizen, E.M., (1994) Regime and mass-energy exchange of subtropical latitude glaciers under monsoon climatic conditions: Gongga Shan, Sichuan China, *Mountain Research and Development* **14**, 2, 101-118.

Alisov, B.P., Drozdov, O.N., Rubinshtein, E.S. (1952) *Course of Climatology*, Hydrometeo-Publishing, Leningrad, USSR, 487 pp. (In Russian.)

Barnett, T.P., Dumenil, L., Schlese, V., Roeckner, E., Latif, M. (1989) The effect of Eurasian snowcover on regional and global climate variations, *J. Atmosphere Sciences* **46**, 5, 666-685.

Barry, R.G. (1984) *Mountain weather and climate*, Hydrometeo-Publishing, Leningrad, USSR, 1984, 311 pp. (Translated into Russian.)

Barry, R.G. (1990) Observational evidence of changes global snow and ice cover, in J.T.Houghton, G.J. Jenkins, and J.J. Ephraums (eds.) *Climate Change: the IPCC Scientific Assessment*, Cambridge University Press, Cambridge, 1.1-1.20.

Budyko, M.I. (1974) *The change of climate*, Works of MGO, Hydrometeo-Publishing, Leningrad, USSR, 279 pp. (In Russian.)

Dzerdzeevsky, B.L. (1974) Some aspects of modern physical and dynamic climatology, in *Investigations of climate genesis*, Moscow, USSR, 6-35.(In Russian.)

Folland, C.K., Karl, T.R., and Vinnikov K.Ya. (1992) Observed Climate Variations and Change, in J.T.Houghton, G.J. Jenkins, and J.J. Ephraums (eds.) *Climate Change: The IPCC Scientific Assessment*, Cambridge University Press, Cambridge.

Gernet, E.S. (1981) *Ice lichens*, Science, Moscow, USSR, 144 pp. (In Russian.)

Glazirin, G.E. (1991) *Mountain glacial systems, their structure and evolution*, Hydrometeo-Publishing, Leningrad. USSR, 110 pp. (In Russian.)

John, B.S. (ed.) (1982) *The winters of the world*, Mir, Moscow, USSR, 31pp. (Translated into Russian.)

Karapetyance, E.M., and Aizen, E.M. (1976) The climatic peculiarities of snowcover spread boundary on the USSR territory, Hydrometeo-Publishing, Leningrad, USSR, Works of MGO, No. 404, 16-29. (In Russian.)

Karapetyance, E.M., Aizen, E.M., and Morozov, V.S. (1983) Correlation of changes of signs of water equivalent anomalies with the changes of circulation condition, *Works of MGO*, Hydrometeo-Publishing, Leningrad, USSR, No. 475, 93-100. (In Russian.)

Kotlyakov, V.M. (1987) Global climatic role of snowcover, in *Interaction between Glaciation, Atmosphere and Ocean*, Nauka, Moscow, USSR, 34-65. (In Russian.)

Krenke, A.N. (1982) Mass exchange in glacier systems on the USSR territory, *Works of MGO*, Hydrometeo-Publishing, Leningrad, USSR, 288 pp. (In Russian.)

Krenke, A.N., and Suslov, V.F. (eds.) (1980) Abramov glacier, Hydrometeo-Publishing, Leningrad, USSR, 205 pp. (In Russian.)

Krenke, A.N., and Bochin, N.A. (ed.) (1984) Tuyuksu glacier,. Hydrometeo-Publishing, Leningrad, USSR, 170 pp. (In Russian.)

Krenke, A.N. (ed.) (1987) Large-scale atmospheric processes in the Northern Hemisphere, *Data of Meteorological Studies*, Moscow, USSR, No.13, 178 pp. (In Russian.)

Kukla, G.J. (1981) Snow covers and climate - *Glaciological Data* (Snow Watch 1980), No. 11, 27-39.

Kumar Bhanu O.S.R.U. (1988) Eurasian snowcover and seasonal forecast of Indian summer monsoon rainfall, *J. des Sciences Hydrologiques* 33, 5, 10/1088, 515-525.

Loktionova, E.M. (1985) Statistical estimate of snowcover extent, *Data of Glaciological Studies*, Moscow, USSR, No. 53, 83-91. (In Russian.)

Matson, M. (1977) Winters snowcover maps of North America and Eurasia from satellite records 1966-76, National Environmental Satellite Service, NOAA, Washington, D.C.

Monin, A.S., and Shishkov, Yu.A. (1979) History of climate, Hydrometeo-Publishing, Leningrad, USSR, 405 pp. (In Russian.)

Pokrovskaya, T.V., and Spirina, L.V (1965) Estimation of snowcover influence on air temperature in spring in the European part of the USSR, *Works of MGO*, Hydrometeo-Publishing, Leningrad, USSR, No. 181, 110-114. (In Russian.)

Robinson, D.A. (1992) Monitoring Northern Hemiphere snow cover, *Proceedings of Snow Watch*, Niagara-on-the-Lake, Ontario.

Vladimir B. Aizen and Elena M. Aizen
Institute for Computational Earth System Science
University of California Santa Barbara, CA 93106, USA

16 GLOBAL WARMING AND THE TREND TOWARD DRYNESS IN THE FRIGID HIGH MOUNTAINS AND PLATEAU OF WESTERN CHINA

LAN-SHENG ZHANG

Abstract

Interior lakes, mountain glaciers, precipitation records and official archives of the frequency of flood and drought are used as evidence to support the following hypotheses. It is evident that temperature in China has been increasing during the past 30 years. The 10-year average air temperature in the 1980s was generally 0.5-1.0°C higher than in the 1950s. Even accounting for variation, the warming in recent years could be considered as the continuation of the warming process that began at the end of the Little Ice Age. This warming process may have been enhanced by the CO_2 greenhouse effect in recent years, but it is difficult to estimate the exact contribution of this effect. Recent research has also provided evidence supporting a trend towards dryness in the cold high mountains and plateau of western China.

16.1. Introduction

Increasing concern has been expressed over global warming and its effects on the environment and sustainable development throughout the world, as well as in China. The purpose of this study is to estmate the impact of global warming on the climate of the cold high mountains and plateau of western China. How will climate change in these regions in the near future? Although it is still too early for scientists to fully answer these questions, some preliminary remarks can be made from the examination of instrumental records and various paleoclimatic proxy data.

16.2. Evidence for a Warming

Increases in temperature in the north and west of China since 1950 have been proven, especially during the winter (Li *et al.*, 1990; Chen *et al.*, 1991). Figure 16.1 shows the decadal average temperature difference between the 1980s and the 1950s. Obvious increases in annual and winter temperatures can be seen in Xingjiang and the Qinghai-Tibetan Plateau, while summer temperatures in northwestern Tibet and southern

J.A.A. Jones et al. (eds.), Regional Hydrological Response to Climate Change, 287–295.
© 1996 *Kluwer Academic Publishers. Printed in the Netherlands.*

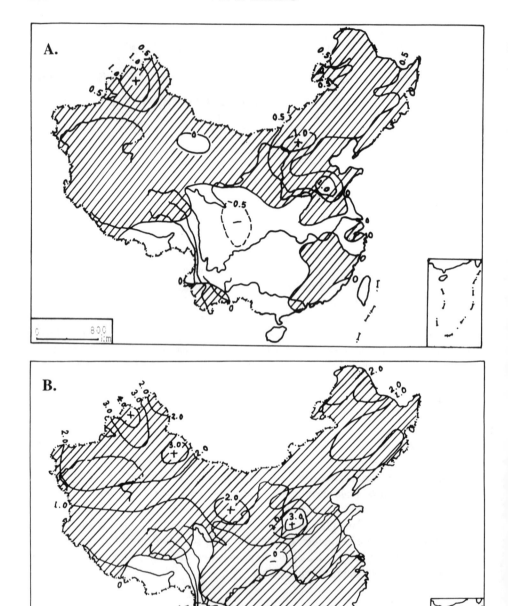

Figure 16.1. The decadal average temperature difference between the 1980s and the 1950s.
a. annual average. b. January. (From Li *et al.*, 1990.)

Xingjiang became a little cooler. In January, the north of Xingjiang was 4.2°C warmer during the 1980s than in the 1950s. This increase is more evident than in other parts of the country.

The temperature increases in the west of China during the last decade could be considered as the continuation of the warming process that began at the close of the Little Ice Age. An oxygen isotope record from the Dunde ice core in the Qilian Mountains shows a continued increase in temperature from the end of the 19th century (Figure 16.2) (Thompson *et al.*, 1993; Yao, 1990).

The regional warming may be partly due to the increasing CO_2 concentration in the atmosphere, though it is difficult at present to estimate the exact contribution of this effect. This conclusion is generally supported by the observation that the warming is more obvious during the winter as well as in higher latitudes within the region, because the same pattern is found in global warming scenarios in GCM simulations.

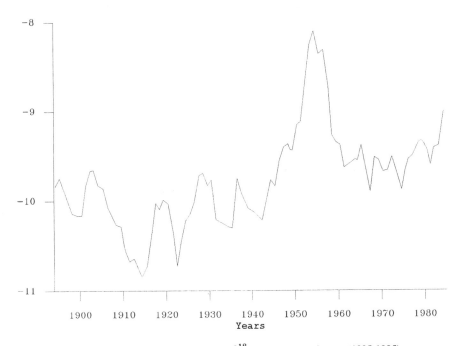

Figure 16.2. Five-year running means of $\delta^{18}O$ from the Dunde core (1895-1985).
(From Thompson *et al.*, 1993.)

16.3. Evidence for a Drying Trend

The drying trend on different scales in the north and west of China was pointed out by Zhang (1984, 1988). Research during the past ten years has provided additional evidence in support of this view.

16.3.1. SHRINKING OF THE INTERIOR LAKES

More than 1000 lakes, most of them salty and with interior drainage, are scattered throughout the Qinghai-Tibetan Plateau. Terraces and lacustrine deposits around these lakes indicate that they were much larger and the lake levels were much higher in the early Holocene. Since then, the lake levels have been dropping continuously and the basins have become closed. For example, Table 16.1 and Figure 16.3 show that in both Qinghai Lake and Bankong Lake the area and lake levels have fluctuated around a declining trend since the early Holocene. Other lakes on the plateau experienced the same phenomenon.

TABLE 16.1. The shrinking of Lake Qinghai.

Year	Altitude of lake surface (m)	Lake area (km²)
1986	3193.80	4304
1956	3196.94	4607
1908	3205.00	4930

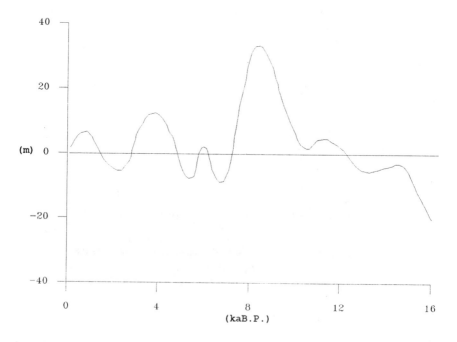

Figure 16.3. The variations in water level on Lake Bankong (33°30'N, 79°50'E, Tibet, China) since 16 ka B.P. The present water level is 4241 m a.s.l. (From Wang, 1992).

The Yangzhouyong Lake provides another example. Being at the foot of the Himalayas, the water level of this lake is 4440 m above sea level, and at present it covers 621 km². Its drainage area is about 6100 km². In the past, the waterflowed northwards into the Yaluzangbu (Brahmaputra) River through the Moqu River. As the climate became drier, the lake decreased in area, closing the outlet (see Liu, this volume, Chapter 21). According to the Tibetan Bureau of Water Resources and Hydrological Survey, the level is falling at the rate of 80 cm/100 years. Extensive beach areas have been formed and some small lakes have been split off from the original large one, suggesting that lake levels are still in decline.

16.3.2. RETREAT OF MOUNTAIN GLACIERS

There are many glaciers in the high mountains of western China. The expansion or retreat of continental glaciers is chiefly dependent on precipitation. It is well known that most of the glaciers in western China have declined steadily during the 20th century (Table 16.2), reflecting a climatic trend towards drier conditions. This retreat is a continuation of a process that began with the close of the Little Ice Age (Table 16.3).

TABLE 16.2. Changes in mountain glaciers in Northwest China during the 20th century.

Mountains	Number	Area (km²)	No. under observation	Changes in glaciers		
				Advancing	Retreating	Stable
Altai	416	293.20	5	1	4	0
Tianshan	8908	9195.98	43	5	29	9
Qilian	2895	1972.50	44	8	25	11
Kunlun	7774	12 482.20	84	31	18	35
Pamir	2112	2992.85	11	2	7	2
Karakoram	1848	4647.17	15	5	7	3
Total	23 917	31 683.90	202	52	90	60

L.-S. ZHANG

TABLE 16. 3. Changes in mountain glaciers in Northwestern China.
(From Wang, 1992.)

Mountains	Maximum area (km^2)	Present area (km^2)	Reduction in last 500 years (km^2)	(%)
Altai	449	293.20	456	36
Tianshan	12 248	9195.48	3052	25
Qilian	3288	1972.50	1316	40
Kunlun	9835	8735.18	1099	11
Pamir	2882	2206.55	676	23
Karakoram	6630	5924.85	705	11
Total	35 332	28 328.26	7004	-

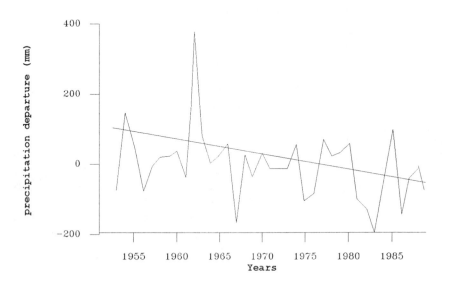

Figure 16.4. Precipitation departures from normal at Lhasa (1951-1989).
(From Lin and Chen, 1993.)

16.3.3. EVIDENCE FROM HISTORICAL DOCUMENTS AND INSTRUMENTAL RECORDS

The records of the weather station at Lhasa show that over the last 30 year period the departure of precipitation from the normal, long-term average has been negative, and in 1983 the departure reached -200 mm. The decrease is illustrated in Figure 16.4.

Data on flood and drought disasters during the last 100 years were analysed using the official archives. The analysis showed that the frequency of drought per 10-year period has been steadily increasing (Figure 16.5). Since 1883, there have been 3 wet periods: 1883-1906 (23 years), 1916-1934 (18 years), and 1947-1962 (15 years). There have also been 3 drought periods: 1907-1915 (8 years), 1935-1946 (11 years) and 1963-1980 (17 years). The wet periods are decreasing in duration and the drought periods are lengthening, which also supports the hypothesis of a trend towards dryness.

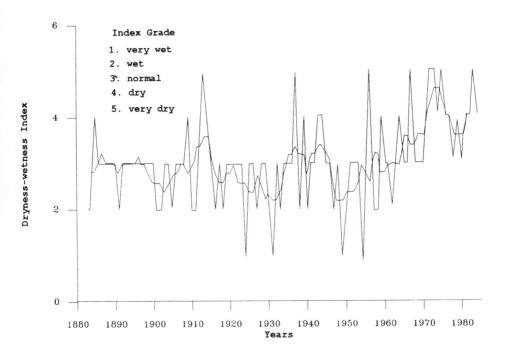

Figure 16.5. Dryness-wetness Index and 5-year running means in Tibet between 1883 and 1984.
(From Lin and Chen, 1993.)

16.4. Conclusions

1. The evident warming since the end of the 19th century in the western high mountains and plateau of China has been observed and corroborated by a number of approaches. The regional warming may, at least in part, be a result of the global greenhouse effect of CO_2, though it is still hard to determine exactly to what extent the greenhouse effect has influenced this trend.

2. The independent evidence from the glaciers, lake levels and weather records consistently indicate both the increasing trend in temperature and the decreasing trend in precipitation are a continuation of processes that have been ongoing since the end of the Little Ice Age.

3. On a longer timescale, the evidence supports the trend towards a drier climate (Zhang, 1988). It coincides with the accelerated uplift in the Qinghai-Tibetan Plateau since the mid-Pleistocene, and it is considered to be caused by the formation of the plateau, which blocked the advection of moisture from the Indian Ocean. However, it is highly unlikely that there is a close relationship between the uplifting of the plateau and the strengthening of the trend towards a drier climate over the last 100 years.

4. It is therefore possible that, if the average global temperature rises due to the effect of greenhouse gases, the western high mountains and plateau of China will tend to be even drier in the future.

References

Li, K., Lin, X., and Wang, W. (1990) The long term variation trend of temperature in China from 1951 to 1988, *Geographic Research* **4**, 26-37. (In Chinese.)

Chen, L., *et al.* (1991) Preliminary analysis of climatic variation during the last 39 years in China, *Advances in Atmospheric Science* **8**, 279-288.

Thompson, G.L., *et al.* (1993) Recent warming: ice evidence from tropical ice cores with emphasis on central Asia, *Global and Planetary Change* **7**, 145-156.

Yao, T. (1990) Climatic records of Little Ice Age in ice cores from Dunde ice cap, Qilian Mountains, in Y. Shi *et al.* (eds.) *Studies on Climatic and Sea Level Changes in China*, vol. 1, China Ocean Press, Beijing, 128-130. (In Chinese.)

Zhang, L. (1984) The major characteristics of environmental changes during the Quaternary period in China, *J. Beijing Normal University (Natural Science)* **4**. (In Chinese.)

Zhang, L. (1988) The trend towards dryness in north and western China since the mid-Pleistocene, in *The Palaeoenvironment of East Asia from the mid-Tertiary*, vol. 1, University of Hongkong, Hongkong.

Shi, Y., *et al.* (1990) Basic features of climate and environments during the Holocene Megathermal in China, in Y. Shi and Kong (eds.) *Climates and Environments during the Holocene Megathermal in China*, China Ocean Press, Beijing, 1-18. (In Chinese.)

Wang, F. (1992) Elementary features of Holocene climate and environment on the Qinghai-Tibetan Plateau, in Y. Shi and Kong (eds.) *Climates and Environments during the Holocene Megathermal in China*, China Ocean Press, Beijing, 179-205. (In Chinese.)

Wang, Z. (1992) Glacier change and its impacts since the Little Ice Age and their future trends in Northwest China, in Y. Shi *et al.* (eds.) *Studies on Climatic and Sea Level Changes in China*, vol. 2, China Ocean Press, Beijing, 121-122. (In Chinese.)

Lin, Z., and Chen, X. (1993) The climate change in the last 200 years in Tibet, in Y. Yang *et al.* (eds.) *Climate Change and its Impact*, China Meteorological Press, Beijing, 50-56. (In Chinese.)

Lan-sheng Zhang
Beijing Normal University
100875 Beijing, China.

SECTION V

THE WATER BALANCE AND
CHANGING REGIONAL RESOURCES

SUMMARY

A water balance methodology for mountainous regions
■ Predictive formulae are derived which combine Turc's formula with standard water balance equations. This enables water balance predictions to be based solely on predicted changes in temperature and precipitation.
■ Turc's formula is found to provide a satisfactory basis for regional-scale predictions and compares well with the Penman approach for Mt. Emei in SW China.
■ Under a 4°C warming with no change in precipitation, evaporation is predicted to rise and runoff to fall at all elevations.

Water resources in Wales
■ GCM-based equilibrium scenarios produce significant reductions in overall resources, especially during the summer.
■ Sources on smaller upland rivers are probably most vulnerable in the summer.
■ Increases in spring rainfall in all scenarios will only be of value to water supply if dams are raised or operational rules are modified.
■ Estimates suggest planning for a 10% reduction in operational yields from reservoirs.
■ This needs to be seen in the context of strategic plans mooted to transfer water from Wales to the water-strapped SE England.
■ An approach to predicting the risks of extreme drought events suggests that return periods could be reduced by 50% under the best estimate scenario.

Water resources in Australia
■ A 19-parameter conceptual daily rainfall-runoff model suggests that regional sensitivities vary widely, although runoff changes always amplify changes in precipitation. Amplification is least in the wet tropical NE (1.5 times) and most in the drier SW coast and South Australian Gulf (2-4 times).
■ Soil moisture changes were generally less than changes in precipitation (0.5 times), except in the drier regions.
■ Temperature changes had little effect on PET in wetter climates, but were an important factor in drier regions.
■ Outputs from 5 GCMs all predict greater warming in the interior and in the south.
■ Regional disagreements on future rainfall regimes amongst the GCMs are resolved by modelling a range of scenarios for each region.
■ These predicted increases in annual runoff of up to 25% in the NE by 2030 and a reduction of up to 35% in the South Australian Gulf.
■ They also showed major disagreements even as to the sign of the response in the temperate SE (+/- 20%) and the West Coast regions (+/- 50%).

■ A stochastic Weather Generator used to predict changes in the timing and frequency of events for a basin in the tropical NE indicated reduced discharges in the summer (-25%, and more in high flows), which suggests problems for public water supply and irrigation. For a basin in the South Australian Gulf it showed a modest 5% increase in summer discharges, rising to 15% in high flow events.

■ Both Weather Generator experiments contradict the general trends predicted from the GCM scenario simulations, and indicate the high degree of current uncertainties.

Water balance in Tibet

■ Levels of an interior drainage lake in S Tibet have been falling at 0.6 m/100 years.

■ Lake levels have fallen particularly over the last 30 years, when precipitation receipts have been consistently below the minimum required to maintain a hydrological balance.

■ Analyses of temperature and precipitation records from 6 meteorological stations indicates that rising temperatures and falling precipitation are the cause.

■ The average period of winter lake ice cover reduced by 10 days during the warming phase of the 1970s and 1980s.

■ Lake levels show a cyclicity with dominant periodicities of 7 to 14 years related to climatic fluctuations.

■ Projected warming of 2-3°C by 2030 will cause lake levels to fall further and this could have a climatic feedback over a wide area.

Satellite data for modelling hydrometeorological change in Tibet

■ Satellites are particularly suited to capturing this high variability in such a remote region. Algorithms use IR and visible band data from the Indian Satellite.

■ Patterns of temperature and albedo follow topographic and biophysical features.

■ High variances are observed, e.g. temperature changes of over 10°C in just 50 km in the centre of the plateau and 20°C on the edge. Albedo varies from 16% to 28%.

■ Radiative processes are especially important at this altitude and latitude for determining water balance components.

■ Desertization will increase albedos and changing General Circulation airflow patterns will alter the cloudiness, which is more important than in general at this elevation and in this position relative to the Indian monsoon.

■ Long-term observations of surface and atmospheric variables, especially surface and subsurface water, are needed to better understand processes and spatial patterns.

Water resources of the Yalong River, China

■ The Langbein, Turc, nonlinear regression and high order grey box models all have low data requirements and are therefore very suitable for remote regions and for predicting changes in riverflow based solely on temperature and precipitation scenarios.

■ The 4 approaches were applied to 3 scenarios - a 2°C warming with +10%, -10% and zero change in precipitation - and displayed a high level of precision and consistency.

■ All but the wettest scenario would cause reductions in discharge, especially during the dry season, with larger changes in runoff than in precipitation receipts.

■ The best estimates suggest that:

   1. A -10% precipitation scenario would result in a 20% reduction in annual discharge and -25% during the dry season.

   2. Zero change would result in 5% lower annual runoff, or -10% in the dry season.

   3. +10% precipitation would increase runoff by 7-8% annually, but only +2% in the dry season.

Water resources in Argentina

■ GCMs suggest a southward shift in the subtropical high pressure zones in the Atlantic and Pacific will cause a drying trend in the north balanced by increased precipitation in the south.

■ There will be a commensurate reduction in the temperature gradient between north and south, with warmer temperatures in the south especially in winter.

■ Snow lines will rise causing an eventual reduction in river resources in rivers with snow-fed headwaters.

■ The reaction of mountain glaciers in Patagonia needs detailed analysis as the response of a number of major rivers will be very sensitive to meltwater patterns.

■ Analyses of precipitation records at 13 stations from the beginning of the century shows fairly consistent increases in the north, but no clear trends elsewhere.

■ This pattern is not consistent with the global warming trends expected from GCMs.

Water balance in Japan

■ Data from Central Japan indicate a rise in PET over the last 100 years, especially during the warm period of the 1980s.

■ A decline in calculated runoff and in the runoff ratio over recent decades, and especially during the 1980s.

■ Runoff changes were much greater than changes in evapotranspiration.

■ No significant long-term trends in precipitation are found.

17 A METHOD TO ASSESS THE EFFECTS OF CLIMATIC WARMING ON THE WATER BALANCE OF MOUNTAINOUS REGIONS

CHANGMING LIU and MING-KO WOO

Abstract

From a regional perspective, systematic vertical variations of the mean annual water balance components (precipitation, evaporation and runoff) are characteristic features of mountainous terrain. For nonglacierized areas, vertical distribution of runoff can be obtained as the difference between measured precipitation and evaporation calculated using such equations as that proposed by Turc. Turc's equation makes use of temperature and precipitation as the input variables, and is well suited to the study of regional water balance impacts due to climatic warming. Using this equation and the water balance relationship, evaporation and runoff from various elevation zones of Mt. Emei in southwestern China were computed. A 4°C temperature rise scenario was applied to the region and this caused a rise in evaporation accompanied by a reduction in runoff at all elevations. The proposed method is simple to use and allows an assessment of the sensitivity of the water balance variables to temperature and precipitation changes.

17.1. Introduction

With an increase in the atmospheric concentration of carbon dioxide and other 'greenhouse' gases, there is a strong possibility of global warming (Houghton *et al.*, 1990). Such climatic change can have notable effects on the hydrology and water resources (Gleick 1987; Waggoner, 1990). Many impact studies make use of climatic change scenarios produced by Global Climate Models (GCMs) to indicate the level of regional warming (Giorgi and Mearns, 1991), particularly for the case when the atmospheric concentration of 'greenhouse' gases is at twice the pre-industrial level. However, different GCMs yield different magnitudes of temperature rise (Schlesinger and Mitchell, 1987), though they generally agree that warming will occur, to be accompanied by precipitation changes.

Mountains are complex in topography, microclimate, vegetation and hydrological behaviour. For convenience, we define mountains as lands where the local relief exceeds 1000 m within an area of 100 km². Significant vertical and horizontal zonation exists within a mountain region (Barry, 1981). Compared with their adjacent lowlands, mountains have lower summer heat and higher total precipitation. Seasonal frost or permafrost, together with varying durations of snow and ice cover, may occur at high elevations. Vertical variation in the climate is also reflected in vegetation zonation

J.A.A. Jones et al. (eds.), Regional Hydrological Response to Climate Change, 301–315.

which, in turn, has important feedback on the climate through interception and evapotranspiration. Given the current state of our knowledge, the study of climatic change impacts on mountain hydrology is feasible only on a regional scale.

This study will explore the possible hydrological responses to climatic change in mountain regions. Several restrictions apply. It is assumed that global warming will also lead to temperature rises in the mountainous areas, accompanied by modifications of the present precipitation regime. As there is little agreement among the GCMs regarding the magnitudes of change at the seasonal time scale, this study will concentrate on average annual values. Although responses to climatic change should take on transient phases, only the equilibrium stage is considered. It is further assumed that most environmental factors (e.g. vegetation, soil and landform) remain little changed so that the manner of hydrological response to climatic forcing will be similar to the present. We further limit the study to a regional level and local variability will not be considered. Thus, altitude and latitude will be the major elements of concern. Finally, we restrict the analysis to non-glacierized areas, using examples from China to provide empirical information. For high mountain zones of northwestern China, Lai and Yeh (1991) have investigated the effects of global warming on runoff from glacierized and permafrost zones.

In view of the scarcity of data and the uncertainties regarding the climatic change forcing, only qualitative interpretation of the results is warranted. One purpose of this work is to stimulate future research so that empiricism can be supplemented by physical understanding, and quantification can be improved to enable better prediction of hydrological responses to climatic change in mountain regions.

17.2. Water Balance Analysis

The average annual water balance, comprising three major elements (precipitation, evaporation and runoff), provides an appropriate framework for analyzing the regional impacts of climatic warming on mountain hydrology. Hydrologically, precipitation (P) and evaporation (E) will be affected directly by climatic change. Climatic change has also a secondary effect on runoff (R) through the water balance $R = P - E$. For a particular geographical region

$$R = f(P, E, Z) \tag{17.1}$$

where Z is a regional factor. For our preliminary investigation, Z for a given location is assumed to be constant under climatic change (i.e. the topography, the soil and vegetation characteristics will remain unaltered). Then

$$R = f(P, E) \tag{17.2}$$

Future changes in precipitation and evaporation will cause runoff to vary, and

$$dR = (\partial R/\partial P)dP + (\partial R/\partial E)dE \tag{17.3}$$

Two situations may arise.

(a)   P and E vary independently with climatic forcing. We obtain from equation 17.3

$$dR/dP = \partial R/\partial P + (\partial R/\partial E)(dE/dP) \qquad (17.4a)$$

and

$$dE/dP = \partial E/\partial P + (\partial R/\partial P)(dP/dE) \qquad (17.4b)$$

(b)   P and E co-vary with climate; and climatic change induces an increment of temperature (T) in the mountain region. Then, applying the chain rule to Eq. 17.2 yields

$$dR/dT = (\partial f/\partial P)(dP/dT) + (\partial f/\partial E)(dE/dT) \qquad (17.5)$$

From equations (4) and (5), all the differentials (dR/dP, dE/dP, dP/dE, dR/dE, dP/dT and dE/dT) and partial differentials ($\partial R/\partial P$, $\partial R/\partial E$, $\partial f/\partial P$ and $\partial f/\partial T$) are physically meaningful indicators of changes in the water balance elements which respond directly or indirectly to climatic forcing. At present, however, not all of the derivatives can be defined deterministically.

17.3. Spatial Variability of Water Balance Components

Both vertical and horizontal variations of hydrological elements are important in mountainous regions. Vertical zonation is especially critical and the water balance components are highly sensitive to altitudinal influences. Horizontal variation is related to locational factors such as latitude and continentality, such that the mean values of P, E and R will differ from one mountainous area to another.

17.3.1. PRECIPITATION

Mountains produce pronounced orographic effects, though the vertical change in precipitation is also influenced by other local factors such as aspect and slope (Johnson *et al.,* 1990). Thus, the relationship between precipitation and elevation (H) varies between geographical locations. The relationship P = f(H) must be specified for particular localities, but it may be expressed by a general form of the polynomial equation

$$P = b_0 + b_1 H + b_2 H^2 + b_3 H^3 + ... \qquad (17.6)$$

where the b's are parameters which can take on positive or negative values, and can be determined by fitting the curve to observed data.

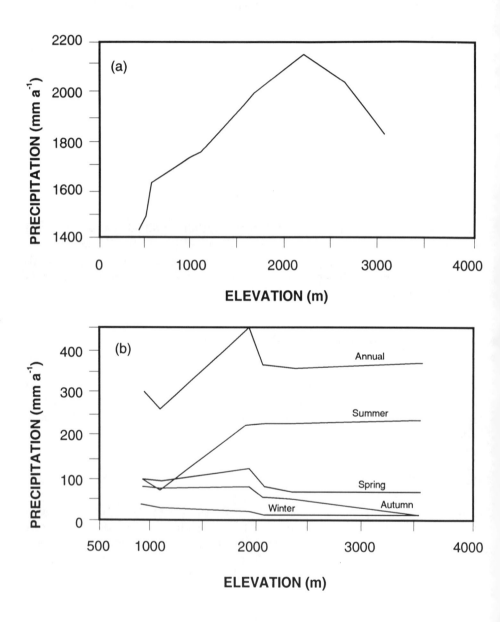

Figure 17.1. Vertical distribution of annual precipitation in selected mountainous regions of China.

Figure 17.1a shows the vertical variation of mean annual precipitation for Mt. Emei in south China. When the polynomial equation is used to fit the precipitation data from ten stations that range in elevation between 400 and 3000 m, the coefficient of determination is 0.97. Another example is from Tianshan in the semi-arid northwestern part of China (Figure 17.1b). The elevation-precipitation relationship changes with the season, and the annual precipitation also varies non-linearly with elevation. In both examples, annual precipitation does not increase monotonically with elevation, though other forms of P-H relationship have been reported. Differences between the examples presented demonstrate the effects of geographical locations superimposed on orography.

17.3.2. EVAPORATION

Evaporation depends on the supply of energy and the availability of moisture, both of which change with altitude, aspect, vegetation and other locational factors. In most areas where moisture is not limiting, energy supply decreases with elevation and so does the mean annual evaporation. This need not be so in the arid zone where water is deficient in the footslopes but the increase in precipitation at higher altitudes allows evaporation rates to increase on mid-slopes. One example is from Tianshan where the potential evaporation (E_0) increases with elevation, but precipitation is limited on the lower slopes (Figure 17.2), thus restricting actual evaporation at the lower altitudes.

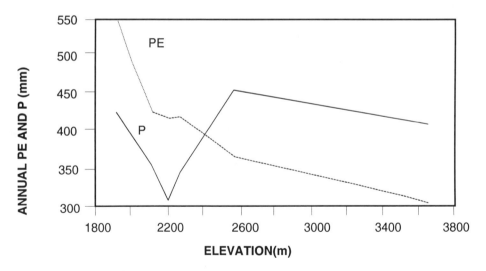

Figure 17.2. Vertical distribution of potential evaporation compared with precipitation, Tianshan.

Polynomial equations can be used to describe the relationship between evaporation and elevation

$$E = c_o + c_1H + c_2H^2 + c_3H^3 + ... \tag{17.7}$$

where the c's are empirical coefficients obtained by curve fitting. One advantage of using this form of equation is that it is compatible with the empirical approximation of precipitation-elevation relationships (Eq. 17.6). Figure 17.3 provides examples of the vertical distribution of evaporation from Henduan Shan in southwestern China, from Qinling in central China and from Qilian Mountain in northwestern China. As with the P-H relationship, spatial differences in the E-H curves reflect the influence of environmental factors other than elevation. The decline in evaporation below 2500 m in Qilian Mountain, for instance, is due to increasing aridity towards the foothill zone which has a semi-desert climate.

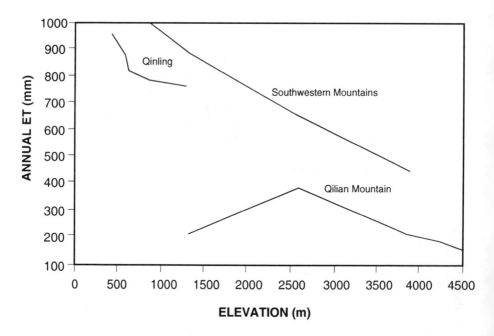

Figure 17.3. Examples of vertical distribution of evaporation in mountainous areas of China.

17.3.3. RUNOFF

Unlike precipitation and evaporation, runoff is seldom obtained for different zones, but as values averaged over entire catchments. Mountain basins tend to traverse a wide altitudinal range, and the areally integrated runoff data do not allow disaggregation of runoff contribution from various elevation zones. Ideally, runoff plots should be set up at different altitudes to provide such information, but to do so is prohibitively expensive. Instead of this approach, runoff will be obtained using the water balance equation

$$R(H) = P(H) - E(H) \qquad (17.8)$$

Combining Eqs. 17.6, 17.7 and 17.8 gives

$$R(H) = (b_0-c_0) + (b_1-c_1)H + (b_2-c_2)H^2 + (b_3-c_3)H^3 + \dots \qquad (17.9)$$

The performance of Eq. 17.9 depends on the reliability of obtaining the relationships of P and E with H.

The elevation of maximum runoff will be located at the elevation where

$$dR(H)/dH = 0 \qquad (17.10)$$

Note again that this analysis refers to nonglacierized catchments where the influence of icemelt is minimal.

17.4. Regional Hydrological Response to Climatic Change

The hydrological responses of mountainous regions to climatic change can be analyzed through the water balance of small to medium size catchments. A number of mountain basins in China were used to establish the spatial differentiation of relationships among precipitation, evaporation and runoff, thus allowing generalizations to be made regarding the regional variability of hydrological responses to possible ranges of climatic changes.

17.4.1. CLIMATIC CHANGE AND EVAPORATION

Following Budyko (1956), mean annual potential evaporation (E_o) for a region is estimated by the energy balance

$$E_o = Q^*/L \qquad (17.11)$$

where Q^* is the energy balance or net radiation, and L is the volumetric latent heat of vaporization. Actual evaporation may be considered to be related to potential evaporation by

$$E = \varepsilon E_o \qquad (17.12)$$

The ratio of actual to potential evaporation ($\varepsilon = E/E_o$) varies according to the supply of moisture (hence precipitation) and available energy which is related to E_o through Eq. 17.11. We propose the following relationship (Liu, 1983, 1985)

$$\varepsilon = 1 - \exp(-P/E_o) \qquad (17.13)$$

Substituting this relationship into Eq. 17.12 yields

$$E = [1 - \exp(-P/E_o)].E_o \qquad (17.14)$$

The ratio E_o/P is Budyko's radiational index of dryness (Sellers, 1965). There are maps showing the spatial distribution of this ratio for China under the present climate (Zhao, 1986).

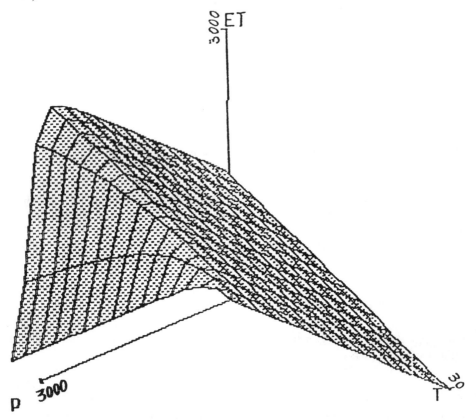

Figure 17.4. Response of evaporation (E) to different magnitudes of mean annual temperature (T) and precipitation (P).

The change in energy balance in a climatic warming scenario will cause an increase in potential evaporation (Eq. 17.11). Precipitation will also change. The partial derivatives $\partial E/\partial E_o$ and $\partial E/\partial P$ will indicate the manner in which evaporation responds to these changes:

$$\partial E/\partial E_o = \varepsilon = 1 - \exp(1-P/E_o) - P/E_o.\exp(P/E_o) \qquad (17.15a)$$

$$\partial E/\partial P = \exp(-P/E_o) \qquad (17.15b)$$

One explicit expression of ε in terms of temperature and precipitation variables is Turc's (1961) formulation

$$E = P/[0.9 + (P/I)^2]^{0.5} \qquad (17.16)$$

where $I = 300 + 25T + 0.05T^3$. This formula was based on a study of data collected from 254 basins located worldwide, and may be suitable for regional characterization of mean annual values. Figure 17.4 shows the response of E to different magnitudes of P and T. As expected, $E = 0$ when $P = 0$, regardless of the magnitude of T. At low temperatures, E is limited even for high P. Large E occurs when both P and T are high.

Using Eq. 17.16, we obtain the derivatives

$$\partial E/\partial T = P^3(25 + 0.15T^2)[0.9 + (P/I)^2]^{-1.5}I^{-3} \qquad (17.17a)$$

and

$$\partial E/\partial P = [0.9 + (P/I)^2]^{0.5} - P^2[0.9 + (P/I)^2]^{-2.5}I^{-2} \qquad (17.17b)$$

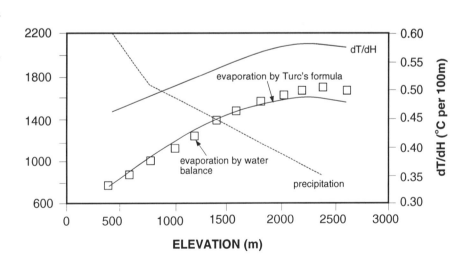

Figure 17.5. Evaporation (E) computed using Turc's equation for precipitation (P) and for temperature gradient (dT/dH) between 0 and 2500 m. Also shown are runoff values derived from water balance (P - E).

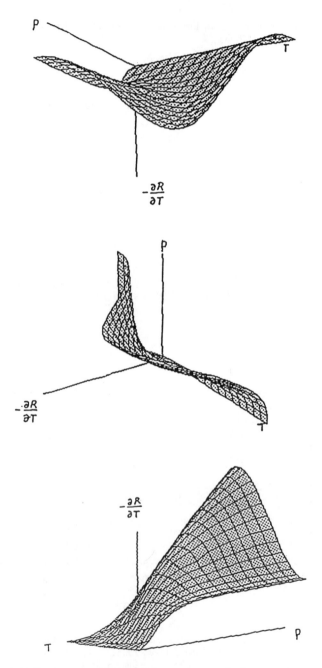

Figure 17.6. Change in runoff with respect to temperature ($\partial R/\partial T$), plotted against mean annual temperature (T) and precipitation (P), as viewed from three different perspectives.

17.4.2. REGIONAL RESPONSE OF RUNOFF TO CLIMATIC CHANGE

Based on the water balance equation and the regional estimation of evaporation, runoff can be obtained as

$$R = P \{1 - [0.9 + (P/I)^2]^{-0.5}]\} \qquad (17.18)$$

Figure 17.5 shows that within a temperature range of $0°$ to $30°C$, and a precipitation domain of 0 to 3000 mm, runoff decreases with T but increases with P.

Taking the derivatives with respect to temperature and precipitation, we have

$$\partial R/\partial T = -P^3(25T + 0.15T^3)(0.9 + P^2/I^2)^{-1.5}I^{-3} \qquad (17.19a)$$

and

$$\partial R/\partial P = 1 - 0.9(0.9 + P^2/I^2)^{-1.5} \qquad (17.19b)$$

Figure 17.6 illustrates the partial differentials $\partial R/\partial T$ from three perspectives. The more the temperature rises, the more runoff is reduced. However, the partial derivative of $\partial R/\partial P$ (Figure 17.7) indicates that runoff increases as precipitation increases, especially when higher precipitation occurs in the zone of low temperatures.

17.5. Climatic Change and Mountain Hydrology

Vertical changes of evaporation and runoff in mountainous terrain under climatic change scenarios can be estimated. The example for Mt. Emei is used to demonstrate the method.

Figure 17.8a shows the vertical distribution of precipitation of Mt. Emei. The mean annual air temperature at the foothill zone is $14.2°C$, but there are no data for the higher elevations and temperatures have to be estimated using lapse rates. At the low elevation, a global average rate of $0.006°C/m$ was assumed. This rate is judged to be too large for higher altitudes because a study for the neighbouring southeastern Tibetan Plateau area showed that temperature decreases at $0.0035°C/m$ (Zhong Shan University and Northwest University, 1979). The lapse rates were therefore linearly reduced from $0.006°C/m$ at 400 m elevation to $0.0035°C/m$ at 2400 m. The precipitation and temperature data thus estimated were substituted into Eqs. 17.16 and 17.18 to calculate the vertical distribution of evaporation and runoff.

Figure 17.8b demonstrates that Turc's equation (Eq. 17.16) provides a good estimation of the vertical distribution of evaporation as the computed values compare favourably with evaporation computed using Penman's equation (Sellers, 1965). Runoff calculated by water balance, using observed precipitation and measured evaporation as inputs, also compares well with the computed runoff obtained by Eq. 17.18 (Figure 17.8c). These results lend confidence to extrapolating the hydrological effects of climatic change in mountainous areas.

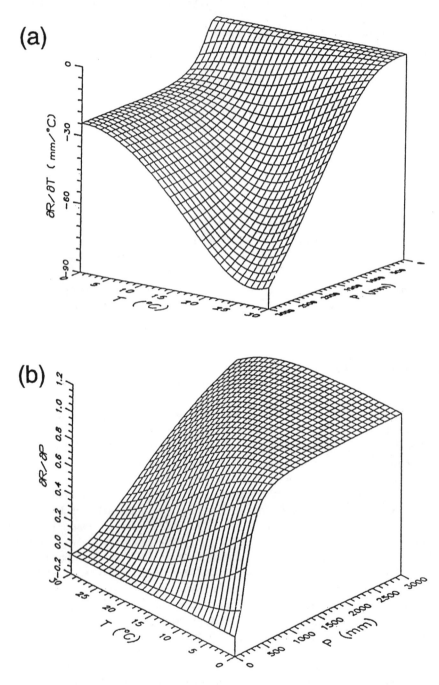

Figure 17.7. Change in runoff with respect to (a) temperature ($\partial R/\partial T$), and (b) precipitation ($\partial R/\partial P$), plotted against mean annual temperature (T) and precipitation (P).

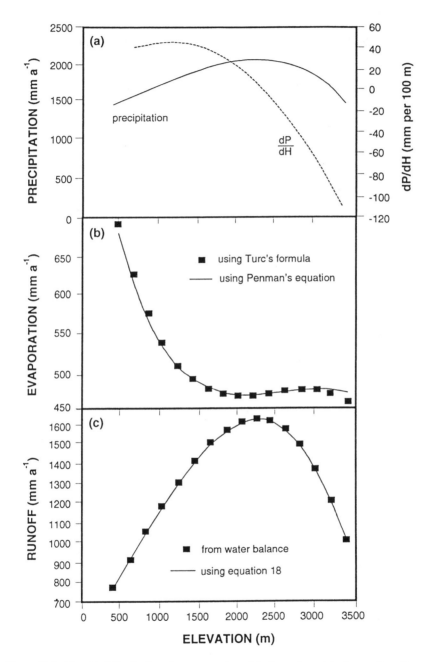

Figure 17.8. (a) vertical distribution of mean annual precipitation and precipitation gradient, Mt. Emei (b) vertical distribution of mean annual evaporation computed using Turc's equation, compared with values calculated using Penman's equation (c) vertical distribution of mean annual runoff computed using equation 18, compared with runoff derived from water balance.

Climatic warming scenarios for the mountainous region where Mt. Emei is located indicate a 3-4°C temperature rise under a doubling of atmospheric concentration of carbon dioxide (Houghton *et al.*, 1990). For our present example, a mean annual temperature rise of 4°C without any precipitation change was assumed. Equations 17.16 and 17.18 were applied to the new climatic forcing and the calculated vertical distributions of evaporation and runoff are presented in Figure 17.9. The computed results show that under the climatic warming scenario, evaporation will increase at all altitudes, accompanied by a decrease in runoff. Such responses may be different if future precipitation is augmented.

The method proposed requires only a limited amount of input data, and is well suited to testing the possible responses of water balance components to climatic change in nonglacierized mountainous terrain.

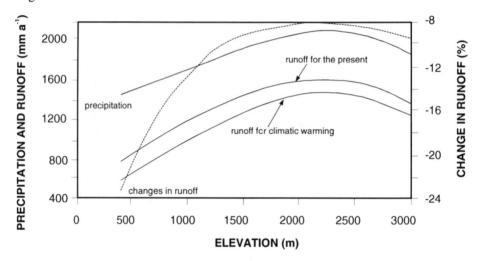

Figure 17.9. Comparison of vertical distribution of runoff for the present and for climatic warming (+4°C) scenario.

Acknowledgements

We wish to acknowledge the Natural Sciences and Engineering Research Council of Canada for financial support towards the completion of this study.

References

Barry, R.G. (1981) *Mountain Weather and Climate*, Methuen, London.
Budyko, M.I. (1956) *The Heat Balance of the Earth's Surface*, English Translation by Stepanova, N.A. (1958) Office of Technical Services, U.S. Department of Commerce, Washington, D.C.
Giorgi, F., and Mearns, L.O. (1991) Approaches to the simulation of regional climate change: a review, *Reviews of Geophysics* **29**, 191-216.

Gleick, P.H. (1987) Regional hydrologic consequences of increase in atmospheric CO_2 and other trace gases, *Climatic Change* **10**, 137-161.

Houghton, J.T., Jenkins, G.J., and Ephraums, J.J. (1990) *Climate Change: the IPCC Scientific Assessment*, Intergovernment Panel on Climate Change, Cambridge University Press, Cambridge, UK.

Johnson, R.C., Blackie, J.R., and Hudson, J.A. (1990) Methods of estimating precipitation inputs to the Balquhidder experimental basins, Scotland, *IAHS Publication No.* 193, 7-14.

Lai, Z.M., and Yeh, B.S. (1991) Water balance model for high and cold mountainous regions, and possible changes in runoff as influenced by climatic warming, *Acta Sci. Sinica* **B 6**, 652-658.

Liu, C.M. (1983) The quantitative features of China's water resources: an overview, *Technical Report No. 38, University of Arizona*, Tucson, Arizona.

Liu, C.M. (1985) An analysis of water balance and water resources in China, *Proceedings of the National Hydrological Conference, Geographical Society of China,* Science Press, Beijing, 113-119.

Schlesinger, M.E., and Mitchell, J.F.B. (1987) Climate model simulatons of the equilibrium climatic response to increased carbon dioxide, *Reviews of Geophysics* **25**, 760-798.

Sellers, W.D. (1965) *Physical Climatology*, University of Chicago Press, Chicago.

Turc, L. (1961) Estimation of irrigation water requirements, potential evapotranspiration: a simple climatic formula evolved up to date. *Annals of Agronomy* **12**, 13-14.

Waggoner, P.E. (1990) *Climate Change and U.S. Water Resources,* Wiley, New York.

Zhao, S.Q. (1986) *Physical Geography of China*, Wiley, New York.

Zhong Shan University and Northwest University (1979) *Meteorology and Climatology*, Education Press, Beijing. (In Chinese.)

Changming Liu
United Research Centre for Water Problems
Chinese Academy of Sciences
Beijing 100101, China.

Ming-Ko Woo
Department of Geography
McMaster University
Hamilton, Ontario, Canada L8S 4K1.

18 SENSITIVITY ANALYSES FOR THE IMPACT OF GLOBAL WARMING ON WATER RESOURCES IN WALES

C.P. HOLT and J.A.A. JONES

Abstract

In order to estimate potential changes in Welsh water resources under a double-CO_2 scenario, a simple methodology is developed based upon Welsh Water yield models and empirical estimates of effective precipitation. The results suggest that even in the best estimate scenario there will be a sizeable reduction in riverflows during summer, and that smaller upland catchments and reservoirs in the lower rainfall areas are most vulnerable. A general increase in spring rainfall may offset these effects somewhat, so that the higher risk of an extreme 5-month drought like 1984 is slightly ameliorated.

18.1. Introduction

The potential implications of global warming for water resources in the UK have become a cause for concern. This concern arises partly from General Circulation Models (GCMs) that suggest increased risk of summer drought (e.g. Hadley Centre, 1992) and from hydrological modelling that suggests an increase in the contrast in water resources between the wetter northwest and drier southeast (Arnell, 1992). It also arises partly from a series of extreme droughts and hot summers during the 1970s and 1980s, with some of the worst droughts on record in 1975-6, 1984 and 1989, and three of the hottest summers this century during the 1980s. Large areas of south east England have been experiencing the worst multiyear drought this century (1988-92), which led the National Rivers Authority to designate 40 rivers as endangered and ban abstractions in spring 1992. This has led to reconsideration of shelved plans for long distance interbasin transfer from the north and west, and could revive plans to expand the Craig Goch reservoir in Wales as a transfer source (National Rivers Authority, 1992, 1994).

The UK Climate Change Impacts Review Group (UKCCIRG, 1991) concluded that the 'best estimate' rainfall scenario for Britain in 2050 AD would be a small increase in annual rainfall mainly due to more winter rainfall, with an 8% increase in winter and no change in summer. However, they acknowledge wide margins of error particularly in summer, with confidence limits of ±16% compared with ±8% in winter. Indeed, there is good reason to believe that summers will be drier, especially in the south and east of

J.A.A. Jones et al. (eds.), Regional Hydrological Response to Climate Change, 317–335.

Britain, as depression tracks shift northwards and evaporative losses increase. This is confirmed in the recent transient run of the new Hadley Centre coupled atmosphere-ocean GCM (AOGCM), which contains the first high resolution window covering Europe (Hadley Centre, 1992). This drying trend is already underway in eastern Britain according to Palutikof's (1987) analysis of recent instrumental records.

Arnell (1992) took the full range of the UKCCIRG (1991) rainfall scenarios (Table 18.1) and concluded that annual runoff could decrease by 10 to 30% in the driest limiting case or increase by 17 to 30% in the wettest scenario. Arnell's calculations emphasise the fact that runoff generating processes will tend to amplify changes in rainfall. This effect is enhanced by regional differences in runoff coefficients, so that a 2.8 times amplification of rainfall changes may be found in runoff in the drier areas of SE England compared with only a 1.2-fold amplification in the wetter northwestern regions of Britain.

TABLE 18.1. Equilibrium climatic change scenarios

	Scenario	Winter	Spring	Summer	Autumn
Temperature (°C)	Best estimate	2.3	2.2	2.1	2.2
Precipitation (% change)	Wettest	+16	+16	+16	+16
	Best estimate	+8	+8	0	+8
	Driest	0	0	-16	0

Using the stochastic model of UK rainfall developed by Gregory *et al.* (1990), Cole *et al.* (1991) have estimated an overall decrease in annual runoff by 4% in northwest Britain and 8% in the southeast by 2030 AD. Putting these runoff changes into reservoir models, Cole *et al.* found yield losses of 4-25% in the wetter, mountainous regions of northwest Britain and confirmed the tendency to amplify changes in rainfall. Law (1989) concluded from a simple yield calculation for a typical reservoir in the Pennine hills that for many British dams even a ±40% change in inflow would only cause a ±20% change in yield.

Important changes are also likely in evapotranspiration and probably in the frequency of extreme events. Actual evapotranspiration losses are proving difficult to estimate, particularly because of their dependence upon soil moisture levels and rainfall patterns (Rowntree, 1990), but the most likely effect will be to reduce summer runoff even when there is no change in summer rainfall. Arnell (1992) used two scenarios for potential evapotranspiration - a 7% and a 15% increase, the latter being based on observations during the hot summer of 1989. His maps for a 15% increase under the 'best estimate' rainfall scenario suggest that any increase in runoff would be less than 4% and limited to the north and west, whilst large areas of south and eastern Britain would have runoff reduced by more than 15%. In contrast, the interim guidelines from the Institute of Hydrology reported in Binnie (1991) suggest planning for a 10% reduction in summer streamflow in the drier southeast and a 20% reduction in Wales and the wetter northwest.

There is greater uncertainty regarding the frequencies of droughts and floods. Hansen *et al.* (1991) estimate that drought events that occurred 5% of the time in 1958 will be a 25% risk by 2020. UKCCIRG (1991) estimate that the risk of events as extreme as the 1976 drought in Britain will increase a hundredfold to reach 10% by 2030 AD. Nevertheless, most projections of drought risk have been concerned with climatic drought, whereas Arnell (1992) points out that hydrological drought tends to be a product not just of dry summers but also of lack of rainfall during the winter recharge period. The widespread agreement that winter rainfall will increase suggests that hydrological droughts could become less frequent.

Again, storminess could increase in the UK due to compression of thermal gradients over the North Atlantic, at least until winter pack ice limits retreat out of the Atlantic into the Arctic Ocean and depression tracks shift further north (Jones, 1993). The Hadley transient run offers muted support for this, though not all GCMs concur (IPCC, 1991). Flood problems could increase as a result of more intense depressions in winter and of increased convective activity in summer.

18.2. Methodology

This paper reports the initial results of a programme undertaken to assess the effects of equilibrium climate change scenarios on representative water resources in Wales. A sensitivity analysis approach has been adopted using the same scenarios as Arnell (1992), as a means of testing a new methodology for assessing water resources in Wales and in anticipation of more reliable regional climatic predictions from the UKHI model (Jones, this volume, Chapter 6). It is proposed to extend the methods developed here to cover the whole of Wales within the next year using UKHI output, when the Hadley Centre is satisfied with the regional output at this scale, and this will be the subject of a subsequent paper.

A wide variety of sources are used for public water supply in Wales (Figure 18.1), which fall into four groups: 1) direct supply (impounding) reservoirs, 2) regulation schemes, and 3) unsupported river abstractions, and 4) groundwater. Each group is likely to have a different degree of climatic sensitivity. Small upland sources on flashy rivers in impermeable basins are more likely to exhibit severe reductions in discharge following 3 or 4 months of hot summer drought, as in 1984 and 1989. Impounding reservoirs also tend to be sensitive to accumulated winter rainfall.

A. PROPORTION OF SOURCE TYPES

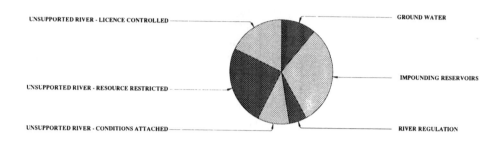

B. PROPORTION OF
YIELDS BASED ON SOURCE TYPE
IN WELSH WATER

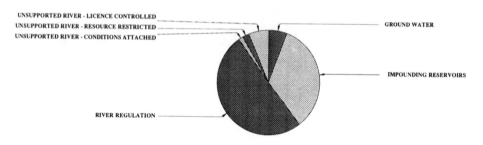

Figure 18.1. Proportional contributions to public water supply in Wales,
a) according to number of source intakes, and b) according to yield.

Three impounding reservoirs, one river regulation scheme and three representative rivers have been selected for the sensitivity analyses (Figure 18.2). The rivers were selected as ones with natural flow regimes and representing a range of catchment types (Table 18.2). The Erch at Pencaenewydd is a small mountain catchment ascending to 564 m above mean sea level (a.s.l.); the Yscir at Pontaryscir is a medium size catchment rising to 474 m a.s.l., and the Taf at Clog-y-Fran is a relatively large lowland basin of 217.3 km^2 with a maximum altitude of 395 m a.s.l.. The three reservoirs also represent a range of locations and sizes: Craig-y-Pistyll in the Cambrian Mountains of mid-Wales with a storage capacity of only 342 Ml storage, Preseli in the Preseli Mountains of southwest Wales (644 Ml storage) and Crai in the Brecon Beacons, south Wales (4204 Ml). The Llyn Brianne scheme in south Wales (62 140 Ml) was selected to represent river regulation.

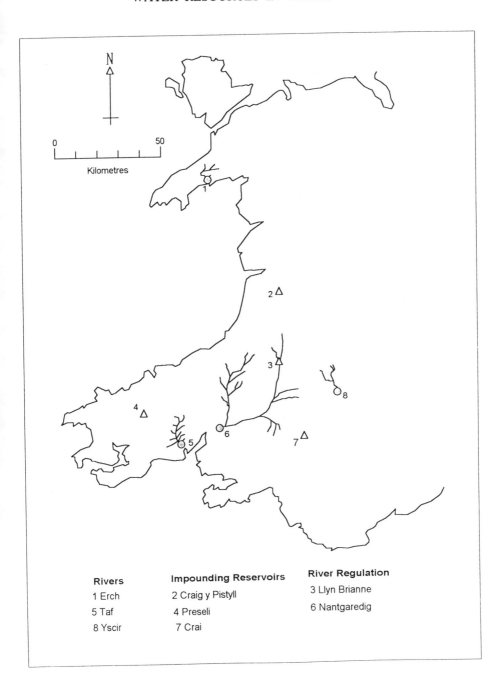

Figure 18.2. Location of selected sites in Wales.

TABLE 18.2. Characteristics of selected catchments.

	Mean annual rainfall (mm)	Catchment area (km^2)	Altitude max. (m)	Storage capacity of reservoir (Ml)
Rivers				
Erch	1407	18.1	564	--
Yscir	1433	62.8	474	--
Taf	1428	217.3	395	--
Impounding reservoir				
Craig-y-Pistyll	1745	5.4		342
Preseli	1203	8.3		644
Crai	1637	11.4		4204
River regulation				
Llyn Brianne	1553	88.0		62 140
Nantgaredig	1503	998.1		

Supercalc spreadsheet models developed by Welsh Water were used to calculate river flows and reservoir yields under each of the UKCCIRG/Arnell scenarios (Table 18.1). The spreadsheets for impounding reservoirs require data to be input on 1) catchment area (km^2), 2) average annual runoff (mm), 3) an inflow factor which is the runoff in a given month as a percentage of the long-term average proportion of annual runoff that normally occurs during that month, 4) the required supply (Ml d^{-1}) and 5) a supply factor which is varied during the progress of a drought as water has to be conserved. The spreadsheets are then driven by calculated or observed inflows. Very few records exist of actual inflows. Similarly, GCM estimates are not available for future runoff and even estimates of evaporation are generally less reliable than for temperature or rainfall. A single method was therefore sought for calculating both past and future flows. Using the same method for each assessment enables the method to be tested and calibrated against existing flow data, where they are available, and should also allow a consistent means of measuring the amount of the change between the present and the future scenarios.

The Meteorological Office Rainfall and Evaporation Calculating System (MORECS) was selected as the basis for these calculations, since it calculates 'hydrologically effective precipitation' which can be used as an indicator of available discharge. Thus, effective precipitation, EP, is calculated as:

$$EP = \text{Rainfall - Actual evapotranspiration - Soil moisture deficit} \qquad (18.1)$$

for 40 x 40 km grid squares. Evapotranspiration values are calculated by a Penman-Monteith formula, using average albedo and vegetation characteristics for each square, and a 2-layer soil moisture model (Wales-Smith, 1975). Unfortunately, this cannot yet be effectively used to simulate future evapotranspirational losses, as too many variables in the equation remain insufficiently predictable (Rowntree, 1990). In fact, only empirical methods have been used thus far, e.g. Bultot *et al.* (1988), Gregory *et al.* (1990), or Arnell (1992). Liu and Woo (this volume, Chapter 17) used Turc's empirical temperature-based formula to calculate potential evapotranspiration.

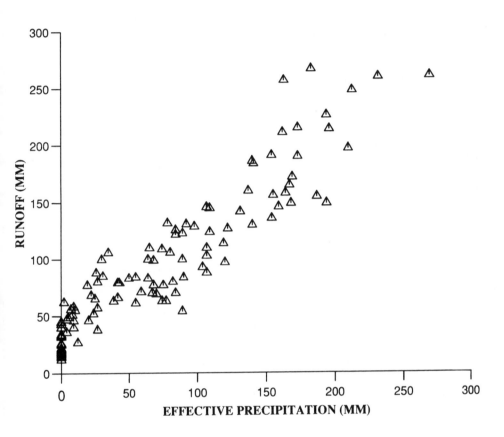

Figure 18.3. Comparison of recorded runoff and effective precipitation.
An example of seasonal data for the River Erch, 1981-1990.

The method adopted was to correlate existing MORECS values for actual evapotranspiration (AET) and for effective precipitation in the grid square(s) covering each site with precipitation and temperature. This was judged to provide the most suitable 'locally-tuned' estimate for actual water loss, and had what turned out to be the critical advantage of being capable of estimating not just AET, but also effective precipitation. The MORECS data set ran from January 1979 to December 1991, and stepwise multiple regressions were performed relating the monthly MORECS values to the precipitation and temperature for that and the previous month. In fact, no statistically significant equations could be established for AET itself on a monthly basis, and on a seasonal basis no successful equations were obtained for summer. However, good 3-month seasonal equations were achieved for effective precipitation. Equations for all sites and seasons were significant at the 1% level, and 80% of them had multiple correlations of over R = 0.9. In effect, the result enabled one step to be cut out of the procedure, since effective precipitation is the required end-product and would have had to be calculated using predicted values for AET. Comparing hindcasted values for the river sites over the 1981-90 period with observed flows recorded in the UK Surface Water Archive tended to show a slight systematic error between the observed and predicted runoff. This could be attributable to different runoff coefficients in the various basins. It was corrected by introducing systematic transfer functions for each site, which were obtained by regressing the observed runoff against the predicted values (Figure 18.3). The regression equations were then incorporated into the spreadsheets for each site. The overall procedure is illustrated by the flowchart in Figure 18.4. Spreadsheets were test run on the current climate, first under average conditions and subsequently under the drought conditions of 1984, which were the worst on record in most of Wales and are now generally used as the design drought.

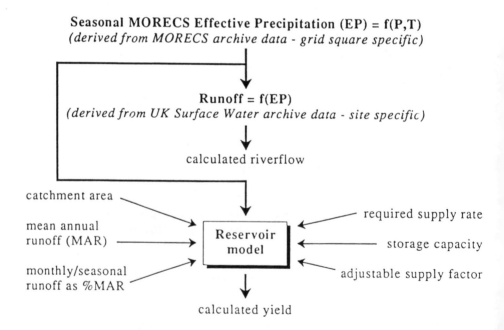

Seasonal MORECS Effective Precipitation (EP) = f(P,T)
(derived from MORECS archive data - grid square specific)

Runoff = f(EP)
(derived from UK Surface Water archive data - site specific)

calculated riverflow

catchment area

mean annual runoff (MAR)

monthly/seasonal runoff as %MAR

Reservoir model

required supply rate

storage capacity

adjustable supply factor

calculated yield

Figure 18.4. Flowchart of method.

18.3. Predictions for a perturbed climate

In order to predict water resources for a 'double-CO_2 scenario, the UKCCIRG (1991) temperatures for summer and winter were entered into the spreadsheet models for each site, together with linear interpolations for spring and autumn, and the models were run for the range of seasonal rainfall scenarios in Table 18.1.

The results are shown in Table 18.3 and Figures 18.5 to 18.7. They suggest that even under the wettest scenario there could be reduced resources during summer and autumn, as a result of increased evapotranspirational losses. The river regulation scheme seems to be the only exception. The best estimate scenario shows an average reduction of 30% in summer discharge in the unsupported rivers. A larger average reduction in reservoir inflows (40%) is mainly due to the Preseli site, which has the lowest effective rainfall of the seven sites in the present climate. However, the regulation scheme also receives 30% less runoff in the best estimate summer. Under the driest scenario even in a normal year summer riverflow is reduced by 40%, or approaching 30% over summer and autumn combined. This compares with a 50% reduction during the 5-month drought in 1984. For the rivers supplying the impounding reservoirs, the best estimate scenario suggests a summer shortfall of nearly 45% or approaching 27% over the 6-month summer-autumn period. Again, the 75% average summer reduction in inflows to the impounding reservoirs under the driest scenario is largely due to the drastic change at Preseli. Craig-y-Pistyll fares best amongst the reservoir inflows and it has twice the mean effective precipitation of Preseli (Table 18.2). Summer inflows to Llyn Brianne are very similar to those of Craig-y-Pistyll and Crai, all very close to the overall average reduction for the three reservoirs of 76% under this scenario. There is also a tendency for about 10% reduction in runoff during winter in the driest scenario as a result of higher evaporation.

There appears to be some geographical variation in the severity of reductions in summer and autumn. Taking the period of reduced flows in summer and autumn as a whole the small mountain catchment of the Erch seems most vulnerable in every scenario. The Erch has the smallest catchment and the lowest rainfall (Table 18.2), and shows the largest net reduction in annual discharge within the group (10%). In contrast, the Yscir, which is intermediate in all respects (size, altitude and effective rainfall) tends to fare slightly better than average. This geographical variation disappears under the driest scenario when all three basins show reductions of between 32% and 36% in the driest six months. Amongst the reservoirs, the smallest, Craig-y-Pistyll, tends to fare best in all scenarios by virtue of its higher altitude and rainfall, whereas Preseli tends to have half the effective rainfall of Craig-y-Pistyll and to suffer most.

TABLE 18.3. Runoff in rivers and inflows to reservoir and river regulation schemes as a percentage of present average, for three future equilibrium scenarios (wettest, best estimate and driest) and during the extreme drought year of 1984.

Scenario	Wettest				Best estimate				Driest				1984			
Season	Win	Spr	Sum	Aut	Win	Spr	Sum	Aut	Win	Spr	Sum	Aut	Win	Spr	Sum	Aut
River flows																
Erch	2.4	12.4	-5.4	-5.4	-3.4	5.1	-25.1	-15.0	-9.2	-2.2	-44.8	-24.7	9.2	-55.5	-63.5	34.6
Yscir	11.7	11.1	0.0	5.3	1.9	5.6	-40.0	0.0	-9.4	-1.1	-56.0	-8.9	19.4	-44.4	-56.0	27.4
Taf	7.9	13.8	5.0	3.3	0.0	6.1	-25.0	-8.2	-7.0	-1.5	-30.0	-18.0	4.7	-50.8	-30.0	49.2
Average	7.2	12.4	-0.1	1.1	-0.5	5.6	-30.0	-7.7	-8.5	-1.6	-43.6	-17.2	11.1	-50.2	-49.8	36.1
Reservoir inflows																
Craig-y-Pistyll	11.2	2.1	1.9	10.0	2.3	-5.4	-30.7	-0.6	-6.7	-12.9	-63.4	-11.3	12.1	-20.7	-33.2	29.5
Preselli	8.2	15.8	-6.8	-6.3	-1.9	5.5	-69.2	-23.8	-11.9	-4.8	-96.6	-41.4	11.8	-88.7	-96.6	104.0
Crai	10.9	18.0	6.0	6.5	1.7	8.7	-31.2	-4.9	-7.4	-0.5	-68.3	-16.4	25.6	-93.2	-90.6	48.6
Average	10.1	23.9	0.4	3.4	0.7	2.9	-43.7	-9.8	-8.7	-6.1	-76.1	-23.0	16.5	-67.5	-73.5	60.7
River regulation flows																
Llyn Brianne	11.5	11.1	8.4	7.1	2.5	3.1	-29.4	-4.0	-6.5	-4.9	-66.1	-15.1	12.1	-20.7	-33.2	29.5
Nantgaredig	15.7	17.2	9.8	4.8	8.1	8.1	-30.8	-8.3	0.0	-1.0	-71.4	-19.2	8.6	-39.9	-85.1	26.6
Average	13.6	14.2	9.1	6.0	5.3	5.6	-30.1	-6.2	-3.3	-3.0	-68.8	-17.2	10.4	-30.3	-39.2	28.1

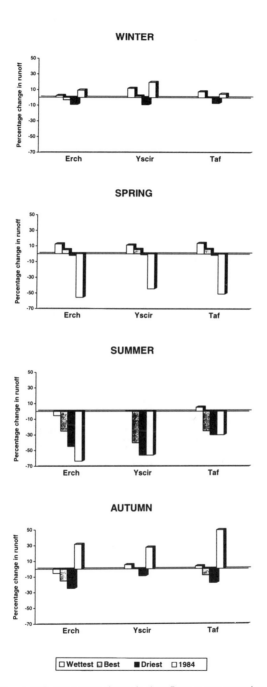

Figure 18.5. Percentage change in river flows at representative
sites under the three equilibrium scenarios.

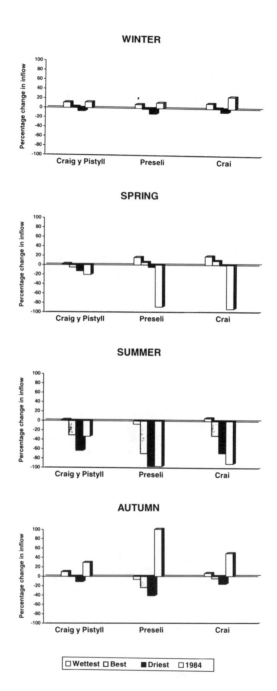

Figure 18.6. Percentage change in reservoir inflows at representative
sites under the three equilibrium scenarios.

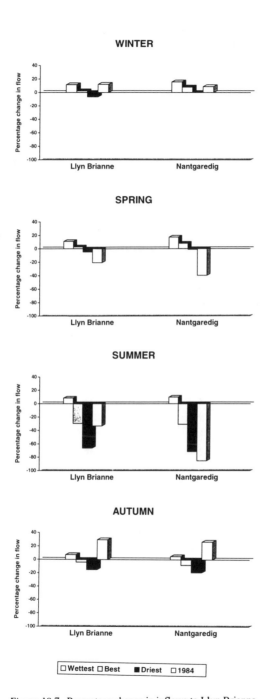

Figure 18.7. Percentage change in inflows to Llyn Brianne
regulating reservoir under the three equilibrium scenarios.

In the final analysis, the vulnerability of a reservoir is a function of its storage capacity and the demand for supply, as well as inflows. The Welsh Water 'operational yield' procedure developed by Jack and Lambert (1992) estimates safe yields by extending the worst summer drought on record through to November, and making commensurate reductions in abstractions for water supply as the drought progresses (Figure 18.8 and Table 18.4).

TABLE 18.4. Demand controls for impounding reservoirs.

Month	Controls	Demand as a % of the average
up to the end of May	• Unrestricted demand	120
June	• Appeals for voluntary restraint	100
	• Hosepipe bans	
	• First cuts in compensation releases	
July	• First bans on non-essential use	100
August	• Second cuts in compensation releases	100
	• First stage of relaxations for river abstractions	
September	• Second ban on non-essential use	90
	• Second stage of relaxations for river abstractions	
October	• Rota cuts	75
Mid-November	• Failure of supply	75

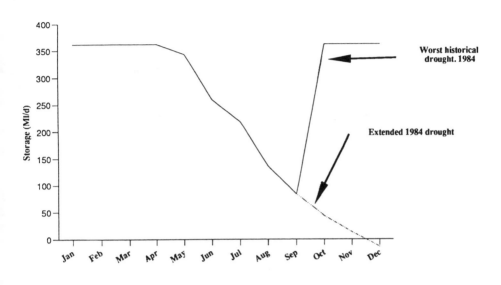

Supply Factors

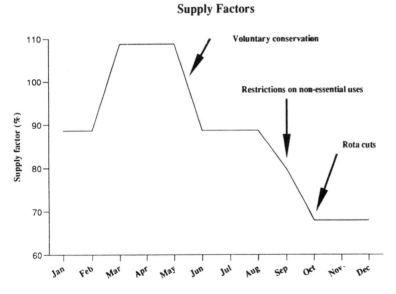

Figure 18.8. The Jack and Lambert (1992) method of calculating operational yield applied to the
Craig y Pistyll reservoir, extending the worst historical drought and controlling supply factors.

Unfortunately, it is still impractical to estimate the progress of extreme droughts in a future climate. However, in order to make an initial stab at predicting the effects on the operational yields of these reservoirs, Table 18.5 has been constructed by comparing average summers under each scenario with the results of running this 'synthetic operational yield' procedure under average present-day conditions. This is only of value if we make the tentative working assumption that droughts have a logarithmic probability distribution, so that the proportional change in mean and extreme events is similar. These synthetic yields suggest an increase in annual resources under the wettest scenario, but net reductions in synthetic operational yield is about 20% in impounding reservoirs and 10% in the Brianne regulation scheme. The smallest reservoir, Craig-y-Pistyll, is most vulnerable in the group, with almost a 50% reduction in yield in the worst scenario.

TABLE 18.5. Synthetic yields for impounding reservoirs and river regulation scheme under the equilibrium scenarios, as a percentage of present values.

	Wettest	Best estimate	Driest
IMPOUNDING RESERVOIRS			
Craig-y-Pistyll	+1.5	-24.4	-50.3
Preseli	-2.5	-33.4	-44.6
Crai	+7.1	-5.4	-17.8
Average	+2.0	-21.1	-37.6
RIVER REGULATION SCHEME			
Llyn Brianne	+11.6	-11.6	-17.9

It is also interesting to speculate on the effects on return periods. Marsh and Lees (1985) plot return periods for the 5-month 1984 design drought at these sites that range from 70 to 100 years. Table 18.6 has been constructed according to Wigley's (1989) method, which assumes a normal distribution of events and relates the shift in mean to the current standard deviation. This suggests an increase in risk under the best scenario, particularly at the two sites with the lowest present rainfall, Taf and Preseli. In general, under the best scenario return periods are reduced by 40% for the river sites, which represent unsupported river abstraction sites, and 50% for the impounding reservoirs: a difference brought about by the strong spatial variation of the 1984 drought severity.

The regulation scheme is intermediate, with a 35% reduction in return period. Under the driest scenario the reductions in return periods are of the order of three to ten-fold.

TABLE 18.6. Estimated return periods under the equilibrium scenarios for five-month droughts equivalent to the current 1984 design event.

	1984[*]	Wettest	Best estimate	Driest
RIVERS				
Erch	70	75	40	20
Yscir	100	165	100	55
Taf	100	165	55	35
IMPOUNDING RESERVOIRS				
Craig-y-Pistyll	70	75	20	10
Preseli	70	110	20	10
Crai	100	280	60	15
RIVER REGULATION SCHEME				
Llyn Brianne	70	100	40	20
Nantgaredig	100	280	70	20

* Based on Institute of Hydrology estimate (Marsh and Lees, 1985).

18.5. Conclusions

The work reported in this paper is part of a continuing programme designed to cover the whole of Wales, which will eventually incorporate the latest 'best estimate' scenario from the Hadley Centre's newly operational high resolution GCM. It uses a relatively simple methodology based upon putting GCM scenarios and empirical estimates of local effective precipitation into models and procedures currently used by Welsh Water.

The results suggest that even in the best estimate scenario there will be significant reductions in overall resources, with higher risks of summer shortfalls. Sources drawing upon the smaller upland rivers are probably most vulnerable. Increases in winter and spring river flows may partly offset drier summers, but reservoirs tend by design and operation to be full in the spring and extra rain is likely to be wasted as overspill unless dams are raised. Very tentative results suggest that water managers should plan for an overall reduction of 20% in the operational yields of reservoirs, although this limited sample of reservoirs reveals a wide range in responses. In the wider, strategic context these results need to be viewed against the current debate in Britain over ways of satisfying projected trends in water demands in southeast England. According to the National Rivers Authority (1992), projected demand in the Thames Region is likely to increase 26% by 2021 AD, reducing the local regional surplus water resource, including planned new sources, to just 1%, and mid-Wales is identified as one of two possible sources for major interbasin transfer schemes.

Acknowledgments

We wish to thank Welsh Water and the University of Wales for financial support, and Logan Jack in particular for data, guidance and unfailing interest.

References

Arnell, N.W. (1992) Impacts of climatic change on river flow regimes in the UK, *J. Institution of Water and Environmental Management* **6**, 432-442.

Binnie, C.J.A. (ed.) (1991) Policy paper on: Securing adequate water supplies in the United Kingdom in the 1990s and beyond, Institution of Civil Engineers, London.

Bultot, F., Coppens, A., Dupriez, G.L., Gellens, D., and Meulenberghs, F. (1988) Repercussions of a CO_2-doubling on the water cycle and on the water balance: a case study from Belgium, *J. Hydrology* **99**, 319- 347.

Cole, J.A., Slade, S., Jones, P.D., and Gregory, J.M. (1991) Reliable yield of reservoirs and possible effects of climate change, *Hydrological Sciences J.* **36**, 579-598.

Gregory, J.M., Jones, P.D., and Wigley, T.M.L. (1990) *Climatic Change and its Potential Effect on UK Water Resources*, Report to the Water Research Centre, Medmenham, UK, Parts 1 and 2.

Hadley Centre (1992) *The Hadley Centre Transient Climate Change Experiment*, Hadley Centre, Meteorological Office, Bracknell, UK.

Hansen, J., Rind, D., Delgenio, A., Lacis, A., Lebedeff, S., Prather, M., and Ruedy, R. (1991) Regional greenhouse climate effects, in M.E.Schlesinger (ed.) *Greenhouse-gas-induced Climatic Change: a Critical Appraisal of Simulations and Observations*, Elsevier, Amsterdam, 211-229.

Intergovernmental Panel on Climate Change (IPCC) (1991) Potential Impacts of Climate Change. WMO/UNEP.

Jack, L., and Lambert, A.O. (1992) Operational yield, in M.N.Parr, J.A. Charles, S.Walker (eds.) *Water Resources and Reservoir Engineering*, Thomas Telford, London, 65-72.

Jones, J.A.A. (1993) Postglacial climatic changes in Western Europe and their implications for agriculture, in D. Wilson, H. Thomas, and K. Pithan (eds.) *Crop Adaptation in Cool, Wet Climates*, European Cooperation in the field of Scientific and Technical Research COST

814, Proceeding of Aberystwyth Conference, Office for Official Publications of the European Communities, Luxembourg, 17-28.

Law, F.M. (1989) Identifying the climate-sensitive segment of British reservoir yield, in Academy of Finland *Conference on Climate and Water*, Helsinki, vol. 2, 177-190.

Marsh, T. and Lees, M. (1985) *The 1984 drought*, Natural Environment Research Council, UK.

National Rivers Authority (NRA) (1992) *Water Resources Development Strategy - a Discussion Document*, N.R.A., Bristol, UK.

National Rivers Authority (NRA) (1994) *Water - nature's precious resource. An environmentally sustainable water resources development strategy for England and Wales*, HMSO, London, 93 pp.

Palutikof, J.P. (1987) Some possible impacts of greenhouse gas induced climatic change on water resources in England and Wales, in S.I. Solomon, M. Beran, and W. Hogg (eds.) *The Influence of Climate Change and Climatic Variability on the Hydrologic Regime and Water Resources*, IAHS Pub. No. 168, 585-596.

Rowntree, P.R. (1990) Estimates of future climatic changes over Britain. Part 2: Results. *Weather* **45**, 79-89.

UK Climate Change Impacts Review Group (UKCCIRG) (1991) *The Potential Effects of Climate Change in the United Kingdom*, First Report, HMSO, London, 124 pp.

Wales-Smith, B.G. (1975) The estimation of irrigation needs, in *Engineering Hydrology Today*, Thomas Telford, London, 45-54.

Wigley, T.M.L. (1989) The effect of changing climate on the frequency of absolute extreme events, *Climate Monitor* **17**, 44-55.

C.P. Holt
School of Environmental Science
Nene College, Boughton Green Road, Northampton, NN2 7AL, UK.

J.A.A. Jones
Institute of Earth Studies
University of Wales, Aberystwyth, SY23 3DB, UK.

19 POTENTIAL HYDROLOGICAL RESPONSES TO CLIMATE CHANGE IN AUSTRALIA

F.H.S. CHIEW, Q.J. WANG, T.A. McMAHON, B.C. BATES, and P.H. WHETTON

Abstract

This chapter describes the use of a conceptual hydrological model to investigate the potential impacts of climate change on runoff and soil moisture in Australia. The potential hydrological responses are estimated by comparing the model simulations driven by historical data and changed input data representing scenarios of climate change. The changed input data series are generated hypothetically as well as using two methods based on results of general circulation model simulations of the climate. The merits and limitations of the methods used here are highlighted and the climate change impact scenarios for Australia based on the current state of science are presented. The results indicate that climate change has the potential to bring about runoff modifications that may require a significant planning response.

19.1. Introduction

It is generally accepted by the scientific community that the increasing concentration of greenhouse gases since pre-industrial time will lead to global warming and changes in precipitation patterns and other climatic variables. These changes will in turn impact on the hydrology and thus the management of land and water resources. For example, larger reservoir spillways and drainage waterways will be required where peak discharges are expected to increase and bigger water supply storages will be needed in areas where runoff decreases. Changes in soil moisture conditions and evapotranspiration will affect irrigation and agricultural crop and land management.

This chapter investigates the potential impacts of climate change on runoff and soil moisture in Australian catchments using a conceptual hydrological model. Measured historical data are used to optimise the model parameters for each catchment. The model is then driven by changed input data representing scenarios of climate change, and the simulations using the historical data and the changed input data are compared. Two methods, based on simulations from atmospheric general circulation models (GCMs), are used to generate the changed input data series. The first method scales the historical

J.A.A. Jones et al. (eds.), Regional Hydrological Response to Climate Change, 337–350.

precipitation by a constant factor and increases the historical temperature by a fixed amount while the second method uses a stochastic weather generator to include the changes in the timing and frequency of the climatic variables as well as their magnitude. The aims of this chapter are to highlight the merits and limitations of the different approaches and to present the climate change impact simulations for selected regions in Australia.

19.2. Data and Rainfall-Runoff Modelling

Eight to 20 years of climate and streamflow data from 28 unregulated catchments, which represent the large range of climatic, physical and flow conditions throughout Australia, are used (Figure 19.1). The hydrological simulations are carried out using the conceptual daily rainfall-runoff model, MODHYDROLOG (Figure 19.2). The model requires daily rainfall and potential evapotranspiration as input data. The potential evapotranspiration is calculated from drybulb temperature, wetbulb temperature and sunshine hours using Morton's (1983) wet environment evapotranspiration procedure. The model has 19 parameters, but in most cases, the calibration of fewer than ten parameters is sufficient to provide satisfactory estimates of runoff (Chiew and McMahon, 1994).

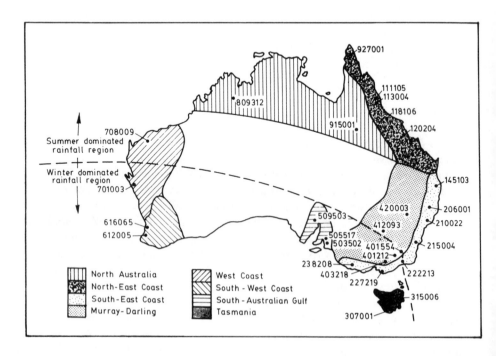

Figure 19.1. Locations of the 28 catchments and catchment regions.

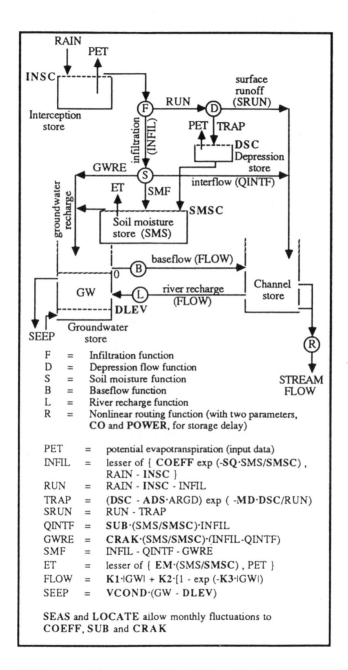

Figure 19.2. Structure of the conceptual daily rainfall-runoff model MODHYDROLOG.
Model parameters are highlighted in bold.

The parameters in MODHYDROLOG are first optimised for each of the 28 catchments. The successful application and calibration of MODHYDROLOG to fit the historical streamflow data are described in Chiew and McMahon (1994). In the following impact simulations, changed input data which reflect the climate change scenarios are used to drive MODHYDROLOG, using the same parameter values obtained earlier in the model calibration against historical data. Results from the simulations using the historical data series and the changed input data are then compared to estimate the impacts on runoff and soil moisture.

The use of the same parameter values assumes that the catchments will continue to behave as they do at present. The potential feedbacks between the surface and the atmosphere are also not taken into account. Nevertheless, although the surface-atmosphere feedbacks and changes in hydrological processes may be significant, there is insufficient understanding as yet to model them adequately.

19.3. Sensitivity Study

Hypothetical scenarios are used to investigate the sensitivity of runoff and soil moisture to changes in precipitation and temperature. The historical precipitation data are scaled by 0%, ± 10%, ± 20%, ± 30% and ± 40% and the drybulb and wetbulb temperatures are increased by 0°C, 2°C and 4°C (resulting also in a slight increase in relative humidity). Sunshine duration (used to compute radiation), which is the remaining data required to calculate potential evapotranspiration, is assumed to be the same. The results are summarised for three populated and agricultural regions in Australia with very different climate conditions.

The tropical north-east coast (Figure 19.1) has an average annual rainfall of more than 1200 mm, 90% of which occurs in the summer-half of the year (November to April). More than 50% of the annual rainfall becomes runoff. Simulations using the hypothetical changes in rainfall indicate that in this wet area, the average annual runoff changes at one and a half times the rate of change in rainfall while the soil wetness (defined as the ratio of the soil moisture level (SMS) to the soil store capacity (SMSC) - Figure 19.2) changes at approximately half the rate of change in rainfall.

The simulations also indicate that the increase in temperature alone does not affect runoff and soil wetness. This is because the model calculates actual evapotranspiration (AET) as the soil wetness multiplied by a parameter which represents the maximum plant-controlled rate of evapotranspiration (see equation for ET in Figure 19.2). However, AET cannot exceed the atmospherically controlled rate of potential evapotranspiration (PET) which is calculated from the climate data. Thus, although higher temperatures increase PET, temperatures (and therefore PET) in this region are generally high throughout the year, and AET is almost always lower than PET. In other words, the change in temperature alone has little effect on the water balance because evapotranspiration here is governed more by the prevailing soil moisture levels than the atmospheric conditions.

In the south-west coast and the South Australian Gulf (Figure 19.1), about 70% of the annual rainfall of 500 to 1000 mm occurs in the winter-half of the year. The streams in these regions generally flow for only 50% of the time, and on average less than 10% of the annual rainfall becomes runoff. It is also common for the total annual runoff to come from only one or two significant flow events during winter. The simulations indicate that the average annual runoff increases at a much faster rate than the corresponding increase in rainfall. A rainfall increase of 10% enhances runoff by 50%, an increase of 20% more than doubles runoff and an increase of 40% results in runoff being almost four times greater. A decrease in rainfall has a potentially more serious consequence as the amount of streamflow drops very quickly. A decrease in rainfall by 20% reduces runoff by one third while a decrease of 40% reduces runoff by 90%. The soil wetness in the winter-half changes at almost twice the rate of the change in rainfall for changes of rainfall of up to 20%. The changes however, approach lower and upper limits with the simulations indicating that it is unlikely for the soil wetness to drop below 10% or increase above 70% even for rainfall changes of more than 50%.

Unlike the tropical region, a temperature increase here affects runoff and soil wetness. This is because a temperature increase would increase PET (and therefore AET on the cold days), resulting in lower runoff and drier soil. The simulations indicate that a temperature increase of 4°C would alone decrease the average annual runoff and soil wetness by almost 5%. Nevertheless, the change in temperature alone has a much smaller effect on the water balance compared to the change in rainfall.

The south-east of Australia has a temperate climate with the average annual rainfall ranging from 500 to 1500 mm. The rain falls fairly uniformly through the year and about 20 to 50% of the annual rainfall becomes runoff. Simulations using the hypothetical scenarios for catchments here indicate that the average annual runoff changes at about twice the rate of change in rainfall and the soil wetness changes at approximately half the rate of change in rainfall.

19.4. Simulations using GCM Scenarios

The sensitivity study in the previous section provides an indication of the hydrological changes in response to potential changes in the climate. Investigating the range of possible changes is important because reliable estimates of climate change are not yet available. In this section, changes in runoff and soil moisture by the year 2030 are simulated using climate change scenarios obtained by analysing the results of enhanced greenhouse experiments from five GCMs. The aim here is to provide an estimate of the range of plausible changes given the current state of science.

The five GCMs used are the CSIRO9 (McGregor et al., 1993), Bureau of Meteorology Research Centre (BMRC) (Coleman et al., 1994), and the three high resolution experiments used by IPCC (Houghton et al., 1990) - United Kingdom Meteorological Office high resolution model (UKMOH), Geophysical Fluid Dynamics Laboratory high resolution model (GFDLH) and Canadian Climate Centre (CCC) GCMs. All these experiments use a simple mixed layer ocean without ocean currents, but with a heat flux correction which implicitly accounts for heat transport due to the present day ocean circulation. The use of a number of GCMs allows for the fact that climate change simulations can differ significantly among GCMs and ensures that the impact scenarios are not highly dependent on the result of a single GCM.

The patterns of temperature increase and percentage change in rainfall were derived by CSIRO (1992) and Whetton *et al.* (1994) by comparing the results of the GCM experiments for 1 x CO_2 and 2 x CO_2 conditions. In deriving the scenarios, CSIRO (1992) ranked the changes simulated by the five GCMs for each grid point from the highest (1) to the lowest (5), and the half-yearly maps of the rank 2 and rank 4 changes were used to define the range of model results at each grid point. The rank 1 and rank 5 changes were not used because they are greatly influenced by some unrealistically extreme results (particularly that of precipitation in arid regions) at some grid points.

The scenarios prepared specifically for this study are shown in Figure 19.3. The ranges given in Figure 19.3 represent warming and percentage rainfall change per degree of global warming. The temperature scenarios reflect a tendency in the GCMs to warm inland regions more than coastal regions and southern areas more than northern areas. The regionalisation of the rainfall scenarios was based on the areas of agreement and disagreement in the sign of rainfall change amongst the five models. In the summer-half (November to April), the GCMs agree on increasing rainfall across Australia, with increases of up to 5% per degree of global warming in the south-east and up to 10% elsewhere. In the winter-half (May to October), three separate regions were identified corresponding to regions where the GCMs agree on an increase (up to 5% per degree global warming in Tasmania), agree on a decrease (up to 5% in central and south Australia) or fail to agree (-5% to +5% in eastern and south-west Australia).

The changes per degree of global warming in Figure 19.3 are then combined with the range of global average warming in Figure 19.4 to provide the scenarios for this study. Figure 19.4 was prepared by the Climate Research Unit in Norwich with the temperature curves calculated using a simple upwelling-diffusion energy balance climate model (see Wigley and Raper, 1992). The broad range of future warming comes from making allowance for two important sources of uncertainty - the range of greenhouse gas emission scenarios considered plausible by the IPCC (Houghton *et al.*, 1992) and the range of uncertainty regarding the sensitivity of the global climate to increased greenhouse forcing. The climate sensitivity is expressed as the global equilibrium warming for a doubling of CO_2 , and has been estimated by the IPCC to range from 1.5°C to 4.5°C. The upper curve in Figure 19.4 thus corresponds to the highest IPCC emission scenario combined with a climate sensitivity of 4.5°C, and the lower curve is based on the lowest emission scenario and a climate sensitivity of 1.5°C. The scenarios derived for this study are therefore time dependent and make some allowance for major sources of uncertainty.

Figure 19.4 shows a global warming of 0.6°C to 1.7°C by the year 2030. Corresponding to the warming of 1.7°C, temperature in the north-east coast will increase by 0.5°C (0.3°C x 1.7) to 1.7°C (1.0°C x 1.7) and summer-half rainfall will increase by 0% to 17% (10% x 1.7) (Figures 19.1 and 19.3). Similarly, corresponding to the global warming of 0.6°C, temperature will increase by 0.2°C to 0.6°C and summer-half rainfall

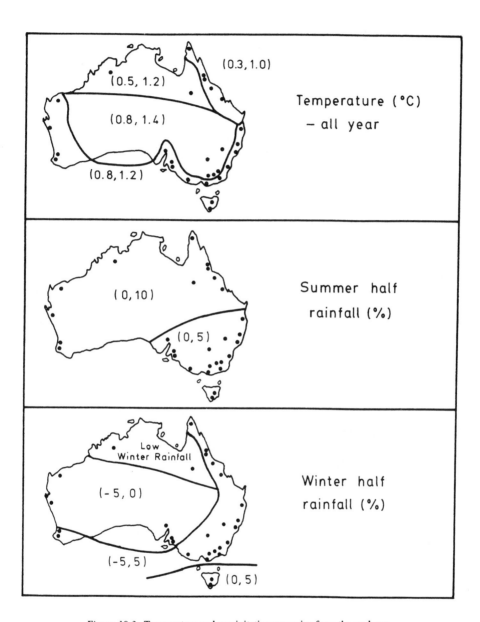

Figure 19.3. Temperature and precipitation scenarios from the analyses
of GCM enhanced greenhouse simulations (numbers in brackets
are limits of changes per degree Celsius of global warming).

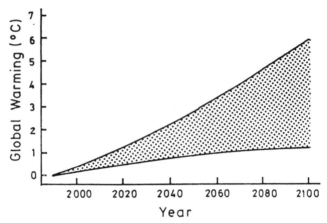

Figure 19.4. Range of future global warming given by Wigley and Raper (1992).

will increase by up to 6%. There are therefore eight possible combinations that can lead to the greatest and least impacts. The eight combinations are used to increase the historical temperature and scale the rainfall data in the 28 catchments. The MODHYDROLOG simulations using these changed input data are then compared with the simulations using the historical data to estimate the range of plausible changes in runoff and soil moisture by the year 2030.

Table 19.1 presents only the generalised changes in the various regions. Detailed results from the simulations and the modelling approach are discussed in Chiew *et al.* (1995). The results indicate increases in annual runoff of up to 25% by the year 2030 in the wet tropical catchments near the north-east coast. The GCMs do not agree on the direction of rainfall change in south-east Australia, and the simulations show runoff changes of up to ± 20% by 2030. For Tasmanian catchments, up to 10% increase in runoff is simulated whereas for catchments in the South Australian Gulf, up to 35% decrease in annual runoff is simulated. Near the western coast of Australia, the simulations using the GCM scenarios show runoff changes of up to ± 50%.

19.5. Simulations using Stochastic Weather Sequences

The simulations in Sections 19.3 and 19.4 ignore potential changes in the temporal distribution and frequency of events. For example, the scaling of historical rainfall assumes that the percentage change in rainfall intensity is the same for every day and that the number of rain days remains the same. As a result of this limitation, there is little point in comparing the simulations of the extreme values and only the average half-yearly and annual impacts are presented above.

In this section, a stochastic weather generator is used to overcome the limitation. The approach relates the characteristics of the historical data at the catchment scale to the

TABLE 19.1. Generalised changes in runoff and soil moisture by 2030.

Geographical Region (see Figure 19.1)	Summer-Half		Winter-Half	
	Runoff	Soil moisture	Runoff	Soil moisture
North-East Coast	0 to +25	0 to +10	*-10 to +20*	*-10 to +10*
South-East Coast				
- northern part	-5 to +25	0 to +10	-20 to +20	-15 to +10
- southern part	*-20 to +20*	*-10 to +10*	-25 to +20	-15 to +10
Murray-Darling				
- northern part	*northern and inland parts are relatively dry*			
- southern part	*-10 to +15*	*-5 to +10*	-20 to +20	-5 to +5
Tasmania	0 to +15	0 to +5	0 to +10	no change
South Australian Gulf	*practically dry*		-35 to -5	-20 to -5
South-West Coast	*practically dry*		-45 to +35	-25 to +15
West Coast				
- northern part	0 to +35	0 to +10	*practically dry*	
- southern part	*practically dry*		-50 to -5	-15 to -5
North Australia	0 to >+100	0 to +10	*practically dry*	

–

Numbers in Table 19.1 are limits of percentage changes relative to current values, simulated by MODHYDROLOG using the GCM scenarios. Italics are used when runoff or soil moisture during that half of the year contributes less than 20% to the annual total or average.

variables simulated by the GCM for the present day climate and uses this relationship to convert the GCM enhanced greenhouse simulations to catchment scale data. The method used here is based on the WGEN generator (Richardson and Wright, 1984), but adapted for generating synthetic series for future climates and for Australian conditions (Wilks, 1992; Bates *et al.*, 1994)

The method is dependent on the choice of appropriate distributions to represent the climate variables and is described in detail by Bates *et al.* (1994, 1995). The weather generator is used with the CSIRO9 GCM (McGregor *et al.*, 1993) to generate 1000 years of daily weather sequences corresponding to the present day and doubled-CO_2 climates for several catchments in Australia. The climate variables analysed using the weather generator are daily precipitation occurrence and amount, maximum and minimum temperatures and solar radiation.

The weather sequences generated for the catchments are used as input data for MODHYDROLOG and Figure 19.5 shows the seasonal runoff characteristics simulated using the two weather sequences for the Broken River (Catchment Number 120204, Figure 19.1) and North Para River (505517) catchments.

The Broken River catchment is dominated by summer-half rainfall and the simulations show that, on average, there will be 25% less runoff in summer and autumn in a doubled-CO_2 environment compared to the present day climate. There is an even greater reduction in the higher runoff values (see higher percentiles in Figure 19.5). The lower catchment yield in summer and autumn has important implications for reservoir management and design while the reduction in winter and spring runoff can affect irrigation and agricultural crop management.

The North Para River catchment is dominated by winter-half rainfall with practically all the runoff occurring in winter and spring. The simulations indicate that although the average annual runoff in a doubled-CO_2 environment will only increase by less than 5%, the higher runoff events will increase by almost 15% (Figure 19.5). The higher runoff will increase the flood risk and bigger spillways and drainage waterways may be required.

As the weather generator approach takes into account changes in the timing, frequency and magnitude of the climate variables, it can be used to study changes in the extremes as well as in the long-term averages. In addition to the above, the simulations can also be used to estimate the potential impacts of climate change on daily peak flows, return periods of floods, reservoir storage estimates, low flow characteristics and the frequency, severity and duration of droughts.

It is interesting to note that the simulations in this section show that runoff in the Broken River catchment in a doubled-CO_2 climate will be lower than the present day runoff while the simulations in the previous section show an increase in runoff in the north-east coast (see Table 19.1). A similar discrepancy is also observed for the North Para River catchment and the South Australian Gulf. The two approaches show opposite results despite the CSIRO9 GCM being one of the five GCMs used to derive the climate change scenarios in the previous section (although results from impact simulations using stochastic weather sequences for most of the other catchments studied are within the range of changes given in Table 19.1). The results highlight the uncertainties in GCM simulations and suggest that estimates of impacts from studies like this should be used with caution.

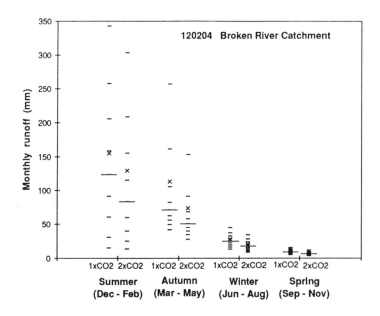

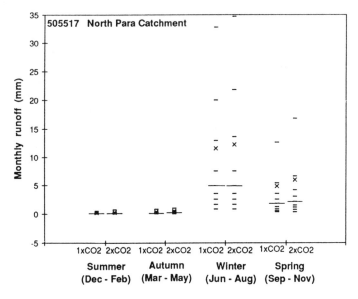

Figure 19.5. Seasonal runoff characteristics simulated using 1000 years of stochastic weather sequences for the present day and doubled CO_2 climates (the average runoff is indicated by the cross while the lines show the 10th to 90th percentiles with the longer line indicating the median).

19.6. Summary and Conclusions

The potential impacts of climate change on runoff and soil moisture in Australia, simulated using the conceptual daily rainfall-runoff model MODHYDROLOG, are presented in this chapter. The model is driven by measured historical data and by changed input data series representing the climate change scenarios. Results from the two simulations are then compared to estimate the hydrological responses to climate change.

The sensitivity analyses using hypothetical scenarios (by scaling historical precipitation and increasing temperatures) indicate that changes in rainfall are amplified in runoff with the amplification factor for runoff being higher in drier catchments. In wet and temperate areas, the percentage change in runoff is about twice as much as the percentage change in rainfall, while in arid areas, large increases in rainfall can enhance runoff by more than five times the change in rainfall. The change in rainfall has little effect on the soil moisture in wet catchments but in drier catchments, the percentage change in soil moisture levels can be greater than the percentage change in rainfall. Compared to precipitation, temperature increases alone have negligible impacts on runoff and soil moisture.

Results from impact simulations using scenarios obtained by analysing enhanced greenhouse experiments from five GCMs are given in Table 19.1. Although the reliability of GCM simulations is questionable, Table 19.1 provides an indication of the range of plausible changes in runoff and soil moisture in various regions of Australia by the year 2030 given the current state of science.

Result from impact simulations using stochastic weather sequences provided by the weather generator (with CSIRO9 GCM) for two catchments are presented in Section 19.5 and Figure 19.5. The weather generator takes into account various characteristics of the climate variables (magnitude, timing and frequency) and can therefore be used not only to study the impacts on the average annual or seasonal quantities, but also the impacts on the extremes (floods and droughts). The simulations indicate that the changes in extreme runoff values (particularly floods) will be much greater than the changes in average values. However, impact simulations of the extremes should be interpreted with caution because although GCMs have some success in simulating the 'average' climate, their simulations of the extremes are notoriously poor.

In any case, the simulations here show the potential for climate change to bring about runoff modifications that may require a significant planning response. They are also indicative of the fact that hydrological impacts affecting water supply and flood studies may be important in considering the cost and benefits of climate change.

References

Bates, B.C., Charles, S.P., Sumner, N.R., and Fleming, P.M. (1994) Climate change and its hydrological implications for South Australia, *Transactions Royal Society South Australia* **118**, 35-43.

Bates, B.C., Jakeman, A.J., Charles, S.P., Sumner, N.R., and Fleming, P.M. (in press) Impact of climate change on Australia's surface water resources, in *Greenhouse 94*, CSIRO Publications, Canberra, Australia.

CSIRO Division of Atmospheric Research (1992) *Climate Change Scenarios for the Australian Region*, Climatic Impacts Group, Report available from the CSIRO Division of Atmospheric Research, Mordialloc 3195, Australia.

Chiew, F.H.S., and McMahon, T.A. (1994) Application of the daily rainfall-runoff model MODHYDROLOG to twenty eight Australian catchments, *J. Hydrology* **153**, 383-416.

Chiew, F.H.S., Whetton, P.H., McMahon, T.A., and Pittock, A.B. (1995) Simulation of the impacts of climate change on runoff and soil moisture in Australian catchments, *J. Hydrology* **167**, 121-147.

Coleman, R.A., McAvaney B.J., and Wetherald, B.T. (1994) Sensitivity of the Australian surface hydrology and energy budgets to a doubling of CO_2, *Australian Meteorological Magazine* **43**, 105-116.

Houghton, J.T., Jenkins, G.J., and Ephraums, J.J. (1990) *Climate Change - The IPCC Scientific Assessment*, Report prepared for IPCC by Working Group 1, Cambridge University Press, Cambridge, UK, 364 pp.

Houghton, J.T., Callander, B.A., and Varney, S.K. (1992) *Climate Change 1992 - The Supplementary report to the IPCC Scientific Assessment*, Working Group 1, Cambridge University Press, Cambridge, UK, 200 pp.

McGregor, J.L., Gordon, H.B., Watterson, I.G., Dix, M.R. and Rotstayn, L.D. (1993) *The CSIRO 9-Level Atmospheric Circulation Model*, CSIRO Division of Atmospheric Research, Technical Paper 26, 89 pp.

Morton, F.I. (1983) Operational estimates of actual evapotranspiration and their significance to the science and practice of hydrology, *J. Hydrology* **66**, 1-76.

Richardson, C.W., and Wright, D.A. (1984) *WGEN: A Model for Generating Daily Weather Variables*, U.S. Department of Agriculture, Agricultural Research Service, Report ARS-8, 83 pp.

Whetton, P.H., Fowler, A.M., Haylock, M.R., and Pittock, A.B. (1994) Implications of climate change due to the enhanced greenhouse effect on floods and droughts in Australia, *Climate Change* **25**, 289-317.

Wigley, T.M.L., and Raper, S.C.B. (1992) Implications for climate and sea level of revised IPCC emissions scenarios, *Nature* **357**, 293-300.

Wilks, D.S. (1992) Adapting stochastic weather generation algorithms for climate change studies, *Climate Change* **22**, 67-84.

F.H.S. Chiew, Q.J. Wang and T.A. McMahon
CRC for Catchment Hydrology
Department of Civil and Environmental Engineering
University of Melbourne
Parkville, Victoria 3052, Australia.

B.C. Bates
CSIRO Division of Water Resources
Private Bag PO
Wembley, Western Australia 6014, Australia.

P.H. Whetton
CSIRO Division of Atmospheric Research
Private Bag 1
Mordialloc, Victoria 3195, Australia.

20 DYNAMICS OF STAGE FLUCTUATION IN YANGZHOUYONGCUO LAKE, TIBETAN PLATEAU

T. LIU

Abstract

Yangzhourongcuo Lake is the largest closed interior lake in the southern part of the Qingzang (Tibetan) Plateau, with a drainage area of 6100 km^2. The elevation of the lake stage is 4440 m and lake storage is about 16 billion m^3. Water recharge mainly depends on rainfall, with meltwater from glaciers accounting for up to 16%. About 2% of the drainage basin is covered by glaciers. The seasonal range of variation in the lake stage is small, usually less than 0.6 m. Highest water level occurs in September or October, not in July or August, the main rainfall period, because of the lake's self-adjustment. Lowest lake level occurs in May, June or July. In wet years the stage fluctuation has clear periodicity; in dry years the lake level tends to remain low throughout the year. Analysis of the patterns of lake level fluctuation during the last 100 years shows that the lake levels have varied by up to 4-5 m, and have been gently descending at a rate of 0.6 m per 100 years.

20.1. The Drainage Basin

Yangzhourongcuo lake is the largest closed interior lake south of the Yaluzangbu (Brahmaputra) River and the north of the Himalayas. Its drainage basin is located between 90°08' and 91°45'E, and 28°27' to 29°12'N. To its north is Yaluzangbu River basin with the narrow Ganbala Mountain as the demarcation line. The shortest distance between the lake and the Yaluzangbu River is only 8 to 10 km, in the Zhamalong region, but the difference in water elevation is up to 840 m. To its east is the Zhegucuo drainage basin, and to its southeast and south are the Himalayas. The lake is contiguous to Pumorongcuo Lake to the south-east and the Nianchu River to west, with the snow-capped Karuola Mountain as its watershed.

To its northwest, Yangzhouyongcuo Lake is only separated by a low hill from the Manqu River, a tributary of Yaluzangbu River. There are several lakelets distributed over the top of this hill. It is estimated that Yangzhouyongcuo was originally a exorheic lake and its water ran into Manqu River through the Yaseya Gap, thence to the Yaluzangbu. Because of a drying climate, the lake level fell, disconnecting it from the Manqu River and so that it became a closed interior lake (Figure 20.1). These

J.A.A. Jones et al. (eds.), Regional Hydrological Response to Climate Change, 351–361.

geographical changes within the lake district clearly reflect the influence of climatic change.

The area of Yangzhouyongcuo Lake drainage basin is 6100 km², with 621 km² of water surface. The elevation of the lake stage is 4440 m above the Huang Sea baselevel. The basin is oddly shaped with a winding lake shoreline and a lake perimeter of about 400 km in length. Lake storage is about 16 billion m³ and it is deepest in the north (55 m). The main feeder rivers enter the lake from the west, southwest and east. The feeder rivers in the north are usually very short, carrying only small quantities of water and most are seasonal.

Yangzhouyongcuo Lake lies within the rain shadow area on the north side of the Himalayas, so it is very dry. Average annual precipitation is only about 350 mm, but average annual evaporation from water surfaces is over 1250 mm. Mean annual temperature is 2.6°C.

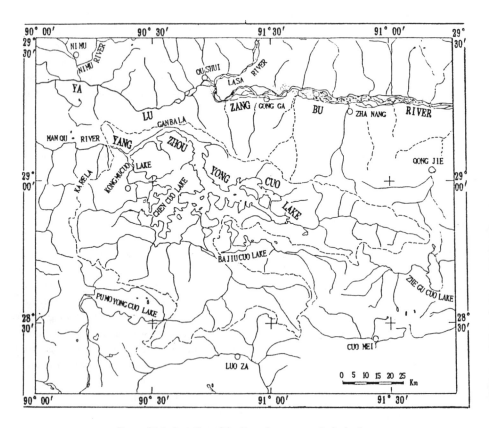

Figure 20.1. Location of the Yangzhouyongcuo Lake basin.

20.2. Sources of Lake Water Recharge

Vapour transfer to Yangzhouyongcuo Lake depends mainly on the warm, humid winds from the Bay of Bengal, which cut over the Himalayas from the Manquya inlet and through the valleys of the Luozhanu and Luoznaxia Rivers south of the lake basin. Vapour transfer cannot reach the northeast of the lake because of continous high mountain ridges, so it is very dry. There are glaciers covering the watershed in the west of the lake with a total area of 110 km^2. There is also a large area of ridges capped with snow throughout the year. These are all sources for lake water recharge, although rainfall is the main water recharge source. Water from the glaciers, which takes up 2% of the total lake drainage area, accounts for 1% of the total recharge water entering the lake.

The meteorological station at Langkazi within the lake basin provides rainfall data from 1961 to 1992. There are seven other hydrological stations within the basin from which data are available from the mid-1970s (Table 20.1).

TABLE 20.1. Summary of rainfall data for stations within the lake basin.

Station	Rainfall (mm)	Duration (years)	Period of record (years)
Langkazi	353.4	31	1961-68, 1970-92
Baidi	411.6	17	1975, 1977-1992
Wengguo	346.6	10	1983-1992
Kadong	352.4	13	1976-78, 1983-92
Rongduo	327.5	10	1983-1992
Quguozhong	326.9	11	1976-78, 1985-92
Dongla	354.4	5	1983-1987
Dui	363.8	9	1976-78, 1985-90

20.3. Patterns of Variation in Lake Stage and the Causal Factors

The recording of stage level in the Yangzhouyongcuo Lake began in 1974; there are 19 years of complete data available up to 1992. The highest water level recorded was 4441.6 m above sea level (in 1980), the lowest was 4437.88 m (in 1990), and the range of fluctuation was about 3.76 m. Historically, the highest water level was 4442.5 m (in 1963), and the largest range of stage fluctuation was 4.26 m.

20.3.1. RECENT ANNUAL PATTERNS OF LAKE STAGE FLUCTUATION AND THEIR CAUSES

20.3.1.1. *Patterns of Fluctuation in Wet and Dry Years*

In wet years, such as 1971 and 1978 (Figure 20.2), lake stage begins slowly rising from May or June, which is the beginning of the rainfall season and of the ice and snow melting period. It then rises quickly during the rainy season of July and Autumn. Because of the lake's self-adjustment, the highest lake level does not appear in July or Autumn, but in September or the first 10 days of October. From then on the lake stage begins falling again until May or June the following year. The period of rising stage is short but with high variance. The period of falling stage is long with low variance. Like river stage, the variation in lake stage is cyclical with a single annual peak, but the range of fluctuation is lower in the lake.

In dry years, such as in 1976, 1982 and 1983, lake level was low all year round, with less water running into the lake because of the reduced rainfall. No lake level rising period occurred during these years. The highest lake level was on January 1st and the lowest was at the end of December. This reflects the special pattern of Yangzhouyongcuo lake stage variation.

Two basic indexes of lake stage variation have been used. One is the difference between the highest and the lowest lake stage within the year. The other is the difference in lake level between the beginning and the end of the year. Table 20.2 shows that the first index ranged between 0.39 to 1.23 m. The average was below 0.6 m; only in one year was the annual range over 1.0 m, which was in 1978 with the value of 1.23 m. Table 20.3 shows that the second index varied between -0.87 m to +0.80 m. All these data show that the fluctuations in lake stage are relatively small.

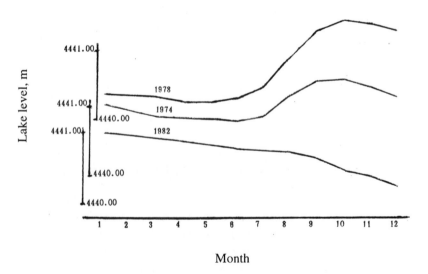

Figure 20.2. Mean monthly lake stage in dry, normal and wet years.

TABLE 20.2. Annual patterns of lake stage fluctuation.

Year	Date of highest level	Date of lowest level	Variation within a year (m)	Variation between years (cm)
1974	28/9	15/6	0.66	-
1975	3/10	19/7	0.40	-11
1976	1/1	31/12	0.56	-34
1977	8/10	19/5	0.68	-33
1978	11/11	3/5	1.23	+47
1979	12/9	16/6	0.68	+53
1980	12/9	13/5	0.49	+13
1981	27/8	25/12	0.40	-13
1982	1/1	30/12	0.87	-54
1983	1/1	30/12	0.67	-80
1984	15/9	24/6	0.45	-60
1985	13/9	1/7	0.40	-26
1986	1/1	30/12	0.40	-37
1987	27/9	30/6	0.81	-15
1988	28/9	14/6	0.58	+4
1989	1/1	19/7	0.42	-21
1990	1/10	19/6	0.72	-17
1991	26/9	2/6	0.72	+15
1992	5/9	11/7	0.39	-3

TABLE 20.3. Variations in Yangzhouyongcuo lake stage over recent years.

Year	Level at the beginning of the year (m)	Level difference ΔH (cm)	Year	Level at the beginning of the year (m)	Level difference ΔH (cm)
1974	4441.04	+3	1984	4439.49	-33
1975	4441.06	-26	1985	4439.15	-28
1976	4440.80	-56	1986	4438.87	-40
1977	4440.25	+17	1987	4438.47	+21
1978	4440.42	+80	1988	4438.68	-9
1979	4441.22	+20	1989	4438.59	-40
1980	4441.41	-5	1990	4438.19	+15
1981	4441.37	-33	1991	4438.34	+10
1982	4441.04	-87	1992	4438.44	-27
1983	4440.16	-67			

20.3.1.2. *The Relationship between Annual Lake Level Fluctuation and Rainfall*
The relationship between lake stage variations and rainfall is shown in Figure 20.3.

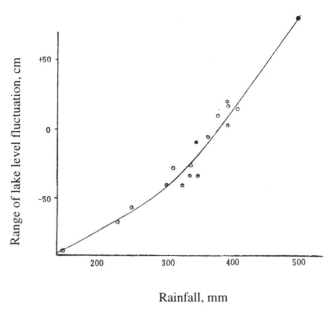

Figure 20.3. Relationship between lake stage fluctuation and precipitation.

Rainfall data were collected from Langkazi Station. In 1978, rainfall was 505.0 mm, which was more than 1.43 times the average rainfall in this area, and the lake stage rose by 1.23 m. In 1982, precipitation was only 150.7 mm, which was only 43% of the average annual rainfall, and lake stage fell by 0.87 m. The difference between the highest and the lowest level was also 0.87 m and lake stage remained low throughout the year.

The curve in Figure 20.3 can be represented by the following equation:

$$\Delta H = 0.27(P - 140)^{1.48} - 90 \qquad\qquad (20.1)$$

in which ΔH is the range of lake level fluctuation (cm) and P is annual rainfall (mm).

This equation and the curve in Figure 20.3 also showed that the lake storage varies with the rainfall. $\Delta H = 0$ when the rainfall is equal to a certain value, P_o, which we call the balance rainfall. If rainfall within the lake basin is above this value, ΔH is positive, otherwise ΔH is negative. P_o was 384 mm from the curve in Figure 20.3 and 380 mm calculated from the equation. This P_o can be used to predict the fluctuations in lake stage. Obviously for each different lake stage, ΔH represents a different change in lake storage, so at each lake level there should be a specific P_o for $\Delta H = 0$. However, this error was ignored in this analysis because the range in lake level fluctuation was relatively small and the period concerned was not very long.

20.3.2. LONG-TERM FLUCTUATIONS IN LAKE LEVEL AND THE CAUSAL FACTORS

Figure 20.4 shows the results of analysing data from the last 100 years. The range of variation was about 4 to 5 m. The longest period with rising lake level was 14 years (from 1923 to 1937), and the longest period of falling levels was also 14 years (from 1963 to 1977). The overall tendency was negative, with lake stage falling slightly at the rate of 0.6 m a century. Based on the following analyses, the author holds that this overall falling tendency was due to the decrease in rainfall and the rise in atmospheric temperature.

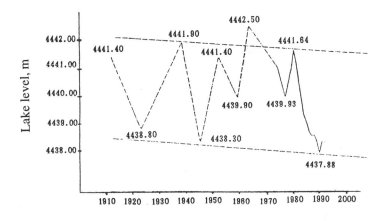

Figure 20.4. Changes in the level of the lake this century.

TABLE 20.4. Range of lake stage fluctuations during different periods.

Order	Year		Range of variation in lake level	Years	Data source
	Beginning	End	(m)		
1	1912	1923	-2.60	11	1912 -1973
2	1923	1937	+3.10	14	data from
3	1937	1945	-3.60	8	other sources
4	1945	1952	+3.10	7	
5	1952	1959	-1.50	7	1974-1992
6	1959	1963	+2.60	4	from direct
7	1963	1977	-2.57	14	measurement
8	1977	1980	+1.71	3	by the Bureau
9	1980	1992	-3.34	12	

There were nine alternating periods of lake level rising and falling between 1912 and 1992. Each period usually lasted 7 to 14 years, although the sixth and the eighth periods only lasted 3 to 4 years. This reflects the fact that the continuously wet periods, i.e. the clusters of wet years, have been becoming shorter. The range of the lake stage fluctuation during those nine periods was 1.5 to 3.6 m.

20.3.2.1. *Long-Term Rainfall Patterns*

Analyses of the rainfall and lake stage records at Yangzhourongcuo, together with analysis of rainfall in the neighbouring Nianchu River valley revealed the following patterns. From 1956 to 1959, Yangzhourongcuo lake fell to the lowest levels, because less rainfall occurred during this period. 1962 and 1963 were wet years, with average P = 441.7 mm; this was above P_0, and the lake stage began rising. In addition, a large amount of water from the burst Pumorongcuo Lake ran to Yangzhouyongcuo Lake, the level rose by 0.77 to 0.8 m and reached the altitude of 4442.5 m a.s.l., the highest in recorded history. From 1964 to 1976, a long period of low rainfall with an average P = 343.2 mm, which was below P_0, the lake stage fell. The lowest level during this period was 4439.93 m a.s.l., which occurred on May 19, 1977. From 1977 to 1979, the average P = 434.7 mm, higher than P_0. In 1978 rainfall amounted to 505.0 mm, the lake level rose again and the highest lake stage was 4441.64 m a.s.l.. From 1981 to 1992, a long period with annual rainfall continuously below P_0 (average P = 321.1 mm), the lowest lake level was 4437.8 m, which was also the lowest in history.

Average rainfall from 1961 to 1992 in Yangzhouyongcuo drainage basin was 353.4 mm, which was below the balance rainfall, P_0. So for the last 30 years annual rainfall has been insufficient to maintain a balanced water budget in the basin and the lake level has fallen. Statistical analysis shows that rainfall during the 1980s was significantly lower, and the falling lake level was more obvious. Rainfall varitions during the last 30 years are shown in Figure 20.5. The curve shows that for most of the years, precipitation was less than P_0. This coincides with the overall decline in lake stage.

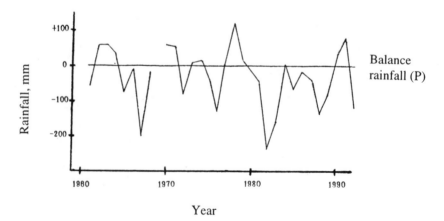

Figure 20.5. Variations in annual rainfall, 1960 to 1992.

20.3.2.2. *Long-Term Temperature Patterns*

Temperature records also reveal that climate change coincides with the declining tendency of the lake levels. During the 30 years of available meteorological data, average annual temperature has been rising at the rate of 0.2 to 0.3°C per decade. Annual minimum temperatures were also rising at a slightly higher rate. Annual maximum temperature was relatively constant.

Further afield, temperatures were also rising at Lhasa and Jiangmei between the 1960s and the 1990s, but at slightly lower rate than that in the Yangzhouyongcuo Lake area. The rising temperature in the Yangzhouyongcuo lake district directly reflected the fact that the climate was becoming warmer and rainfall was declining, both of which were consistent with the falling lake stage. Figure 20.6 shows the pattern of variation in mean annual air temperatures (from 1961 to 1990) at six stations surrounding the Lake Yangzhouyongcuo area. They are located to the northeast (Lhasa), the north (Dangxiong), the west (Rikaze), southwest (Longzi), and the east (Zedang) of the lake district, and Langkazi Station is within it. The results of these analyses show that the mean annual air temperature has been rising at an average rate of 0.1-0.6°C/decade, and the maximum rate at a single station was 0.25-0.4°C/decade. The greatest temperature rise occurred between October and March, while air temperature remained relatively stable from June to September. The air temperature during May dropped at most of the stations.

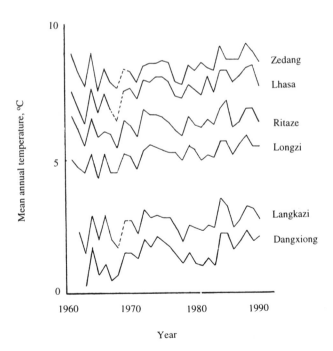

Figure 20.6. Fluctuations in mean annual temperature at representative stations.

T. LIU

As a result of increased air temperatures in wintertime, the dates of early freezing on the lake were delayed and spring breakup dates were advanced. According to the statistics of the freeze-up date in various parts of the lake surface from the 1970s to 1980s, the ice period shortened by about 10 days. This obviously reflected the close relationship between the shrinkage of lake water level and climatic warming.

TABLE 20.5. Temperature at the Langkazi Meteorological Station.

Year	1962	1963	1964	1965	1966	1967	1968	1969	1970	1971
Mean	2.3	1.5	2.9	2.0	2.0	2.0	1.7		2.7	2.2
Max.	19.5	18.7	21.8	21.2	21.6	19.9	19.7		21.8	20.7
Min.	-24.9	-24.3	-21.5	-20.5	-24.6	-20.2	-25.0		-21.5	-20.8

Year	1972	1973	1974	1975	1976	1977	1978	1979	1980	1981
Mean	3.1	2.8	2.9	2.8	2.8	2.4	1.9	2.5	2.4	2.4
Max.	22.5	20.1	19.3	20.0	19.4	18.8	19.8	21.4	19.3	20.5
Min.	-19.9	-20.2	-23.4	-21.6	-22.9	-22.9	-23.2	-22.7	-21.9	-22.7

Year	1982	1983	1984	1985	1986	1987	1988	1989	1990	1991
Mean	2.5	2.4	3.5	2.7	2.4	2.7	3.2	3.1	2.7	2.9
Max.	20.9	21.8	20.5	21.1	21.5	22.6	21.5	21.6	21.6	19.0
Min.	-20.2	-22.4	-18.6	-20.5	-20.1	-21.4	-17.3	-19.4	-19.4	-19.8

20.4. Conclusions

There are three important characteristics in the fluctuation of Yangzhuoyongcuo lake stage:

1. Variations in lake stage are closely related to rainfall, and show clear periodicities within the year and persistence over a number of years.

2. Because of the lake's self-adjustment, the range of fluctuation in lake level is smaller than in the feeder rivers, and the dates of highest and lowest level and range of long-term fluctuation were all different from the rivers.

3. Based on analyses of data from the last 30 and the last 100 years, it can be concluded that there is an overall tendency of the lake level to fall at a rate of 0.6 m per 100 years.

Yangzhouyongcuo Lake is the largest closed interior lake in the southern part of the Qingzang (Tibetan) Plateau. The fluctuation of the lake stage is controlled by climate. Lake level is falling due to rising temperatures and declining rainfall. The average temperature in Qingzang Plateau in 2030 is forecast to rise by 2 to 3°C. This would undoubtedly aggravate the falling tendency of the lake level. More attention should be concentrated on changes in the lake level, since it also has major feedback effects upon the climate of the Qingzang Plateau and, indeed, over a large area of Asia.

References

Chinese Academy of Sciences Comprehensive Survey of Qingzang Plateau (1984) *Rivers and Lakes in Tibet*, Science Publishing House, Beijing, 159-168.
Yang Zhenniang (1991) *Glacier Water Resources in China*, Gansau Science and Technology Publishing House, Lanzhuo.

Liu Tianchou
Bureau of Hydrological Water Resource Survey of Tibet
Lhasa, China.

21 DERIVATION OF SURFACE TEMPERATURE, ALBEDO, AND RADIATIVE FLUXES OVER THE TIBETAN PLATEAU BASED ON SATELLITE MEASUREMENT

LEI SHI

Abstract

To gain a better understanding of the regional hydrometeorology in relation to the unique climate and topography of the Tibetan Plateau, algorithms are developed to estimate the surface temperature, surface albedo, cloud properties, and surface radiative fluxes based on Indian Satellite infrared and visible measurements. The spatial distribution patterns of temperature and albedo are found to closely correspond to surface topographical and biophysical features. Temperatures are much lower over the elevated plateau than over the surrounding regions. The variations of temperature can exceed 10°C within a distance of 50 km in the central plateau region and 20°C across the boundary of the plateau during daytime. The surface albedo field also exhibits non-smooth features with the smallest values less than 16% over the eastern plateau and the largest values exceeding 28% over the northern plateau. The horizontal distributions of surface fluxes are largely influenced by both topography and distribution of cloudiness. Following the topographical feature, longwave upwelling fluxes generally decrease with increasing elevation; and the largest values of surface downwelling shortwave flux are found over the western plateau where the surface elevation is highest. Under the influence of the cloudiness patterns, the downwelling shortwave flux decreases toward the south as there are larger cloud amounts and greater moisture contents over the southern plateau and the region off the southern plateau. In these regions the frequent occurrences of deep-convective clouds sngwave upwelling fluxes generally decrease with increasing elevation; and the largest values of surface downwelling shortwave flux are found over the western plateau where the surface elevation is highest. Under the influence of the cloudiness patterns, the downwelling shortwave flux decreases toward the south as there are larger cloud amounts and greater moisture contents over the southern plateau and the region off the southern plateau. In these regions the frequent occurrences of deep-convective clouds significantly decreases the transparency of the atmosphere and thus reduces the solar radiation reaching the surface.

J.A.A. Jones et al. (eds.), Regional Hydrological Response to Climate Change, 363–380.
© *1996 Kluwer Academic Publishers. Printed in the Netherlands.*

21.1. Introduction

The studies of the impact of climatic change on regional hydrology require long term observations of surface and atmospheric variables. Many of these essential observations can be accomplished through the use of satellite measurements. For example, Hall et al. (1986) and Allen and Mosher (1986) used Nimbus-7 and GOES measurements to monitor snowpack conditions. Miller (1986) applied LANDSAT images to determine snowcover areas. Robinson (1986) studied spring snowmelt over Arctic lands based on surface albedo and surface temperature derived from NOAA AVHRR and DMSP measurements. A thorough review of the progress made in the remote detection of hydrological phenomena was provided by Engman and Gurney (1991).

In this study, Indian satellite (INSAT) infrared and visible measurements during summer, 1988 are used to retrieve the clear-sky temperatures, albedos, cloud amounts and cloud-top pressure distributions. Then, in conjunction with estimates of surface temperature, surface albedo, and cloud optical properties, medium-resolution infrared and solar radiative transfer models are used to calculate radiative fluxes and surface heat budget. Descriptions of the models can be found in Smith and Shi (1992) and Shi (1994).

Recent observational studies on the climatological, meteorological, and surface heating properties of the Tibetan Plateau can be found after several field experiments. In the summer of 1979, the Academia Sinica and the National Meteorology Administration of China organized the Qinghai-Xizang Plateau Meteorological Experiment (QXPMEX-79) (Tao et al., 1986). A number of radiative process studies were carried out based on local observations from this experiment. For example, Chen and Yang (1984) examined the radiation processes in Linzi (Xizang province); Weng et al. (1984a, 1984b) studied solar radiation over the Lhasa River Valley; and Xie et al. (1984) investigated the radiation characteristics over Golmud region. Ji et al. (1984) found that the value of surface net longwave radiation in the western plateau was higher than in other parts of the plateau, and was considerably higher than in the low-elevation plain areas. Further examinations of local radiative processes were also performed after another meteorological field experiment designed for obtaining heating properties during the winter season of 1982-1983. Detailed analyses can be found in Jiang et al. (1985), Yao et al. (1985), Yang and Shui (1985), and Yuan (1985). These studies revealed the importance of regional differences on atmospheric radiation processes over the plateau due to complicated topographical and biophysical surface conditions.

Surface temperature, albedo and heat budget are closely related to climatic and hydrological conditions of the region. At the surface, the magnitude of temperature greatly depends on soil heat capacity, which is a function of soil moisture. The albedo is determined by the biophysics of the surface. The distributions of vegetation patterns and albedo values are indications of horizontal soil moisture distribution in the region. In the atmosphere, the water vapor and clouds enhance downward infrared flux, while reduce solar radiation to the surface. The distribution patterns of surface temperature, albedo, infrared flux and solar flux reflect the variations of hydrometeorological factors across the region. In the modeling studies of snowpack and snowmelt, surface temperature, albedo, and radiative fluxes are important variables required by the models (Robinson, 1986; and Hall *et al.*, 1986).

21.2. Estimations of Surface Temperature and Albedo

A method for determining clear-sky radiances from satellite measurements has been discussed by Rossow *et al.* (1985). The method assumes that clear-sky and cloudy radiances form a monotonic distribution, with the clear-sky radiance as an extreme (minimum reflectivity or maximum brightness temperature). This assumption is also made in the present study. In order to avoid spectral inconsistencies in defining clear-sky planetary properties caused by using separate maximum temperature and minimum albedo time composites, a 'combined ranking method' has been used. A twenty-one day moving time frame, centered on the analysis day, is used to obtain the best estimate of cloud free planetary temperature and albedo. Each pixel is first ranked according to the temperature compared to the other pixels at the same location for the time window. The pixel at the day with highest equivalent black body temperature (EBBT) in the twenty-one day frame is ranked '1'. The pixel at the day with second highest EBBT is ranked '2', etc., until every pixel for every day is ranked. This yields the '1' to '21' EBBT ranking numbers. Next, each pixel location at each day is ranked again in a similar way but according to the planetary broad-band albedo compared to those on the other days. The lowest albedo is ranked '1' whereas the highest is ranked '21'. The temperature and albedo ranking numbers at each pixel location are then added to yield a combined ranking number. Finally, the temperature and albedo of each pixel location at the day with the lowest combined ranking number are chosen to represent the estimated clear-sky planetary temperature and albedo. In order to reduce residual cloudiness effects, a low-pass Fourier filter is then applied to the clear-sky planetary temperature and albedo time series. The filtered, composite temperatures and albedos are then considered to be the optimum representations of cloud-free conditions. Due to the slight opacity produced by the water vapor continuum, ozone and aerosol absorption, the clear-sky planetary temperature measured by the satellite is lower than the actual ground skin temperature. Cogan and Willand (1976) developed a simple parametric formula in terms of precipitable water to derive skin temperature. This scheme is applied in this study to obtain surface temperature.

The distribution patterns of surface temperature for June and July, 1988 are shown in Figures 21.1 and 21.2, respectively. The temperature retrievals are first averaged over 2.5° x 2.5° grid areas to match the spatial resolution of the ECMWF data. The upper panels in the Figures show the monthly mean surface temperature, whereas the middle and lower panels show the monthly averaged surface temperature at 6 GMT (local noontime) and 18 GMT (local midnight), respectively. Before the onset of the summer monsoon over the plateau in mid-June, the surface and lower atmosphere are dry and warm. The aridity conditions are more severe over the western plateau than over the eastern plateau in this period in proximity to the summer solstice. The smaller heat capacity of the western plateau surface leads to much larger surface temperatures than

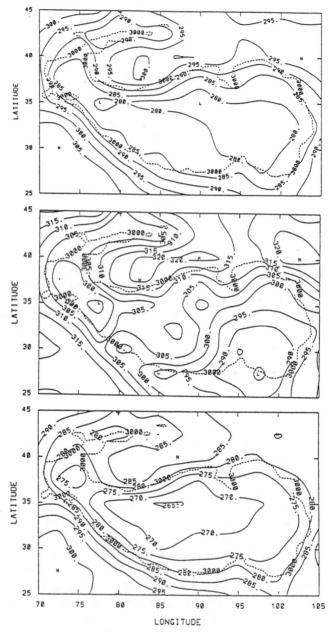

Figure 21.1. Distribution of surface temperature (degree K) for June. The upper panel shows the monthly mean surface temperature; the middle and lower panels show the monthly averaged surface temperature at 6 and 18 GMT respectively. The surface topography contour for 3000 m is plotted with dashed lines.

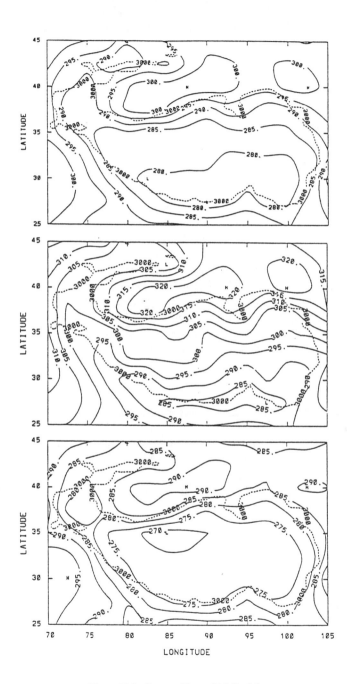

Figure 21.2. Same as Figure 21.1 for July.

over the eastern plateau during daytime as shown in the middle panel of Figure 21.1. After the intense precipitation in July, the water content of the soil rises. Therefore, the vegetation over the plateau increases, especially in the southern portion of the plateau. The surface temperatures between the southwestern and southeastern plateau thus tend to be more homogeneous. However, there are notable surface temperature differences between the northwestern and northeastern plateau. These temperature differences mainly result from differences in soil moisture and in surface biophysics between the two regions, i.e., mountainous dry-land in the west and grass land in the east. Nighttime surface temperatures do not vary much with time (lower panels of Figures 21.1 and 21.2). The lowest surface temperatures are always located in the central-western plateau because this region has the highest grid-averaged elevation.

In order to obtain surface broad-band albedo, an algorithm is developed based on the use of Xie's (1984) surface broad-band albedo distribution map for July, 1979. These albedos were measured during the Qinghai-Xizang Plateau Meteorological Experiment (QXPMEX-79), when several temporary ground stations were established over the western plateau to support the First GARP Global Experiment (FGGE). The mean clear-sky planetary broad-band albedos of July are first averaged over each $2.5° \times 2.5°$ grid point corresponding to the ECMWF grid. Then the grid-averaged planetary albedo values are compared with Xie's (1984) surface albedo values at the same location. In order to avoid uncertain values generated through contour drawing for the points far from measured sites in Xie's (1984) map, only the albedo values closed to the ground stations are chosen in the comparison. A linear regression line is then computed from the planetary and surface albedo data sets. The linear regression is expressed as

$$y = -7.5824 + 1.4485x \qquad (21.1)$$

where 'x' depicts the planetary clear-sky broad-band albedo, and 'y' depicts the surface broad-band albedo.

Figures 21.3a-b show the distribution pattern of surface albedo for June and July respectively. The smallest albedo values (16-18%) are found in the eastern plateau (mainly grass-land surface). The albedo becomes larger in the northwestern plateau (mountainous dry land). There is a high value region of albedo (28-30%) in the region off the northern plateau (desert/gobi surface).

21.3. Estimation of Cloud Properties

A bispectral (infrared and visible) algorithm has been implemented for a better estimation of fractional cloud cover. The cloud detection is based on the infrared and visible threshold method discussed by Rossow *et al.* (1985) and Rossow (1989). Each pixel is classified as clear or cloudy according to whether its planetary EBBT or planetary albedo differs from the clear-sky values by more than a threshold amount. Within each $2.5° \times 2.5°$ grid area, the pixels are tested individually for being cloudy or cloud-free. The percentage of cloudy pixels within the grid area are counted to yield

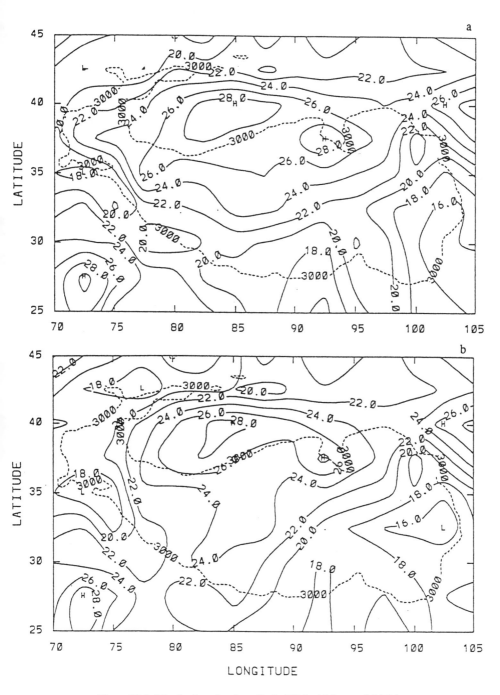

Figure 21.3. Distribution of surface albedo (%) for (a) June and (b) July.

the fractional amount of clouds. In the present study, a threshold value of 6°C is applied for temperature, whereas a value of 3% is applied for albedo.

The evolution patterns of cloud cover over the plateau during June-July are shown in Figures 21.4a-b. As thin cloudiness over the plateau region does not vary significantly with time, the temporal variation of total cloud cover tends to follow that of thick cloud cover. The largest variation of cloud amount occurs over the southern plateau. There is an approximately 20% increase of cloud cover for both thick clouds and total clouds from June to July in this region. The distribution pattern of cloudiness is closely associated with monsoon activities. Within the elevated plateau region, organized large scale cyclone systems occur most frequently over the southern portion of the plateau during the summer monsoon season. Thus the cloudiness over the southern plateau increases dramatically after the commencement of the southwest monsoon.

21.4. Heat Budgets at the Surface

The heat budget at the surface can be expressed as:

$$Q^* = Q_H + Q_E + Q_G \qquad (21.2)$$

where Q^* stands for the net radiation, Q_H for the upward sensible heat flux, Q_E for the upward latent heat flux, and Q_G for the downward molecular heat flux. The net radiation can also be expressed as:

$$Q^* = L_u + L_d + S_u + S_d \qquad (21.3)$$

where L_u and L_d represent the upward and downward longwave radiation fluxes respectively; and S_u and S_d denote the upward and downward shortwave radiation fluxes respectively. In this section, the horizontal distributions of monthly averaged longwave (L_u and L_d) and shortwave (S_u and S_d) surface fluxes for the month of July are examined in detail.

The longwave and shortwave radiative transfer models are used for computations in conjunction with INSAT satellite data with a time resolution of eight times per day. The ECMWF data sets of pressure, temperature and relative humidity at 1000, 850, 700, 500, 300, 200 and 100 hPa for 0 and 12 GMT are used to construct atmospheric profiles. For the levels above 100 hPa, the temperature and mixing ratio values are obtained from the climatological profiles of McClatchey et al. (1972). The atmospheric profiles for the times between 0 and 12 GMT are obtained by linear interpolation with respect to time. The horizontal resolution of the results are mapped at the spatial resolution of the ECMWF data, that is at 2.5° x 2.5°. The concentrations of carbon dioxide and oxygen are set uniformly at 345 ppm and 2.095×10^5 ppm respectively. The mid-latitude summer ozone profile of McClatchey et al. (1972) is used to account for ozone effects. For longwave calculations, the surface emissivity is set to 0.95, which represents an average value for emissivities of soils, desert, grass, agricultural crops,

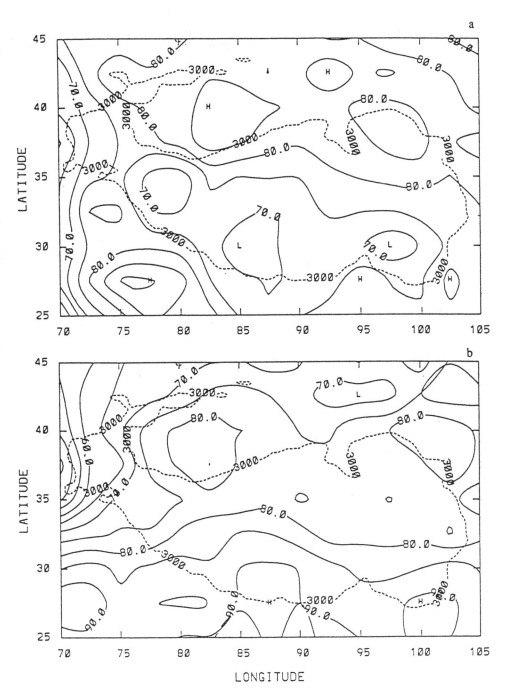

Figure 21.4. Monthly mean cloud cover (%) for (a) June and (b) July.

forest, water, snow and ice as given in table 1 of Oke (1978). The surface temperatures, surface visible and infrared albedos, cloud types, cloud top and base heights, cloud water contents and effective radii of cloud drops are determined by the methods described in the previous section.

Each 2.5 x 2.5 degree grid area may contain cloud-free and cloudy regions of various cloud types. The directional fluxes at each level z for a grid area are calculated by a combination of cloud-free and cloudy fluxes weighted by the corresponding fractional area coverage:

$$F(z) = (1 - w_{cld}) F_{clr}(z) + \Sigma w_{cldi} F_{cldi}(z) \tag{21.4}$$

where F represents the upward or downward flux for the grid area; subscripts 'clr' and 'cld' denote clear or cloudy situations; w_{cld} represents total cloud cover; and w_{cldi} represents the fractional cloud coverage for each separate type of cloud. A total of five types of cloud are considered, i.e., thick clouds with high, middle and low cloud-tops, and thin clouds with high and middle cloud-tops.

The surface net longwave flux reflects the amount of energy lost by the surface due to longwave radiative divergence. The distribution of surface upward longwave flux is a function of surface temperature and surface emissivity, while the distribution of downward longwave flux is determined primarily by the atmospheric temperature, moisture and cloudiness.

The July mean longwave upward, downward and net fluxes at the surface are shown in Figure 21.5. As depicted in the upper panel of Figure 21.5, large areas of the southern plateau have longwave upward fluxes of less than 360 W m^{-2}. The upward fluxes increase progressively towards the north. An east-west belt with upward fluxes larger than 440 W m^{-2} is formed over the desert-type surface along 40°N latitude. The distribution of monthly mean longwave downward fluxes closely follows the large scale surface elevation features (middle panel of Figure 21.5). Smaller values (320 W m^{-2}) are found over higher elevated plateau areas, while larger values (> 380 W m^{-2}) are distributed over the regions off the plateau. The isolines of longwave net flux at the surface are northeast-southwest aligned (lowest panel of Figure 21.5). The net fluxes increase from less than 20 Wm^{-2} over the southern plateau to larger than 60 W m^{-2} over the northern plateau.

The surface net longwave flux pattern is determined by the influences from the surface and from the atmosphere. The distribution of surface temperature determines the distribution of the upward flux. The smallest values of the upward fluxes over the southern plateau result from the surface temperature as shown in the upper panel of Figure 21.2. The high surface temperature over the desert region north of the plateau leads to the large values of longwave upward flux along 40°N.

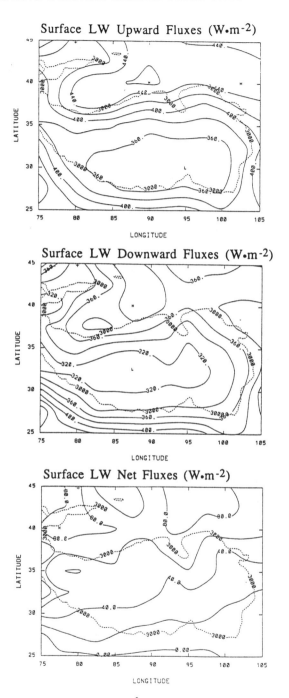

Figure 21.5. July monthly mean values (W m^{-2}) for the surface longwave upward flux, downward flux, and net flux. The surface topography contour for 3000 m is plotted with dashed lines.

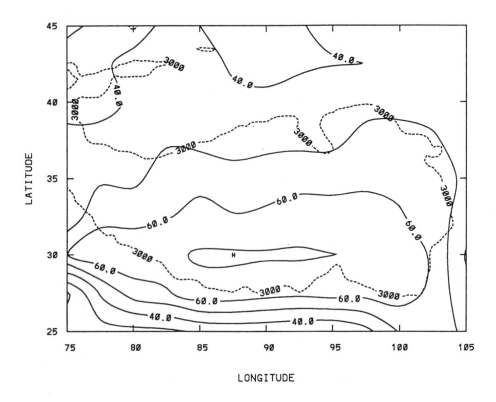

Figure 21.6. Cloud effect on the July surface longwave downward flux (W m^{-2}).

The distribution of surface longwave downward flux is determined primarily by the effects from atmospheric water vapor and clouds, along with surface elevation. The smaller downward fluxes over the elevated plateau can be explained by the thinner atmosphere above this region. It can be noted that the downward fluxes over the western plateau are smaller than over the eastern plateau, due to the drier atmosphere over the western plateau. Clouds play an important role modifying the distribution of downward flux. In order to examine cloud effects, the cloud forcing on the longwave downward flux is calculated and shown in Figure 21.6. In this study, cloud forcing is defined as the difference between the radiative downward flux which occurs in the presence of clouds, and that which would occur if the clouds were removed. Figure 21.6 shows that cloud forcing is largest over the southern plateau. In July, deep convective systems develop frequently over the southern plateau region, which result in more than 70% of monthly mean thick cloud amount as shown in the middle panel of Figure 21.4b. The cloud forcing over the southern plateau is as large as 22% of the surface longwave flux.

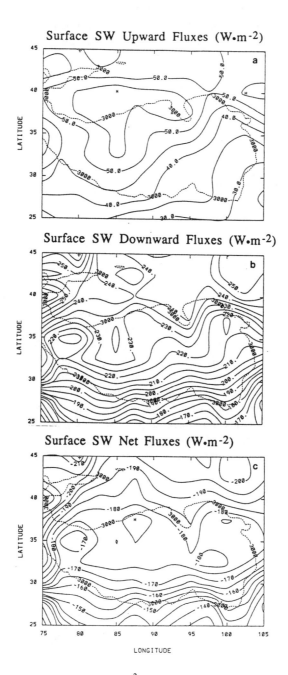

Figure 21.7. July monthly mean values (W m^{-2}) for the surface shortwave upward flux, downward flux and net flux. The surface topography contour for 3000 m is plotted with dashed lines.

Figure 21.7 shows the July mean shortwave upward, downward and net fluxes at the surface. Cloudiness is a major factor influencing the distribution pattern of shortwave downward flux. Under the influence of the Bengal wet current, there are larger amounts of cloud and greater moisture contents which decrease the transparency of the atmosphere to solar radiation over the southern plateau. Therefore, the downward fluxes increase progressively from 180 W m^{-2} over the southern plateau to 240 W m^{-2} over the northern plateau (middle panel of Figure 21.7). For the same reason, the downward shortwave flux values over the western plateau are larger than over the eastern plateau.

Another important factor influencing the distribution pattern of shortwave downward flux is the large scale surface elevation. The downward fluxes are much higher over the plateau region in comparison to the region off the southern plateau. This is due to the high elevation and unique climatological conditions of the plateau. The elevation of the plateau region is above 4000 m; the density of air, the concentration of moisture, and the average thickness of clouds are all considerably smaller over the plateau than the surrounding regions. All of these factors lead to the atmosphere being more transmissive to shortwave radiation, and thus to enhancement of the solar insolation at the surface. This downward solar flux pattern as well as the distribution of surface albedo determine the distribution pattern of shortwave upward flux as shown in the upper panel of Figure 21.7, with smaller upward flux values (30 W m^{-2}) over the southern plateau and larger values (55 W m^{-2}) over the northern plateau. Previous studies of Ye and Gao (1979), Luo and Yanai (1984), Chen et al. (1985), and Yanai et al. 1992) have shown that there are large differences of sensible heating and latent heating between the eastern and western plateau due to the differences in surface properties of these two areas. In the surface radiation fields, the east-west contrasts due to the variations in surface temperature, albedo, and topography are also indicated in Figures 21.5 and 21.7.

The distribution of net shortwave flux is shown in the lowest panel of Figure 21.7. The values of net shortwave flux increase progressively from south to north. Although the surface albedo values of the southern plateau are generally smaller than those of the northern plateau, the cloud effect is more dominant in determining the magnitude of net shortwave flux.

The cloud forcing for the surface shortwave flux is shown in Figure 21.8. Following the distribution of cloud amounts (as depicted in Figure 21.4b), the cloud forcing increases progressively from 120 W m^{-2} over the northern plateau to 150 W m^{-2} over the southern plateau. Cloud forcing decreases the shortwave downward flux through absorbing and reflecting solar radiation. Over the southern border of the plateau, the strong attenuation of solar radiation is due to clouds associated with large-scale convective activity.

21.5. Summary and Conclusion

The use of satellite measurements for the Tibet region shows promise in that the high resolution measurements are capable of presenting the detailed structure of the highly variable surface features which cannot be observed by existing ground stations. Based on the satellite measurements, the spatial distributions of clear-sky planetary temperature and albedo are found to closely correspond to the surface topography and biophysical features. Temperatures are much smaller over the elevated large-scale plateau compared to the surrounding regions. The temperature fields show that the variation of temperature can be more than 10°C within a distance of 50 km in the central Tibetan region. Temperature variations across the boundary of the plateau for the same distance can be as large as 20°C due to rapidly changing topography. The surface albedo field also exhibits non-smooth features which result from complicated surface biophysics. The smallest albedo values (< 16%) are found in the eastern plateau, while the largest values (> 28%) are found in the northern plateau.

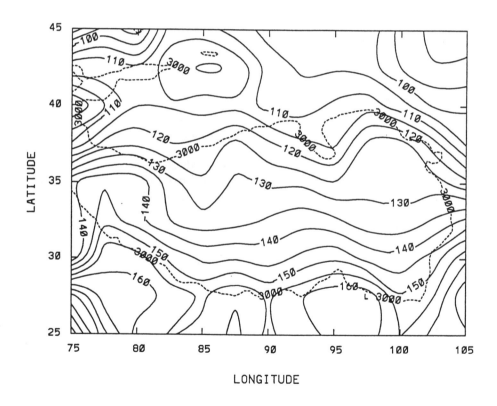

Figure 21.8. Cloud effect on the July surface shortwave downward flux (W m^{-2}).

generally decrease with increasing height. The downward longwave flux values over the plateau region are always smaller than those over the surrounding regions due to the smaller amounts of water vapor and cloud water content in the thinner atmosphere above the elevated plateau. The net longwave flux varies from ~ 20 W m^{-2} in the southeastern plateau to ~ 60-70 W m^{-2} in the northern plateau. There is large differentiation of surface shortwave downward flux in the region analyzed. Larger monthly mean values for July (240-250 W m^{-2}) are found north of the plateau, while smaller values (170-180 W m^{-2}) are found south of the plateau. This distribution pattern primarily results from cloud effects because of the fact that under the influence of the Bengal wet current, there are larger cloud amount and greater moisture contents which decrease the transparency of the atmosphere.

Global warming impacts hydrological condition through the changes in surface and atmospheric conditions. From a hydrometeorological point of view, an increase in surface temperature may increase evaporation of the surface layer and upward infrared flux. A possible regional desertization associated with global warming may result in larger value of surface albedo and increase upward shortwave flux. The change in atmospheric circulation due to global warming processes can result in redistribution of cloudiness. For a region with increasing cloudiness there will be significant reduction of solar downward radiation at the surface, and vice versa. The impact of cloudiness change on the surface heat budget of Tibetan Plateau region is more remarkable than over other surrounding regions due to the high elevation of the plateau. To monitor large scale and long term variations of hydrological and meteorological variables, satellites provide useful measurements especially over the remote areas. The present study focuses on the application of satellite remote sensing technique in obtaining surface temperature, albedo, radiative fluxes, and cloud properties. Further studies on various aspects of remote sensing application, including the detection of precipitation, runoff, snow hydrology, evapotranspiration, soil moisture, ground water, and water quality, are necessary in order to obtain a better understanding of hydrological cycle and the impact of global change upon it.

Acknowledgments

The author wishes to thank Eric Smith for guidance throughout the study, V. Ramanathan for useful suggestions, and Robert Bernstein for encouragement. The author has been supported by the NSF grant ATM-8812411 to Eric Smith, the U.S. Department of Energy - Atmospheric Radiation Measurement Program through DE-FG03-91ER61198 to V. Ramanathan and Tim Barnett, and the SeaSpace Corporation.

References

Allen, M. W., and Mosher, F. R. (1986) Operational demonstration of monitoring snowpack conditions utilizing digital geostationary satellite data on an interactive computer system, in D.L. Kane (ed.) *Proceedings of the Symposium: Cold Regions Hydrology*, University of Alaska-Fairbanks, Fairbanks, Alaska, 531-540.

Cogan, J. L., and Willand, J. H. (1976) Measurement of sea surface temperature by the NOAA 2 satellite, *J. Applied Meteorology* **31**, 173-180.

Chen, M., and Yang, H. (1984) Radiation in Linzi (Xizang), *Collected Works of QXPMEX (1)*, Science Press, Beijing, 95-103. (In Chinese.)

Engman, E. T., and Gurney, R. J. (1991) *Remote Sensing in Hydrology*, Chapman and Hall, London and New York, 225 pp.

Hall, D. K., Chang, A. T. C., and Foster, J. L. (1986) Seasonal and interannual observations and modeling of the snowpack on the Arctic coastal plain of Alaska using satellite data, in D.L. Kane (ed.) *Proceedings of the Symposium: Cold Regions Hydrology*, University of Alaska-Fairbanks, Fairbanks, Alaska, 521-530.

Ji, G., Yuan, F., Shui, D., Chen, Y., and Wang, W. (1984) Radiation in the western plateau. *Collected Works of QXPMEX (1)*, Science Press, Beijing, 10-22. (In Chinese.)

Jiang, J., G. Ji, and Wang, J. (1985) A study of the radiation Properties at Lhasa. *Plateau Meteorology* **4**, 67-79. (In Chinese.)

McClatchey, R. A., Fenn, R. W., Selby, J. E. A., Volz, F. E., and Garing, J. S. (1972) Optical properties of the atmosphere (Third Edition), *Environmental Research Papers*, No. 411, AFCRL, Bedford, MA, 108 pp.

Miller, W. (1986) Applying a snowmelt-runoff model which utilizes Landsat data in Utah's Wasatch mountains, Proceedings of the Symposium: Cold Regions Hydrology. Edited by D. L. Kane, University of Alaska-Fairbanks, Fairbanks, Alaska, 541-546.

Oke, T. R. (1978) *Boundary Layer Climates*, Methuen, London, 372 pp.

Robinson, D. A., (1986) Initiation of spring snowmelt over Arctic lands, in D.L. Kane (ed.) *Proceedings of the Symposium: Cold Regions Hydrology*, University of Alaska-Fairbanks, Fairbanks, Alaska, 547-556.

Rossow, W. B. (1989) Measuring cloud properties from space: a review, *J. Climate* **2**, 210-213.

Rossow, W. B., Mosher, F., Kinsella, E., Arking, A., Desbois, M., Harrison, E., Minnis, P., Ruprecht, E., Seze, G., Simmer, C., and Smith, E. A. (1985) ISCCP cloud algorithm intercomparison, *J. Climatology and Applied Meteorology* **24**, 877-903.

Shi, L., (1994) Cloud radiative forcing on surface shortwave fluxes: a case study based on Cloud Lidar and Radar Exploratory Test, *J. Geophysical Research* **99**, 25,909-25,919.

Smith, E. A., and Shi, L. (1992) Surface forcing of the infrared cooling profile over Tibetan Plateau. Part I: influence of relative longwave radiative heating at high altitude, *J. Atmospheric Science* **49**, 805-822.

Tao, S., Luo, S., and Zhang, H. (1986) The Qinghai-Xizang Meteorological Experiment (QXPMEX) (May-August) 1979, *Proceedings of International Symposium on the Qinghai-Xizang Plateau and Mountain Meteorology*, Chinese Meteorological Society and American Meteorological Society, Science Press, Beijing, 3-13.

Weng, D., Chen, W., and. Chen, L (1984a) The study on solar radiation over Lhasa river valley
    (2), *Proceedings of QXPMEX (1)*, Science Press, Beijing, 82-94. (In Chinese.)

Weng, D., Chen, W., Shui, G. Ji, and Shui, D. (1984b) Computation methods for 10 day and 30
    day period effective radiation in Qinghai-Xizang Plateau, *Collected Works of QXPMEX (2)*,
    Science Press, Beijing, 12-16. (In Chinese.)

Xie, X. (1984) The distributive character of the surface albedo over the Qinghai-Xizang Plateau
    in summer, *Kexue Tongbao* **29**, 365-367.

Xie, X., Y. Zhou, Y. Xiang, Z. Xu and Y. Ma (1984) The radiation characteristics over Gulmu
    region during May-August, 1979. *Proceedings of QXPMEX (1)*, Science Press, Beijing, 48-
    60. (In Chinese.)

Yang, H., and Shui, D. (1985) The characteristics of the radiation balance in Garze over the
    eastern part of Qinghai-Xizang Plateau, *Plateau Meteorology*, **4**, 80-93. (In Chinese.)

Yao, L., Jing, F., Chen, Y., and Wang, W. (1985) The characteristics of solar radiation in the
    three-river valley, the northern Tibetan Plateau and the Himalayas mountain area over the
    Qinghai-Xizang Plateau, *Plateau Meteorology* **4**, 94-111. (In Chinese.)

Yuan, F. (1985) The characteristics of the radiation over the Gerze region in the Qinghai-Xizang
    Plateau, *Plateau Meteorology* **4**, 36-49. (In Chinese.)

Lei Shi
SeaSpace Corporation
San Diego, California, USA.

22 CLIMATIC WARMING AND ITS IMPACT ON THE WATER RESOURCES OF THE YALONG RIVER, CHINA

DENG YUREN and HOU YUGUANG

Abstract

The water resources of the Yalong River basin may be affected by variations in air temperature and precipitation. In this paper, the impact has been estimated by using the Langbein model, the Turc model, a non-linear regression model and a higher order graybox dynamic model. The results are quite good in the optimization of model parameters, the model assessment and forecast verification. The average annual discharge will probably decrease if the air temperature rises 2°C in the future. The rate of decrease is greatest in the dry season.

22.1. Introduction

Climatic warming of the Earth from increases concentration of CO_2 and other greenhouse gases in the air may affect water resources. The impact is principally due to changes in air temperature and precipitation.

The calculated effect of climatic warming was derived by numerical analysis based on GCM output. A forecast of potential change is that if the concentration of CO_2 doubles, the average annual air temperature of the Earth will increase 1-2°C after A.D. 2030 (Second NCCC, 1990). The degree of climatic change in China was forecasted using a climatic model and a synthetic study. We concluded that the annual air temperature will probably increase 2-3°C in South-West China, but the change in precipitation is not clear. This paper uses statistical hydrometeorological models to estimate the impact climatic warming on the Yalong River.

22.2. Yalong River Basin

Yalong River is one of the main branches of the ChangJiang river. It joins from the north in the upper reaches of ChangJiang river and is located in the eastern part of the Qinghai-Xizang (Tibet) Plateau of China. It lies between latitude 26°32'-33°58'N and longitude 96°52'-102°48'E. The form of the basin resembles a long rectangle. The

J.A.A. Jones et al. (eds.), Regional Hydrological Response to Climate Change, 381–388.
© 1996 *Kluwer Academic Publishers. Printed in the Netherlands.*

elevation range of the watershed is about 2000-4000 m. The main river is 1500 km long and the area of the basin is about 130 000 km². The Ertan hydro-electric Project (3.3 x 10^6 kW), one of the big water power stations of China, is now under construction in the lower reach of the river.

The climate of Yalong river basin is affected by west wind circulation and the south-west monsoon. The contrast between dry and wet seasons is very marked. Every hydrological year, the wet season is considered to be from May to October and the dry season from November to next April. The precipitation is not evenly distributed. About 73% of annual precipitation occurs during the wet season and 27% during the dry season. In this basin the population is small and the impact on nature by human activities is very small.

The air temperature of this region may increase because of climatic warming. This change will cause an increase in evapotranspiration within the basin, a decrease of snow accumulation, and changes in the temporal distribution of snowmelt, thus changing the quantity and time distribution of streamflow. These indirect effects will be in addition to the direct effect of changes in precipitation on the quantity and time distribution of streamflow.

An important question is whether a change in quantity of water resources that may result from climatic warming will seriously impede the normal operation of the Ertan power station?

We used hydrometeorological data from the Yalong river for the period 1959 to 1990 in this paper. The calculated periods in this study are considered as hydrological year (May-next April), flood season (May-October) and dry season (November to next April). The basin-wide mean of precipitation and air temperature are estimated arithmetically, based on the best observed data available.

22.3. Models Used

Four statistical models were used to estimate the impact on the water resources of Yalong river. The four models are empirical relationship models. Hydrometric data were obtained from the Xiaodeshi station in the basin, for which 32 years of records were available.

The four statistical models are detailed below.

22.3.1. LANGBEIN MODEL

The model is based on the single-valued relation of P/F(t), and Y/F(t), where P is precipitation (cm), Y is runoff (cm), and F(t) is a factor of air temperature, such that $K(t) = 10^{A_1 t + A_2}$ and A_1, A_2 are parameters determined by optimization (WMO, 1983). We tested linear, single or double value and twin-logarithmic relationships for optimization and found the twin-logarithmic relationships were the best.

According to our calculation, we can establish the model as follows:

Hydrological year

$$Y = 10^{-0.690 + 0.028t + 1.436\log\left(P/10^{0.780 + 0.028t}\right)}$$ (22.1)

Flood season

$$Y = 10^{-0.766 + 0.027t + 1.520\log\left(P/10^{0.799 + 0.027t}\right)}$$ (22.2)

The results of model assessment and forecast verification are presented in Table 22.1. The model assessment involved checking the modeling results against the original observed data, which is used to construct the model. The success rate is the frequency with which the model results fall within ± 20% of observed values. Forecast Verification consisted of checking the predicted values against a predetermined subset of the observed data. The success rate is the same as above.

22.3.2. TURC MODEL

The Turc model is used to estimate the long-term loss of water, D, from the basin by using precipitation, P, and air temperature, T, over a long period (WMO, 1983). That is

$$D = P\left(a_1 + \frac{P^2}{L^2}\right)^{-1/2}$$ (22.3)

where, $L = a_2 + a_3T + a_4T^3$.
The runoff depth can be calculated by following formula:

$$Y = P - D = P - P\left[a_1 + P^2 / \left(a_2 + a_3T + a_4T^3\right)^2\right]^{-1/2}$$ (22.4)

Parameters a_1, a_2, a_3, a_4 can be determined by optimization. The resultant model is as follows:

Hydrological year

$$Y = P - P\left[0.19 + P^2 / \left(62.1 + 7.0T + 0.2T^3\right)^2\right]^{-1/2}$$ (22.5)

Flood season

$$Y = P - P\left[0.14 + P^2 / \left(45.4 + 3.81T + 0.08T^3\right)^2\right]^{-1/2}$$ (22.6)

where D, Y and P are measured in mm. The model assessment and the forecast verification are presented in the Table 22.1.

22.3.3. NON-LINEAR REGRESSION MODEL

This regression model uses precipitation and air temperature parameters that have been transformed in a variety of non-linear ways (Fu and Liu, 1991). The best transform type for runoff depth was determined by optimization. Then, we can determine the parameters of the model by a least squares non-linear model. The model can be written as follows:

Hydrological year

$$Y = P - 38 \times 10^{7.9t/199 + t} - 271 \tag{22.7}$$

Flood season

$$Y = P + 35 \times 10^{7.4t/209 + t} - 501 \tag{22.8}$$

where P and Y are measured in mm.

The model assessment and the forecast verification are also presented in Table 22.1.

22.3.4. HIGHER ORDER GRAY-BOX DYNAMIC MODEL

The model proposed by Deng Julong (1986) is:

$$\hat{Y}(k + 1) = \left[Y(1) - \frac{B_1}{A}\hat{P}(k + 1) - \frac{B_2}{A}\hat{T}(k + 1) \right] \exp(-Ak) + \frac{B_1}{A}\hat{P}(k + 1)$$

$$- \frac{B_2}{A}\hat{T}(k + 1) \qquad\qquad k = 1, 2,.... \tag{22.9}$$

where $\hat{Y}$, $\hat{P}$ and $\hat{T}$ are respectively the sum of statistical series of runoff depth, precipitation and air temperature. $Y(1)$ is the initial value of runoff depth; A, B_1 and B_2 are the parameters of the model. We estimated the parameters of the model and the formula after reordering as follows:

Hydrological year

$$\hat{Y}(k + 1) = \left[Y(1) - 1.47\hat{P}(k + 1) - 59.67\hat{T}(k + 1) \right] \exp(-0.11k) + 1.47\hat{P}(k + 1)$$

$$- 59.67\hat{T}(k + 1) \qquad\qquad k = 1, 2,.... \tag{22.10}$$

Flood season

$$\hat{Y}(k + 1) = \left[Y(1) - 1.17\hat{P}(k + 1) - 29.52\hat{T}(k + 1)\right]\exp(-0.43k) + 1.17\hat{P}(k + 1)$$
$$- 29.52\hat{T}(k + 1) \qquad\qquad k = 1, 2,.... \qquad\qquad (22.11)$$

The result of the model are not as good as the previous models, and the model assessment and forecast verification are not presented in Table 22.1.

TABLE 22.1. The results of model assessment and forecast verification.

Model	Period	Success rate (%)	
		Model assessment	Forecast verification
Langbein	Hydrologic year	100	85.7
	Flood season	100	85.7
	Dry season	95.8	71.4
Turc	Hydrologic year	87.5	85.7
	Flood season	100	85.7
	Dry season	58.3	71.4
Non-linear regression	Hydrologic year	100	100
	Flood season	85.8	85.7
	Dry season	91.7	71.4

Note: The success rate is the frequency with which model results fall within ± 20% of observed values.

22.4. Analysis of Models

The above models have represented the quantitative relationship between runoff depth, precipitation and air temperature. The parameters have been tested using observed data from the basin. Most models display a relatively high precision.

The construction of the models is simple. The application of the models is convenient and requires relatively little data. They are particularly suitable for regions where hydrological and meteorological data are deficient because of a lack of observation stations. Thus, it is hardly necessary to use a full water balance method.

These models show there are non-linear relations between the runoff depth and air temperature. The Turc model is most sensitive to the change in air temperature and the non-linear regression model is slow in reacting. In general, the relation is also non-linear hydrological and meteorological data are deficient because of a lack of observation stations. Thus, it is hardly necessary to use a full water balance method.

These models show there are non-linear relations between the runoff depth and air temperature. The Turc model is most sensitive to the change in air temperature and the non-linear regression model is slow in reacting. In general, the relation is also non-linear between the runoff depth and precipitation. Changes in precipitation are decisive for the impact on runoff depth.

As shown in Table 22.1, if the permissible error is not over ±20%, the success rate in the model assessment can be more than 95% for both the hydrological year and the flood season. The success rates of forecast verification are more than 85%. If we put the permissible error at ±15%, the success rate of model assessment and forecast verification are generally more than 83%, but is lower in the dry season. Since the results are very good for the hydrological year and the flood season, we can make up for a deficiency of the models during the dry season by using a water balance approach.

22.5. Prediction

According to the environmental scenario presented in the introduction, if climatic warming increases air temperature about 2°C above present in the Yalong river basin in the future, the precipitation may increase (e.g. +10% of annual mean precipitation), decrease (e.g. -10%) or even stay the same. We predicted the runoff depth from the models for these three cases. Table 22.2 shows the predicted results.

TABLE 22.2. The predicted change in runoff at the Xiaodeshi hydrometric station, Yalong River.

Period		Year		Flood season		Dry season	
Average runoff		346.3		252.0		94.3	
Change		ΔY (mm)	Δc (%)	ΔY (mm)	Δc (%)	ΔY (mm)	Δc (%)
	-10%	-87.3	-20.0	-63.4	-18.0	-23.9	-25.3
P	0%	-24.0	-5.5	-14.1	-4.0	-9.9	-10.5
	+10%	+30.5	+7.0	+28.2	+8.0	+2.3	+2.4

Note: ΔY and Δc are respectively the change in runoff in milllimetres relative to the present average and the percentage change.

According to Zhao (1989), analyses based on the same assumption adopted in this paper, i.e. a 2°C increase in temperature and a 10% reduction in annual precipitation, suggest reductions in annual runoff ranging from -35% to -17% in at least ten basins. The results in Table 22.2 fall within this range and appear to be reasonable, given the physical environment of the Yalong basin.

From Table 22.2, we can see that if the air temperature rises by 2°C, but the precipitation remains unchanged, water resources will decrease, but only by a small amount. However, if the precipitation decreases as well, the amount of water resources will clearly decrease, especially during the dry season. Runoff can fall to about 1/4 of the average for the dry season. If, on the other hand, precipitation increases, the change in water resources is quite small, because the loss of water resources by the increase in air temperature will counterbalance the increase due to greater precipitation.

In conclusion, the impact of climatic warming of 2°C on the flow of the Yalong river does not appear to be very great, except in the scenario in which precipitation is reduced by at least 10%. In this scenario the more obvious impact will be during the dry season. Otherwise the predicted amount of change seems to lie roughly within the expected degree of error of the procedure used for prediction.

Acknowledgments
This research is supported by National Nature Science Fund of China. No. 49070063.

References

Deng, Julong (1986) *Gray-Box Prediction and Policy Decision*, Publishing House of HUSE. (In Chinese.)

Fong, Yaohung (1991) Application of Non-linear Regression in Weather Forecast, *Meteorology* **16**. (In Chinese.)

Fu, Guobin, and Liu, Changming (1991) Estimation of Regional Water Resources Response to Global Warming - A case study of Wan Quan Basin, *Acta Geographica Sinica* **46**.

Second Group of NCCC (China) (1990) The Impact Assessment for Climatic Change Caused by Human Activities in China, *Environmental Science of China* **10**.

WMO (1983) *Guide to Hydrological Practices,* Vol.2, WMO Pub No 168, WMO, Geneva.

Zhao, Zhong-Ci (1989) Modeling the Hot-house Effect to Impact on the Change of Climate, *Meteorology* **15**, 3.

Deng Yuren and Hou Yuguang
Department of Hydraulic Engineering
Chengdu University of Science and Technology
Chengdu 610065, China.

23 THE PROBABLE IMPACT OF GLOBAL CHANGE ON THE WATER RESOURCES OF PATAGONIA, ARGENTINA

R.M. QUINTELA, O.E. SCARPATI, L.B. SPESCHA and A.D. CAPRIOLO

Abstract

Very little work has been undertaken on the hydrological impact of global warming in South America. In this study, some initial estimates are made using IPCC and Hadley Centre scenarios, supported by an analysis of long-term trends in rainfall patterns. Under the global warming scenarios, we expect a drying trend in the north of Argentina and a wetter climate in the south, accompanied by a reduced meridional temperature gradient. Snowlines are likely to rise, although the overall response of glaciers needs detailed investigation. However, analyses of rainfall records suggest that rainfall has been increasing in the north, contrary to the expected global warming pattern.

23.1. Introduction

Little detailed work has been undertaken on the potential effects of global warming in South America. However, there is natural concern over the likely physical and socio-economic impacts.

The work described in this chapter uses the report of the Intergovernmental Panel on Climate Change (IPCC, 1990) together with the paper by Mitchell and Gregory (1992) as the main bases for prediction. Mitchell and Gregory (1992) developed four scenarios for future emissions of greenhouse gases, which were used to estimate the resulting rate and magnitude of climate change.

Figure 23.1 shows the simulated changes in global mean temperature with 'high', 'best estimate' and 'low climate sensitivities' (4.5, 2.5 and 1.5°C) and in Figure 23.2 we can observe changes of mean global temperature assuming a 'best estimate' climate sensitivity of 2.5°C for the respective 1990 and 1992 scenarios. Using the 'best estimate' climate sensitivity, these scenarios give a range of warming from 1.5°C to 3.5°C for the year 2100 (mean = 2.8°C) and 1.5°C for the year 2050 (IPCC, 1992). The increase in sea level will be 0.20 m and 0.65 m in 2030 and 2100 respectively.

Although these values are the 'official' ones, many authors have given different values for global temperature change. There is even greater difficulty in deriving these values for a regional analysis (Waggoner, 1990).

J.A.A. Jones et al. (eds.), Regional Hydrological Response to Climate Change, 389–407.
© 1996 *Kluwer Academic Publishers. Printed in the Netherlands.*

With reference to the Southern Hemisphere, several authors (Pittock *et al.*, 1988; Nuñez, 1990; Burgos *et al.*, 1991) applied global circulations models (GCMs), such as GISS, NCAR and others, and have obtained different results. Burgos *et al.* (1991) and Nuñez (1990) found a transition zone about 40°S. The former authors concluded that for the year 2010, to the North the temperature will be increased, the humidity of the air will be lower and the radiation increased because of diminished cloudiness, whereas to the South of this transition zone the air humidity will be greater, the surficial temperature will not change, but the tropospheric temperature will be increased by the heat exchange. For the year 2050, the process will be more intense, but with the same characteristics. Nuñez (1990) predicted very high ΔT values, but then reduced them in line with other researchers.

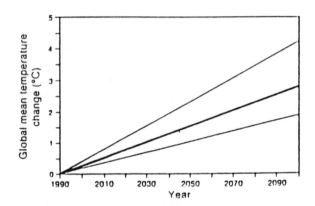

Figure 23.1. Simulated changes in global mean temperature after 1990 due to doubling CO_2 concentrations, assuming high (4.5°C), 'best-estimate' (2.5°C) and low (1.5°C) climate sensivities (IPCC, 1992).

On the other hand, Russian researchers have developed analogue techniques based on the analysis of paleoclimates (Budyko *et al.*, 1994). These results shall be discussed in a later part of this analysis.

Novelli *et al.* (1995) have referred to recent changes in CO_2 concentrations and their implications on global changes. The impact on the climate of the recent changes in CO_2, CH_4 and CO is difficult to predict. The latest measurements of atmospheric concentrations by NOAA/LCDC show that they have begun to increase markedly.

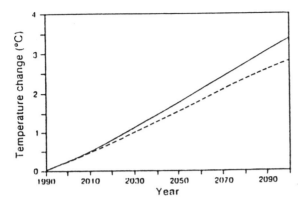

Figure 23.2. Changes of mean global temperature after 1990 assuming a 'best estimate' climate sensitivity of 2.5°C (estimates from IPCC (1992) = dashed curve; from IPCC (1990) = solid curve).

23.2. Present Hydrological and Climatic Characteristics of Patagonia

Patagonia is an arid to semiarid region. It grades westwards into a narrow and wetter climatic region dominated by forests and lakes (Figure 23.3), which is the source area for many of its more important rivers.

It extends approximately between 38°S and 55°S to the south of the Colorado River, and it covers an area of 700 000 km². The principal soils, according to the Soil Taxonomy (USDA, 1975) are: aridisols (50-60%), entisols (18%), molisols (10-20%), and alfisols, histosols and inceptisols in smaller percentages (INTA, 1990) (Figure 23.4).

Patagonia's water resources are contained in important rivers that flow from west to east through the Patagonian plateau, and in numerous lakes, glaciers and a huge mass of ice in the South Andes piedmont. Some of these drain to the Atlantic Ocean and others into the Pacific Ocean (Table 23.1) (Ferrari Bono, 1990). The river regimes are characterized by two annual peaks, one caused by snow and rainfall in winter, and the other produced by the snowmelt from the Andes mountain chain in spring, as shown on the Limay River illustrated in Figure 23.5. The most important characteristics of some rivers that we must mention are:

1) the Neuquen and Limay Rivers that form the Negro River, which has the largest discharge (1000 m³ s⁻¹ of mean flow). These two tributaries have the greatest potential for hydroelectric power. They provide almost 1/4 of the total energy installed in Argentina (26 400 GWh/year)

2) the Futaleufu dam provides energy (448 MW) to the Aluar aluminium factory near the shores of the Atlantic Ocean, and the Ameghino dam on the Chubut River feeds 22 000 ha of irrigation area.

3) south of the Andes, the Patagonian Continental Icefield, which is a huge mass of ice extending 280 km long between 47°30'S and 50°50'S.

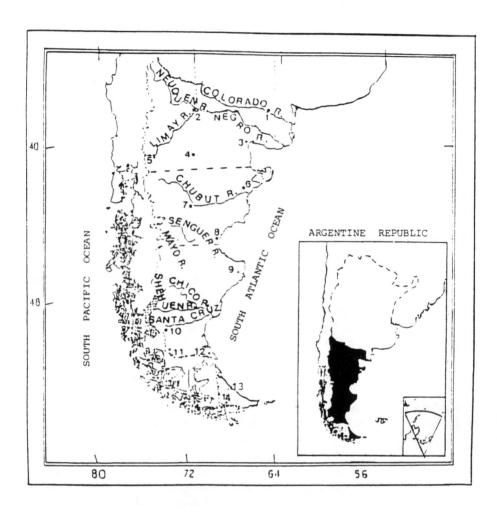

Figure 23.3. Map of Patagonia, with the most important rivers and pluviometric stations.

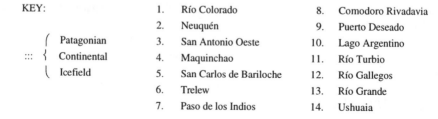

KEY:		1.	Río Colorado	8.	Comodoro Rivadavia
		2.	Neuquén	9.	Puerto Deseado
	Patagonian	3.	San Antonio Oeste	10.	Lago Argentino
:::	Continental	4.	Maquinchao	11.	Río Turbio
	Icefield	5.	San Carlos de Bariloche	12.	Río Gallegos
		6.	Trelew	13.	Río Grande
		7.	Paso de los Indios	14.	Ushuaia

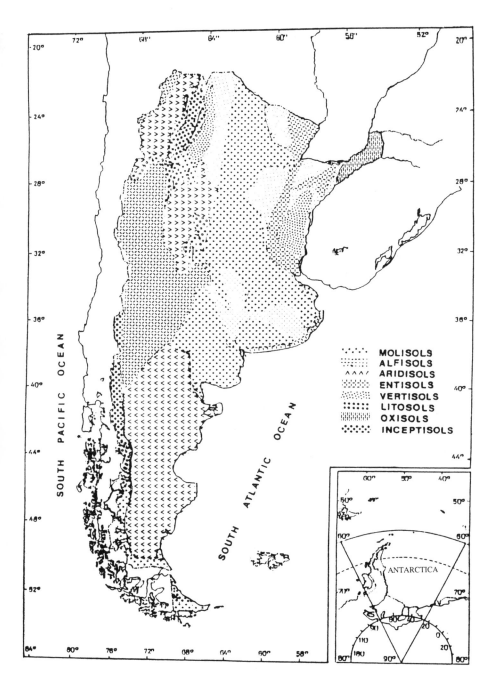

Figure 23.4. Argentine soils taxonomy.

TABLE 23.1. Characteristics of the principal rivers.
(From Ferrari Bono, 1990.)

Rivers	Basin area (km²)	Mean flow (m³ s⁻¹)	Length (km)	Generating potential (GWh/year)	Natural lakes	
					Number	km²
a) Basins with Atlantic drainage						
Colorado	69 000	133	977	3700	3	37
Neuquén	32 450	316	541	5000	15	205
Limay	56 185	733	804	18 600	29	1113
Negro	95 800	1014	635	5666	--	--
Senguer	28 025	50	360	1540	--	--
Chubut	29 400	49	867	1000	--	--
Deseado	14 450	5	615	--	9	48
Sheuen-Chico	16 800	30	2	--	2	74
Santa Cruz	24 510	750	382	6576	13	2632
Coyle	14 600	5	350	--	2	14
Gallegos	5200	15	360	--	22	39
Chico	800	2	75	--	--	--
Mayer	6500	235	250	2600	14	764
b) Basins with Pacific drainage						
They are eleven rivers. One of the principal rivers is:						
Manso	2810	132	70	2500	16	115

The rivers which drain into Atlantic Ocean have a total discharge of 1920 m³ s⁻¹ and those that flow into Pacific 1240 m³ s⁻¹; together they represent approximately 3000 m³ s⁻¹ of surficial resources. Many basins have a water resources potential that are not totally used (Table 23.1). Figure 23.6 shows the principal dams built and projected in the hydroelectric complex of the Neuquen and Limay Rivers (installed power: 2700 MW).

The climate is dominated by position of the Subtropical anticyclones in the Atlantic and Pacific oceans. Figures 23.7a and b show atmospheric pressure at sea level for summer and winter in South America (Schwerdtfeger, 1976). We can deduce that in the summer the thermo-orographic low pressure is intensified over the higher regions in the west of Patagonia (Lichtenstein, 1990) and the South Pacific high pressure centre is situated at 32°S 90°W. In winter, the isobars shift toward a north-south direction; the

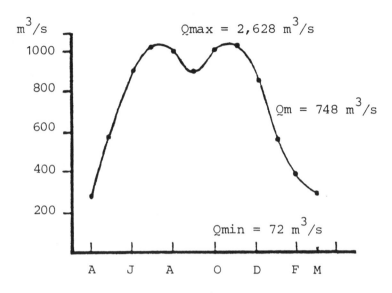

Figure 23.5. Mean annual discharge of the Limay River (1903-1980).

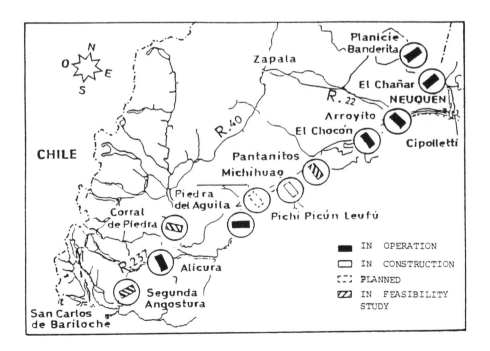

Figure 23.6. Principal dams functioning and planned in the Negro River basin.

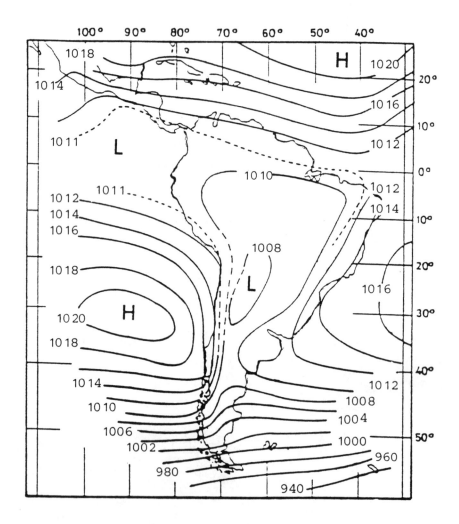

Figure 23.7A. Atmospheric pressure at sea level in South America during the summer.

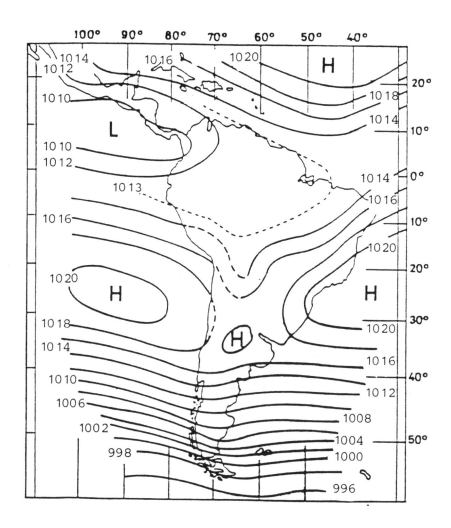

Figure 23.7B. Atmospheric pressure at sea level in South America during the winter.

notorious westerlies are situated along 40°S and in the region immediately polewards, and the two semi-permanent anticyclones are in a similar position.

The north-south orientation of the mountain chain, the wind circulation from west to east and the rising of humid air masses from the Pacific Ocean cause more abundant precipitation in west Patagonia than in east Patagonia. The heavy rainfall supports forests, peat-bogs and tundra, on organic, wet and acid soils in the piedmont zone. On the other hand, in the central plateau and in the Atlantic riparian plain, a xerophytic steppe has been formed on mineralized soils (Burgos, 1985; Endeicher and Santana, 1988). To the west, over the Andes mountain chain, the rainfall are more than 4000 mm.

The water balance calculated by the methods of Budyko (1948) or Thornthwaite and Mather (1955), shows a very high water deficit. El Turbio, (51°40'S 72°02'W) has an annual deficit of 87 mm and Rio Gallegos (51°35'S 69°00'W) has a deficit of 314 mm (Burgos, 1995). The most important climatological values for some Patagonian stations are given in Table 23.2.

Hoffmann (1990) has studied the trends in air temperature, comparing the periods 1941-50 and 1981-90, with the following results: Rio Gallegos, +1.2°C; Lago Argentino, +0.4°C; San Julian, +0.7°C (49°19'S 67°42'W); Comodoro Rivadavia, +0.2°C and Trelew, +0.1°C. It is observed that the decadal changes increase from north to south, coinciding with the results of the GCM models for global warming.

23.3.   Possible Changes in Patagonian Water Resources Induced by Global Warming

Two global atmospheric circulation models (GCM) were chosen, CSIRO9 from the Division of Atmospheric Research, Commonwealth Scientific and Industrial Research Organization, Australia, and UKHI from the United Kingdom Meteorological Office, following the results obtained by Boer (1992) and Labraga (in press).

A comprehensive assessment of GCM performance over the Southern Hemisphere as a whole is still pending and would be very valuable. The procedure used to evaluate the capability of models to simulate the contemporary earth climate was as follows. Selected sets of climate features in the observed MSL pressure, surface temperature and precipitation fields, relevant to the present climate, were highlighted. Qualitative evaluation was then undertaken of the model's capability to simulate each of these selected climate characteristics, and statistical measures of global performance were computed.

Double CO_2 equilibrium experiments were compared, and some consistent patterns of climate trends were detected in the selected surface variables. Attention was focussed on surface data only, because any significant change in this variables could seriously affect the evolution of natural ecosystems and the output of many essential human activities.

The resolution of the transform grids used for calculating the model physics and number of vertical levels of each model are indicated in Table 23.3.

TABLE 23.2. Climatic characteristics of selected stations, 1981-90.
(Servicio Meteorológico Nacional Estadísticas climatológicas.)

		S. C. de Bariloche	Catedral 2000	Paso de los Indios	Rio Gallegos	Ushuaia
Location		41°09' S	41°15' S	43°49' S	51°37' S	54°48' S
and		71°10' W	71°37' W	68°53' W	69°17' W	68°19' W
Altitude		840 m	1955 m	460 m	19 m	14 m
Mean	S*	15.0	8.1	19.1	14.1	10.3
Temperature	W	2.1	-2.5	3.4	1.2	1.6
(°C)	A	8.4	--	11.3	8.0	--
Mean Min.	S	6.7	4.1	9.6	8.1	5.7
Temperature	W	-1.5	-4.7	-2.6	-1.9	-1.4
(°C)	A	2.3	-0.8	3.8	3.3	--
Mean Max.	S	22.2	14.5	27.1	20.1	15.0
Temperature	W	6.6	0.5	8.7	4.6	4.5
(°C)	A	14.6	6.9	18.0	13.3	--
Precipitation	S	16.4	--	9.1	34.6	30.7
(mm)	W	105.7	152.0	20.0	19.3	46.2
	A	714.5	--	200.0	274.0	--
Main Wind	S	W	W	W	W	W
Direction	W	W	W	W	W	W
	A	W	W	W	W	--
Wind velocity	S	26.9	33.1	16.6	32.9	16.9
(km h^{-1})	W	17.8	41.1	10.9	23.3	10.8
	A	22.6	--	15.3	29.2	--

* S - January (Summer); W - July (Winter); A - Annual.

TABLE 23.3. GCM horizontal and vertical resolution.

	CSIRO9	UKHI
Horizontal resolution (Lat. x Long.)	3.2°x5.6°	2.5°x3.75°
Number of grid points	3584	6912
Number of vertical levels	9	11

R.M. QUINTELA et al.

Latitudinal pressure gradients in the southern tip of the continent calculated by these GCMs are shown in Table 23.4. There is general agreement among the model results about a possible southward shift and intensification of the summer continental low pressure system in the 2 x CO_2 equilibrium climate.

TABLE 23.4. Latitudinal pressure gradient between 45° and 55° S, at 70°W (hPa / ° Lat.).

Season	Observed	CSIRO9 prediction	UKHI prediction
Summer	1.4	1.4	1.2
Winter	1.2	1.7	1.8

The computed RMS errors between observed and model MSL pressure, surface temperature and precipitation rate are presented in Table 23.5. This provides an objective measure of model performance. Calculations were carried out after interpolating model output onto a mesh equal to the observed data (3.18°Latitude x 5.63°Longitude). A weighting function was used to take into account the latitudinal variation of the area represented by grid points.

TABLE 23.5. Annual and seasonal pattern correlation and RMS error (in brackets) of the MSL pressure, surface temperature and precipitation fields.

	Annual	DJF	MAM	JJA	SON
Pressure (hPa)					
CSIRO9	0.96 (2.5)	0.92 (3.7)	0.96 (2.5)	0.96 (2.8)	0.96 (3.1)
UKHI	0.95 (3.0)	0.89 (4.8)	0.95 (2.9)	0.95 (2.9)	0.95 (3.3)
Temperature (°C)					
CSIRO9	0.88 (2.8)	0.76 (2.9)	0.88 (2.7)	0.93 (3.1)	0.87 (3.3)
UKHI	0.95 (2.4)	0.91 (2.3)	0.96 (2.2)	0.96 (3.2)	0.93 (3.5)
Precipitation (mm)					
CSIRO9	0.68 (1.7)	0.66 (2.5)	0.65 (2.5)	0.61 (2.5)	0.73 (2.1)
UKHI	0.67 (2.0)	0.75 (2.6)	0.67 (2.8)	0.71 (2.6)	0.66 (2.2)

MSL pressure difference fields between the 2 x CO_2 and 1 x CO_2 numerical experiments were analysed. No consistent change can be discerned in the latitudinal pressure gradient at the southern end of the continent, between 45° and 55°S. Both CSIRO9 and UKHI have predicted for Patagonia an increase of summer surface temperature of about 4-5°C and for winter, 3-5°C. The mean precipitation differences between the 2 x CO_2 and 1 x CO_2 are not significant.

Budyko *et al.* (1994), working with analogous models and paleoclimates, arrived at a forecast for temperature change in South America of 1°C by the year 2000, 2-3°C for 2025 and 3-4°C about the middle of the 21st century. Furthermore, they calculated the latitudinal differences of the mean air temperature for the different latitudinal belts in both hemispheres and their results for Patagonia are summarised in Table 23.6. Mean annual rainfall in the centre-oceanic region will decrease by 50 mm by the year 2000; by the year 2025 they calculate an increase of 100 mm for the northern zone and a 50 mm decrease for the extreme south and Tierra del Fuego. By the year 2050, increases of 100 and 50 mm in the centre-north and in the centre-south respectively are indicated.

TABLE 23.6. Latitudinal differences for mean air temperature (°C).

(From: Budyko *et al.*, 1994.)

Latitude (South)	Year 2000		Year 2020		Year 2050	
	Winter	Summer	Winter	Summer	Winter	Summer
40-50°	1.4	1.6	2.5	2.6	3.2	4.0
50-60°	1.7	2.6	3.0	3.2	5.1	6.2

Comparison of the regional changes in temperature determined by the two climate models with those derived from the paleoclimatic data permits a forecast for the winter air temperature anomalies, especially in middle and high latitudes. The numerical climate models and the paleoclimatic reconstruction all provide similar values on the temperature changes, with correlation coefficients among them of between $r = 0.8$ to 0.9. It is more difficult to predict the anomalies for summer temperature, because these changes are of smaller magnitude (Budyko *et al.*, 1992).

Changes have already been observed in precipitation. Ten-year moving averages and rainfall trends with a fourth degree equation, have been calculated for Patagonian stations and are shown in Figure 23.8. A positive trend is observed in the two stations near to the Andes piedmont, i.e., Bariloche and Esquel (42°54'S 71°22'W) (Quintela *et al.*, 1993; Scarpati and Faggi, in press).

Most Patagonian water resources are held in its rivers, lakes and snow mass, located in the piedmont region and high mountain chain. Patagonia's surficial water resources have experienced a disminution during the last three decades, in contrast with what has

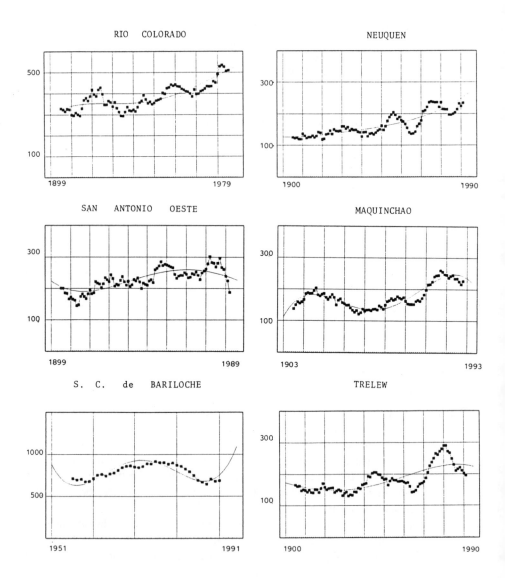

Figure 23.8. Moving averages (■) and trends (—) for mean annual precipitation.
Y axis is annual precipitation in millimetres. See Figure 23.3 for locations.

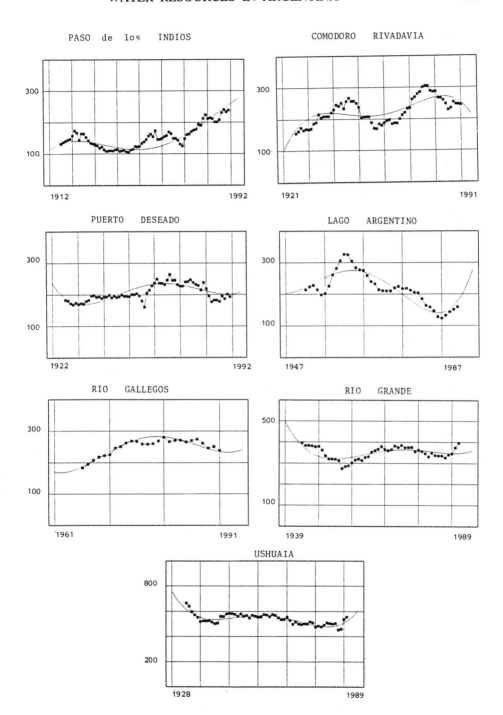

occurred in the pampean region, where there has been a water excess (Quintela *et al.*, 1989; Quintela and Scarpati, in press).

In Patagonia, the Andes mountain chain introduces a very important morphologic element. It can be conjectured that an increase of 20-50% in rainfall would produce an improvement in water resources in the Andes piedmont subzone (Quintela *et al.*, 1993; Scarpati and Faggi, in press). A large part of Patagonia has been suffering a growing deterioration from eolian and water erosion, due to mismanagement of the soils and over-grazing; 40% of soils has suffered serious deterioration (Prego, 1988). Soriano and Movia (1988) noted that the Patagonian semidesert and neighboring steppes are areas where the advancement of dunes, covering fields and settlements, the formation of active gullies and ravines, the cutting of roads, the filling up of river channels and the formation of wide rocky pavements, have been occurring with increasing intensity.

The main change below an altitude of 2000 m is likely to be the temperature increase. According to the soil characteristics of the different Patagonian regions (Figure 23.4), it would be expected the variations produced on them.

The temperature increase would act as a severe limiting factor for the snowline level. According to Australian researchers, it would be raised by 100 m for each °C of warming. The glaciers would also suffer a disquieting impact. In Bolivia, icemelt is already occurring on the high summits, for example the Chacaltaya glacier (5343 m) near La Paz, where B. Francou has recently reported measurements confirming losses of 6 m height of ice thickness in one year. Kellog (1993) reports similar situations occurring in other places of the world related to the greenhouse effect. According to Shepard, the mountains of this nation and of Peru show a rise in the snowline.

Finally, we must consider an important physico-chemical phenomenon, the disminution in the ozone layer in Antarctica. This has decreased the temperature in the south of Argentina during recent decades. Models confirm that the loss of stratospheric ozone is capable of reducing temperatures by up to 0.3-0.4°C per decade, according to Bojkov (1995). The great sensitivity of stratospheric temperatures to ozone radiation shows the necessity to analyze the climatic implications. The same applies to CO and CH_4.

23.5. Conclusions

The preceding analyses suggest that global warming will have the following effects on Patagonian water resources.

1) The temperature differences between the north and south of the region will be reduced. Temperatures will rise most in the south and during the winter. This will be associated with a southward shift in the average location of the Subtropical anticyclones in both the Atlantic and Pacific Oceans.

2) The altitude of the snowline will rise with increased temperatures, which will eventually produce a decrease in snowmelt resources feeding the headwater rivers. Rivers that currently receive major contributions from snowmelt, like the Senguer and Mayo will be most affected. Those that are fed by large lakes or from ice fields, like the

Santa Cruz, will be least affected.

3) The response of the Patagonian Continental Icefield is more difficult to predict and should be the subject of a special study. It could gradually increase the levels of the Argentino and Viedma lakes and eventually enhance the relatively small discharges of the Rivers Coyle and Sheuen.

4) A positive rainfall trend is observed in the region near the Andes piedmont and, if continued, this would produce an improvement in the land resources in this zone, depending on the prevailing soil type.

5) It remains very difficult to predict the response of Patagonian water resources to global warming with certainty. Some zones could benefit, like the rivers of Santa Cruz province. Others may experience reduced resources, like the Negro River basin. However, we see no reason to share the despair of Lauria (1995):'A realistic global model exceeds the actual computation and data communication capacities', and we expect a sustained improvement in the reliability of regional predictions from the new generation of GCMs.

Acknowledgements

We thank *in memoriam* the important climatological contribution given by Ing. R.J. Broqua, before his death. We also thank P. Fernández, M.A. Soria Nóbile and A. Quintela for their technical assistance.

References

Boer, G.I., Mc Farlane, N.A., and Lazare, M. (1992) Greenhouse gas-induced climate change simulated with the CCC second-generation general circulation model, *J. Climat.* **5**, 1045-1077.

Budyko, M.I., Borzenkova, I.I., Menzhulin, G.V., and Selyakov, K.I. (1992) Forthcoming Climate Change, *Izvestia RAN Sr Georgia* **4**, 36-52.

Budyko, M.I., Borzenkova, I.I., Menzhulin, G.V., and Shiklomanov, I.A. (1994) *Cambios antropogénicos del clima en América del Sur*, Serie de la Academia Nacional de Agronomía y Veterinaria de la República Argentina, No. 19.

Bruniard, E. (1986) Aspectos geográficos de las precipitaciones nivales en la República Argentina, *Boletín de Estudios Geográficos* (National University of Cuyo, Mendoza, Argentina) **22**, 82-83.

Burgos, J.J. (1985) Clima del extremo sur de Sudamérica, in O.Boelcke, D.M. Moore and G.R. Roig (eds.) *Transecta Botánica de la Patagonia Austral*, Argentina.

Burgos, J.J., Fuenzalida Ponce, H., and Molion, L. (1991) Climate Change Predictions for South America, *Climate Change* **18**, 223-239.

Bojkov, R.D. (1995) *La evaluación internacional del ozono, WMO Bulletin* **44**, 1, 42-50.

CSIRO (1988) *Greenhouse. Planning for climate change*, Commonwealth Scientific and Industrial Research Organisation, Australia.

Endeicher, W., and Santana Aguila, A. (1988) El clima del sur de la Patagonia y sus aspectos ecológicos. *Anales del Instituto de La Patagonia.* Serie Ciencias Naturales. Univ. de Magallanes, Punta Arenas, Chile **18**, 58-85.

Ferrari Bono, B.V. (1989) La potencialidad del agua, Recursos Hídricos continentales de la Patagonia, Argentina, *Ciencia Hoy* **2**, 7.

Instituto Nacional de Tecnología Agropecuaria (INTA) (1990) *Atlas de suelos de la República Argentina, Escala 1:500.000 y 1:1.000.000, 1990,* Secretaría de Agricultura, Ganadería y Pesca. Proyecto PNUD ARG 85-019.

Intergovernmental Panel on Climate Change (IPCC) (1990) *Climate Change. The IPCC Scientific Assessment,* J.T. Houghton, G.J. Jenkins and J J Ephraums (eds.), Cambridge University Press Cambridge.

Intergovernmental Panel on Climate Change (IPCC) (1992) *Climate Change 1992. The Supplementary Report to the IPCC Assessment,* J.T. Houghton, B.A. Callander, and S.K. Varney (eds.), Cambridge University Press, Cambridge.

Hoffmann, J.A. (1990) De las variaciones de la temperatura del aire en la Argentina y estaciones de la zona subantártica adyacente, desde 1903 hasta 1989, *Actas de la Primera Conferencia Latinoamericana sobre Geofísica, Geodesia e Investigación Espacial Antárticas,* Buenos Aires, Argentina.

Labraga, J. C. (in press) The climate change in South America due to a doubling in the CO_2 concentration: intercomparison of general circulation models equilibrium experiments, *CENPAT-CONICET,* Argentina.

Lauría, E.H. (1995) Los grandes desafíos, *La Nación,* Buenos Aires, Argentina, 20th May.

Lichtenstein, E. (1990) *La depresión termo-orográfica del noroeste argentino,* Unpublished Ph.D Thesis.

Novelli, P.C., Conwait, T.J., Dlugokenck, E.J., and Tans, P. (1995) Cambios recientes en el dióxido de carbono y el metano y sus implicancias en el cambio climático mundial, *WMO Bulletin* **44**, 32-37.

Nuñez, M. (1990) Cambio climático en Sudamérica. Uso de modelos de circulación general, *Geofísica* **32**, 47-64.

Prego, A. (1988) *El deterioro del ambiente en la Argentina.* Fundación para la Educación, la Ciencia y la Cultura (FECIC), Buenos Aires, Argentina.

Quintela, R.M., Forte Lay, J.A., and Scarpati, O.E. (1989) Modification of the water resources characteristic of the pampean subhumid-dry region, *Sixth Conference on Applied Climatology of the American Meteorology Society. 19th Conference Agricultural and Forest Meteorology and 9th Conference Biometeorology and Aerobiology,* J-30-J-35.

Quintela, R.M., Broqua, R.J., and Scarpati, O.E. (1993) Posible impacto del cambio global en los recursos hídricos del Comahue, Argentina, *X Simposio Brasileiro de Recursos Hídricos, I Simposio de Recursos Hídricos do Cone Sul,* Gramado, Brasil, ANAIS **3**, 320-329.

Quintela, R.M., and Scarpati, O.E. (in press) Incidencia del Cambio Global sobre los recursos hídricos del sur de la Patagonia, Argentina, *Geofísica* **39**, IPGH, México.

Scarpati, O.E., and Faggi, A.M. (in press) Relaciones entre parámetros climáticos y bosques en el Parque y Reserva Nacional Lago Puelo. *Revista Facultad de Agronomía. Universidad de Buenos Aires.* Argentina.

Schwerdfeger, W. (1976) The atmospheric circulation over Central and South America, Elsevier

Scientific Publishing Company, *World Survey of Climatology* **12**, 1-12.

United States Department of Agriculture (USDA) (1975) Soil taxonomy. A basic system of soil classification for working and interpreting soil surveys, *Agriculture Handbook No. 436*, Soils Conservation Service, USA.

Waggoner, P.E. (1990) *Climate Change and U.S. Water Resources*, John Wiley and Sons, USA.

R.M. Quintela and A.D. Capriolo
Biometeorological Research Center
National Scientific and Technical Research Council (CIBIOM -CONICET)
Serrano 669 (1414)
Buenos Aires, Argentina.

O.E. Scarpati
Humanity and Education Sciences Faculty
La Plata National University, La Plata, Argentina.

L.B. Spescha
Agronomy Faculty
University of Buenos Aires, Argentina.

24 LONG TERM TRENDS IN THE WATER BALANCE OF CENTRAL JAPAN

K. MORI

Abstract

Changes in the components of the hydroclimatological water balance over the last 100 years were investigated in Tsu City in the central part of Japan as a case study. Annual potential evapotranspiration has increased over a long period of time. Smoothed secular changes in the difference between the annual values of precipitation and potential evapotranspiration are analogous to those in the annual precipitation. The difference between annual precipitation and annual potential evapotranspiration is inclined to be below the average since the latter half of the 1920s, but the imbalance has increased over the last few years. A trend curve of the long-range changes in the annual runoff ratio is dominated by annual precipitation. The average annual runoff ratio for each ten years has a tendency to decrease during the last few decades.

24.1. Introduction

Global warming caused by increasing concentrations of carbon dioxide has been examined for different climatic zones by means of various models, especially using General Circulation Models (GCMs). The repercussions of climatic change on hydrological processes are, however, still uncertain.

Gleick (1986) discussed the approaches for evaluating the regional hydrologic impacts of climatic change. The effect of changes in climate on water availability was investigated in the Sacramento Basin, California, by using water balance methods (Gleick, 1987). The results suggested that information on the regional hydrologic effects of climatic change is extremely important for long-range water resources planning. The implications of climatic change for the hydrologic cycle and for future water management were reviewed by Gleick (1989). An increase of streamflow induced by the doubling of atmospheric carbon dioxide was simulated for an experimental catchment area in New South Wales, Australia (Aston, 1984). Bultot *et al.* (1988) carried out an investigation into the impact of climatic change on the hydrologic system in three drainage basins in Belgium. In addition, quantitative changes in the flow regime of the Tone River, Japan, induced by the so-called 'greenhouse effect' were estimated by Masukura *et al.* (1991, 1992).

J.A.A. Jones et al. (eds.), Regional Hydrological Response to Climate Change, 409–416.

final goal of this study is to elucidate the quantitative features of alterations in runoff in the humid temperate regions as induced by global warming.

24.2. General Descriptions of the Runoff Pattern in the Study Area

In order to calculate the annual runoff ratio based on the water balance, meteorological data from Tsu City on the Pacific side of central Japan were analyzed. The study area was selected as typical of the region with humid temperate climate. As shown in Figure 24.1, the annual runoff pattern of Japanese rivers is divided into three types (e.g. Ichikawa *et al.*, 1967; Arai, 1977). The study area is located in the regional division 'A' in Figure 24.1, where the maximum discharge appears in early summer or in autumn and the minimum occurs in winter.

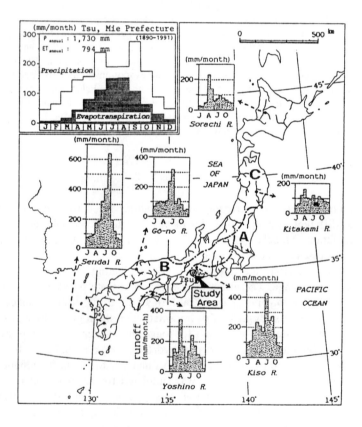

Figure 24.1. Map showing the regional division of annual runoff patterns in Japan, including annual variations in precipitation and potential evapotranspiration of the study area.

The arithmetic mean annual precipitation for the last 102 years in the study area is 1730 mm (Figure 24.1). It seems reasonable to assume that a monthly surplus of water storage occurs throughout the year except during extreme drought years.

24.3. Method

If a sufficiently long-term period is taken into consideration, changes in water storage can be neglected. Consequently, the annual runoff in a given area can be estimated by subtracting the annual potential evapotranspiration from the annual precipitation. Kayane and Takeuchi (1971) pointed out that in comparison to the observed values of annual runoff for Japanese rivers such water balance estimates are generally overestimates. However, Mori (1989) found that the annual runoff for major drainage basins in the central part of Japan, calculated as the difference between annual precipitation and annual potential evapotranspiration, was much more reliable than the observed values.

The data on monthly precipitation and mean air temperature were collected at the Tsu Local Meteorological Observatory for the last 102 years, 1890-1991. Because of the limitations of available meteorological data over the last 100 years, annual potential evapotranspiration for each year was calculated by Thornthwaite's method (Thornthwaite, 1948). The Thornthwaite equation is based primarily on air temperature, and is unable to take account of variations in such controls as wind speed, humidity, solar radiation and soil moisture. As already indicated in Figure 24.1, however, significant monthly deficits of water storage in the study area do not develop throughout the year. Hence, as far as annual value is concerned, it is assumed that the values of potential evapotranspiration calculated using the Thornthwaite formula are numerically equivalent to values of actual evapotranspiration. Based on the assumption mentioned above, annual runoff totals were calculated as the difference between annual precipitation and annual potential evapotranspiration.

24.4. Changes in Annual Evapotranspiration as Basic Data for the Calculation of Annual Runoff

Long-term temporal changes in the annual evapotranspiration in Tsu are shown in Figure 24.2. In this figure, the record of annual evapotranspiration has also been characterized by a 5-year iterative running mean, in order to represent the general trend of long-term changes more clearly. The annual evapotranspiration in the study area ranges from 758 mm to 880 mm, and its arithmetic mean is 796 mm.

The smoothed trend curve indicates that the annual evapotranspiration has increased over the long term. Up to the first half of the 1910s, annual evapotranspiration was below average. During the 40 years from the latter half of the 1910s to the mid-1950s, annual evapotranspiration was fairly constant, showing a slightly lower value than the arithmetic mean. Since the mid-1950s, however, annual evapotranspiration has been inclined to increase with a peak in the early 1960s. This was followed by a substantial increase in the last ten years. It is pointed out that the key influence on the trend of annual evapotranspiration values shown on Figure 24.2 primarily reflects increased air temperatures.

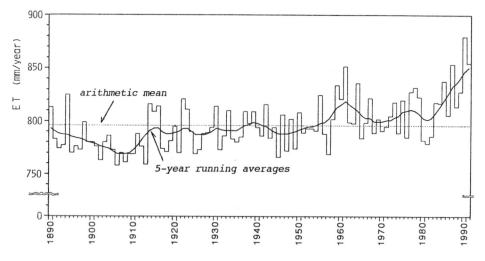

Figure 24.2. Secular changes in the annual evapotranspiration in Tsu.

24.5. Changes in Annual Runoff Based on the Water Balance

24.5.1. ANNUAL RUNOFF

Under the assumption that there is little or no storage carry-over between years, annual runoff was calculated as the difference between annual precipitation and annual evapotranspiration. The long-range changes in annual runoff and precipitation in the study area are shown in Figure 24.3. In this figure, a long-term trend in both elements is also indicated by 5-year running averages. On the basis of water balance. the arithmetic mean of annual runoff in the study area over 102 years is calculated to be 934 mm. It is obvious from the Figure that the trend of secular changes in annual runoff is analogous to that in annual precipitation. The temporal changes both in annual runoff and in annual precipitation are characterized by several pairs of repeated peaks and troughs of the trend curve. The annual runoff in the study area is apt to be below the average since the latter half of the 1920s, but runoff shows a tendency to increase in the last few years.

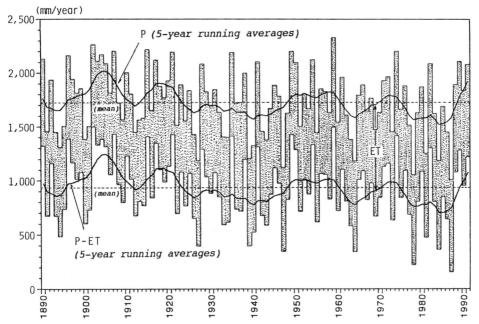

Figure 24.3. Secular changes in the annual precipitation (P) and annual 'runoff' (P-ET) in Tsu.

24.5.2. ANNUAL RUNOFF RATIO

Using the value of calculated annual runoff for each year, yearly annual runoff ratios were obtained for the study area. The annual runoff ratio can be obtained as a quotient of the difference between annual precipitation and annual evapotranspiration divided by the annual precipitation. Figure 24.4 shows the change in the annual runoff ratio in the study area, including 5-year running averages. The arithmetic mean as indicated by the horizontal broken line in the Figure is 0.52. By comparing Figure 24.4 with Figure 24.3, it is clear that the trend curve of long-term changes in the annual runoff ratio is strongly influenced by annual precipitation. From the late 1970s to the mid-1980s, the annual runoff ratio is remarkably low. Since this period, the annual runoff ratio has begun to increase again.

In order to establish the trend of changes in annual runoff ratio more clearly, the average ratios were calculated for each decade from 1890 to 1989. The results are plotted in Figure 24.5. In this figure, annual runoff (P - ET) is expressed by the parallel broken lines. A similar exercise was undertaken using data from a number of experimental basins by Kayane and Takeuchi (1971) and for major rivers in the central part of Japan by Mori (1989). From Figure 24.5, it is evident that the range of changes in annual runoff is much larger than that in annual evapotranspiration. The Figure shows that the average annual runoff ratio for each decade has tended to decrease during the last few decades.

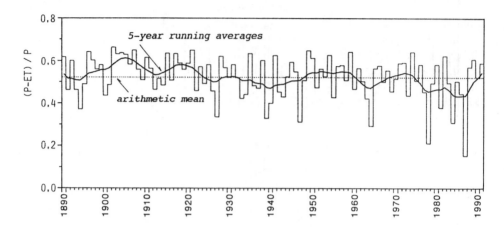

Figure 24.4. Secular changes in the annual 'runoff ratio' [(P - ET)/P] in Tsu.

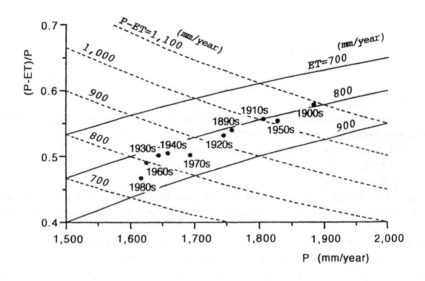

Figure 24.5. Relationship between annual 'runoff ratio' and annual precipitation in
Tsu for the average value of each decade over the period 1890-1989.

24.6. Concluding Remarks

Long-term changes in annual runoff and in the runoff ratio were investigated in the central part of Japan based on the water balance method. The most striking feature during the last 100 years has been the very low values of both annual runoff and annual runoff ratio immediately before and after 1980. Both values, however, have continued to increase since the latter half of the 1980s. Although comparison of the results presented for central Japan with those for different climatic regions is left for further studies, the trends identified provide a meaningful perspective on the variability of the components of the water balance, and may be useful as basic information to evaluate the regional hydrological response to global warming.

Acknowledgments

The author is indebted to Professor Isamu Kayane of the University of Tsukuba for drawing his attention to this research topic. He is also grateful to Dr. J.A.A. Jones of University of Wales for a critical reading of the manuscript.

References

Arai, T. (1977) Water balance, in E. Fukui (ed.) *The Climate of Japan -Developments in Atmospheric Science* (8), Kodansha, Tokyo, 167-198.

Aston, A.R. (1984) The effect of doubling atmospheric CO_2 on streamflow: A simulation, *J. Hydrology* **67**, 273-280.

Bultot, F., Coppens, A., Dupriez, G.L., Gellens, D., and Meulenberghs, F. (1988) Repercussions of a CO_2 doubling on the water cycle and on the water balance - A case study for Belgium, *J. Hydrology* **99**, 319-347.

Gleick, P.H. (1986) Methods for evaluating the regional hydrologic impacts of global climatic changes, *J. Hydrology* **88**, 97-116.

Gleick, P.H. (1987) The development and testing of a water balance model for climate impact assessment: Modeling the Sacramento Basin, *Water Resources Research* **23**, 1049-1061.

Gleick, P.H. (1989) Climate change, hydrology, and water resources, *Reviews Geophysics* **27**, 329-344.

Ichikawa, M., Sugai, A., Fusuki, K., Maejima, I., and Yamana, S. (1967) *Explanatory Diagrams of Geography*, Kokon-shoin, Tokyo, 115 pp. (In Japanese.)

Kayane, I., and Takeuchi, A. (1971) On the annual runoff ratio of Japanese rivers, *Geographical Review Japan* **44**, 347-355. (In Japanese with English summary.)

Masukura, K., Yoshitani, J., Yoshino, F., and Kavvas, M.L. (1991) Possible changes of hydrological cycle by greenhouse effect in Japanese rivers, *Proceedings International Conference Climatic Impacts on the Environment and Society (CIES)*, Tsukuba, 7 pp.

Masukura, K., Yoshitani, J., and Watanabe, Y. (1992) Evaluation of possible precipitation change over Japan and discharge variation in the Tone River, *Proceedings Japan-United States Committee on Hydrology, Water Resources and Global Climatic Change*, Tsukuba, 122 pp.

Mori, K. (1989) Runoff characteristics of the Kiso, Nagara and Ibi Rivers as inferred from distributions of precipitation and evapotranspiration, *Reports Environmental Science, Mie University* **13**, 27-36. (In Japanese with English summary.)

Thornthwaite, C.W. (1948) An approach toward a rational classification of climate, *Geographical Review* **38**, 55-94.

K. Mori
Department of Geography
Mie University, Kamihama, Tsu, Mie 514, Japan.

CONCLUSIONS

25 THE IMPACT OF GLOBAL WARMING ON REGIONAL HYDROLOGY AND FUTURE RESEARCH PRIORITIES

J.A.A. JONES

The present status of the global hydrological cycle is merely a snapshot in a long continuum of change. The geological record embodies periods when the balance of the cycle was very different, from the Carboniferous swamps to the Triassic deserts. Changes in the distribution of land masses have wrought fundamental shifts in the cycle: continental drift and orogeny as well as the regular shifts in the Earth's orbit have all contributed to great range of activity in the hydrological cycle that is revealed in the geological record. The sensitivity of the hydrological cycle is illustrated by the fact that a 10% change in global continental precipitation only requires a 2% change in oceanic evaporation. The creation of new deserts requires an even smaller change, 0.2%, in the overall cycle (Kayane, Section I).

Although there remain considerable uncertainties about the patterns of climate change under gobal warming scenarios, especially in terms of hydrological components, this is no argument for not pursuing research in this field. The potential implications are huge for both humanity and the biosphere, and the timescales of change are now clearly set within the range of prudent planning. The issue is firmly and rightly within the remit of world governments, as the Rio Earth Summit (UN, 1992) and the continued work of the Intergovernmental Panel on Climate Change show.

Quite apart from these policy issues, however, global warming research has become a major impetus to bringing together the various environmental sciences, to identifying gaps in scientific knowledge and to begin to link models developed in different sciences and at different scales in time and space (cp. Bass et al., Section I). Perhaps no other 'crisis' could have mobilised the funds and the will to succeed in this extremely important venture. The ultimate aim of environmental science to predict the interlinked responses of the Earth-atmosphere system will be greatly advanced by it.

25.1. Main Conclusions

The preceding chapters have attempted to provide the first global coverage of the probable hydrological consequences of global warming. There are still areas of uncertainty and even significant disagreement in both the nature of regional climatological change and the hydrological results. Nevertheless, these are probably

J.A.A. Jones et al. (eds.), Regional Hydrological Response to Climate Change, 419–422.
© *1996 Kluwer Academic Publishers. Printed in the Netherlands.*

outweighed by the considerable degree of agreement on both the methods and the results of prediction.

The main conclusions to be drawn are:

1. Global warming is likely to reduce water resources in the lower mid-latitudes, approximately equatorwards of 50°N and 35-40°S.

2. Resources are likely to increase in the higher mid-latitudes, roughly polewards of 55°N and 35°S.

3. Even where there is a net increase in annual water balance, there are still likely to be many areas that will experience seasonal shortages, mainly in the summer and autumn.

4. Most reductions in water resources are likely to be due to reduced precipitation, although increased evapotranspiration rates generally have an important secondary role.

5. Increases in annual precipitation will mostly be in the form of rain. The amount and/or proportion of precipitation falling as snow will generally reduce, even in mid-latitudes north of 60°.

6. The durations of snowcover, freshwater ice cover and frozen ground will decrease.

7. Nonpolar glaciation will generally decrease.

8. The risk of extreme events may well increase, and many areas will suffer from parallel increases in the frequency of both floods and droughts.

9. Meltwater floods may increase in many but not all areas, but over a period of a few decades these will eventually decrease to less than the present.

10. High resolution GCM output, especially from nested HRLAMs are beginning to provide predictions at a spatial resolution commensurate with the requirements of regional hydrology.

11. Downscaling techniques currently in use are beginning to provide useful input to catchment simulation models.

12. Correlations between precipitation and General Circulation patterns appear to have an important potential use in downscaling, as too do Stochastic Weather Generators, which may have particular potential for studying the return periods of extreme events.

13. There is still more variation in GCM predictions for precipitation than there is for temperature, and hydrological add to the climatological uncertainties to create quite wide margins of error.

14. Many analyses of recent instrumental records indicate trends that are consistent with global warming.

15. Even so, there are few regions that show widespread and consistent trends in river discharge, and there is no evidence of any increase in total global runoff (Chiew *et al.*, Section I); but neither is there evidence of the decrease that would be consistent with decreasing global precipitation and increasing temperature (Kayane, Section I).

25.2. Priorities for Future Research

25.2.1. CLIMATOLOGICAL PRIORITIES

Advances in GCMs continue at a rapid pace. The recent successful simulation runs of the UK Hadley Centre High Resolution model have included the effects of dust and sulphate aerosols, which create a small but significant reduction in the amount of global warming (Mitchell *et al.*, 1995; Hadley Centre, 1995), represent a major advance. So too does progress with AOGCMs, models capable of running in the transient mode, and with the European Window in UKHI and HRLAMs in other GCMs.

We can identify the following priorities for hydrological applications:

1. Improved prediction of the key hydrological variables of precipitation and soil moisture, as well as the factors that control evaporation, relative humidity, plus windspeeds and insolation.

2. This also requires continued improvements in the prediction of the type and amount of regional cloud cover, including continuation of research in cloud physics, especially on the processes and extent of ice crystal formation and retention within the clouds.

3. More realistic parameterisation is needed of biospheric feedbacks, including the response of terrestrial plants to CO_2-fertilisation and heat and drought stress, plus an urgent need to understand the dynamics of marine phytoplankton as a carbon sink and the controls on their distribution.

4. Further progress is urgently needed in operationalising the upscaling of SVAT models and the parameterisation of variability in meso- and micro- scale land surface exchanges in order to link with global-scale climate models.

25.2.2. HYDROLOGICAL PRIORITIES

1. More information is needed on variability in precipitation and evaporation, including durations and frequency-magnitude relations. It will clearly be many years before GCMs can provide enough high resolution runs to enable statistical estimates to be made directly from GCM output. In the meantime, indirect approaches such as perturbing ideomorphic probability density functions that fit current distributions, or statistical links between events and atmospheric circulation patterns, seem to be valuable routes forward. So too do Stochastic Weather Generators.

2. More work is needed to develop physically-based hydrological simulation models for large basins, as currently underway on the Rhine (Parmet and Mann, 1993) and Mississippi (NOAA, 1994), that are spatially commensurate with the resolution of GCM output.

3. This must be matched by linking between GCMs, GIS models of local topography and hydrological simulation models. The UK LINK and Core Model Programmes are active in this general area of downscaling.

J.A.A. JONES

4. Hydrological simulation has to date generally side-stepped the important question of changing land surface response to climate change. Research papers normally begin with the explicit assumption that these remain unchanged. Yet a body of evidence is emerging of likely changes in evapotranspirational responses from vegetation (cf. Jones, Section III). Changes are also to be expected following increased exposure to drought, both from the vegetation and from the soil. Depending on the soil type, increased desiccation will increase cracking, hydrophobic reactions and runoff ratios.

5. Research on hydrological responses to climate change also highlights the need to improve hydrological modelling of the effects of agriculture and landuse changes (cp. Jones, 1996). Examples of these effects are reviewed in Jones (Section II) and many of them are estimated to be of the same order as projected climate changes, but there are also large uncertainties as to the response in many cases.

6. Finally, there is a very clear need for economic analyses of the probable effects upon water resources. Initial attempts have been made recently to put a monetary value on these effects. Fankhauser (1995) estimates the global 'welfare loss in the water sector' at nearly $47 billion, of which the USA and the EU each account for about $14 billion. But these estimates only cover the cost of lost water. Titus (1992) also looked at the costs of increased water pollution in the USA and estimated these alone to amount to $33 billion under a 4°C warming.

References

Fankhauser, S. (1995) *Valuing Climate Change*, Earthscan, London, 180 pp.
Hadley Centre (1995) *Modelling Climate Change 1860-2050*, Meteorological Office, Bracknell, 13 pp.
Jones, J.A.A. (1996) *Global Hydrology: Processes, Resources and Environmental Management*, Addison Wesley Longman, Harlow.
Mitchell, J.F.B., Johns, T.C., Gregory, J.M., and Tett, S.F.B. (1995) Climate response to increasing levels of greenhouse gases and sulphate aerosols, *Nature* **376**, 501-504.
NOAA (1994) *The great flood of 1993*, US Department of Commerce, National Disaster Report.
Parmet, B.W.A.H., and Mann, M.A.M. (1993) Influence of climate change on the discharge of the River Rhine - a model for the lowland area, in *Exchange Processes at the Land Surface for a Range of Space and Time Scales*, International Association of Hydrological Sciences, Publication No. 212, 469-477.
Titus, J. (1992) The cost of climate change to the United States, in S.K. Majumbar, L.S. Kalkstein, B. Yarnal, E.W. Miller, and L.M. Rosenfeld (eds.), *Global Climate Change: Implications, Challenges and Mitigating Measures*, Pennsylvania Academy of Science, Pennsylvania.
UN Conference on Environment and Development (1992) *The Earth Summit, Rio de Janiero 1992*, Regency Press Corporation.

J.A.A. Jones
Institute of Earth Studies, University of Wales, Aberystwyth, SY23 3DB, UK.

INDEX

Acidification, 118
Aerosols, 11, 92, 421
Africa, 2, 32, 35, 41, 65
Agriculture, see also Landuse interactions and Vegetation, 2, 120, 337, 404, 422
Analogues, temporal/spatial, 71, 115
ANNs (Artificial Neural Networks), 154, 173-179
Antarctic, 15, 404
AOGCMs (Coupled Atmosphere-Ocean GCMs), 8, 90, 421
Aquifers, see Groundwater
Arabian peninsula, 2
Aral Sea, 238, 259-267
Arctic, 3, 14, 15, 70, 73-86
Arid lands, see Drylands
Asia, 2, 16, 32, 63-68, 133-151, 197-211, 238-295, 298-315, 351-381, 409-416
Australia/Australasia, 32, 51, 64, 65, 298, 337-350

BAHC (Biospheric Aspects of the Hydrological Cycle), 39-62
BEST (Bare Essentials of Surface Transfer) model, 42
Biospheric interactions, see also Vegetation and Landuse, 4, 22, 25, 36, 39-62, 88, 108, 364, 368, 419, 421
BMRC (Bureau of Meteorological Research GCM), 39, 341-350
BOD, 117
Breakup, see Freshwater ice
Budyko's formulae, 94-98, 265, 307, 308, 398

CCC (Canadian Climate Centre GCM), 341-350
China, 2, 4, 65, 70, 71, 133-151, 198-211, 239, 241-257, 287-295, 297-299, 301-315, 351-388
Closed basins, see Interior drainage
Cloud cover, 14, 70, 78-80, 247, 299, 363-380, 421
Cloud properties, 90-92, 105, 106, 109, 363-380
Cloud type, see Cloud properties and cover
Cloudiness, see Cloud cover and properties
Core Model Programme, 88, 421
Costs, to water industry, 123, 319, 333, 334, 422
Coupled Atmosphere-Ocean GCMs, see AOGCMs
CSIRO9 (Commonwealth Scientific and Industrial Organisation GCM), 341-350, 398-401

Dams, see Hydroelectricity, River regulation and Reservoirs

423

The GeoJournal Library

KLUWER ACADEMIC PUBLISHERS – DORDRECHT / BOSTON / LONDON